Modeling and Simulation in Science, Engineering and Technology

Series Editor
Nicola Bellomo
Politecnico di Torino
Italy

Advisory Editorial Board

M. Avellaneda (Modeling in Economics)
Courant Institute of Mathematical Sciences
New York University
251 Mercer Street
New York, NY 10012, USA
avellaneda@cims.nyu.edu

K.J. Bathe (Solid Mechanics)
Department of Mechanical Engineering
Massachusetts Institute of Technology
Cambridge, MA 02139, USA
kjb@mit.edu

P. Degond (Semiconductor
  & Transport Modeling)
Mathématiques pour l'Industrie et la Physique
Université P. Sabatier Toulouse 3
118 Route de Narbonne
31062 Toulouse Cedex, France
degond@mip.ups-tlse.fr

M.A. Herrero Garcia (Mathematical Methods)
Departamento de Matematica Aplicada
Universidad Complutense de Madrid
Avenida Complutense s/n
28040 Madrid, Spain
herrero@sunma4.mat.ucm.es

W. Kliemann (Stochastic Modeling)
Department of Mathematics
Iowa State University
400 Carver Hall
Ames, IA 50011, USA
kliemann@iastate.edu

H.G. Othmer (Mathematical Biology)
Department of Mathematics
University of Minnesota
270A Vincent Hall
Minneapolis, MN 55455, USA
othmer@math.umn.edu

L. Preziosi (Industrial Mathematics)
Dipartimento di Matematica
Politecnico di Torino
Corso Duca degli Abruzzi 24
10129 Torino, Italy
preziosi@polito.it

V. Protopopescu (Competitive Systems,
  Epidemiology)
CSMD
Oak Ridge National Laboratory
Oak Ridge, TN 37831-6363, USA
vvp@epmnas.epm.ornl.gov

K.R. Rajagopal (Multiphase Flows)
Department of Mechanical Engineering
Texas A&M University
College Station, TX 77843, USA
KRajagopal@mengr.tamu.edu

Y. Sone (Fluid Dynamics in Engineering
  Sciences)
Professor Emeritus
Kyoto University
230-133 Iwakura-Nagatani-cho
Sakyo-ku Kyoto 606-0026, Japan
sone@yoshio.mbox.media.kyoto-u.ac.jp

Antonio Romano
Renato Lancellotta
Addolorata Marasco

Continuum Mechanics using *Mathematica*®

Fundamentals, Applications and Scientific Computing

Birkhäuser
Boston • Basel • Berlin

Antonio Romano
Dipartimento di Matematica
 e Applicazioni "R. Caccioppoli"
Università degli Studi di Napoli "Federico II"
via Cintia
80126 Napoli
Italy

Renato Lancellotta
Dipartimento di Ingegneria
 Strutturale e Geotecnica
Politecnico di Torino
Corso Duca degli Abruzzi, 24
10129 Torino
Italy

Addolorata Marasco
Dipartimento di Matematica
 e Applicazioni "R. Caccioppoli"
Università degli Studi di Napoli "Federico II"
via Cintia
80126 Napoli
Italy

AMS Subject Classification: 74-xx, 74Axx, 74Bxx, 74Jxx, 76-xx

Library of Congress Cataloging-in-Publication Data
Romano, Antonio, 1941
  Continuum mechanics using Mathematica : fundamentals, applications, and scientific computing / Antonio Romano, Renato Lancellotta, Addolorata Marasco.
    p. cm — (Modeling and simulation in science, engineering and technology)
  Includes bibliographical references and index.
  ISBN 0-8176-3240-9 (alk. paper)
    1. Continuum mechanics–Data processing. 2. Mathematica (Computer file) I. Lancellotta, Renato. II. Marasco, Addolorata. III. Title. IV. Modeling and simulation in science, engineering & technology.

QA808.2.R66 2004
531–dc22   2004047748

ISBN-10 0-8176-3240-9   eISBN 0-8176-4458-X   Printed on acid-free paper.
ISBN-13 978-0-8176-3240-3

©2006 Birkhäuser Boston

All rights reserved. This work may not be translated or copied in whole or in part without the written permission of the publisher (Birkhäuser Boston, c/o Springer Science+Business Media Inc., 233 Spring Street, New York, NY 10013, USA), except for brief excerpts in connection with reviews or scholarly analysis. Use in connection with any form of information storage and retrieval, electronic adaptation, computer software, or by similar or dissimilar methodology now known or hereafter developed is forbidden.
The use in this publication of trade names, trademarks, service marks and similar terms, even if they are not identified as such, is not to be taken as an expression of opinion as to whether or not they are subject to proprietary rights.

Printed in the United States of America.   (LAP/SB)

9 8 7 6 5 4 3 2 1

www.birkhauser.com

Contents

Preface		ix
1	**Elements of Linear Algebra**	**1**
1.1	Motivation to Study Linear Algebra	1
1.2	Vector Spaces and Bases	2
1.3	Euclidean Vector Space	5
1.4	Base Changes	9
1.5	Vector Product	11
1.6	Mixed Product	13
1.7	Elements of Tensor Algebra	14
1.8	Eigenvalues and Eigenvectors of a Euclidean Second-Order Tensor	20
1.9	Orthogonal Tensors	24
1.10	Cauchy's Polar Decomposition Theorem	28
1.11	Higher Order Tensors	29
1.12	Euclidean Point Space	30
1.13	Exercises	32
1.14	The Program VectorSys	36
1.15	The Program EigenSystemAG	41
2	**Vector Analysis**	**45**
2.1	Curvilinear Coordinates	45
2.2	Examples of Curvilinear Coordinates	48
2.3	Differentiation of Vector Fields	50
2.4	The Stokes and Gauss Theorems	55
2.5	Singular Surfaces	56
2.6	Useful Formulae	60
2.7	Some Curvilinear Coordinates	62
	2.7.1 Generalized Polar Coordinates	63
	2.7.2 Cylindrical Coordinates	64
	2.7.3 Spherical Coordinates	65
	2.7.4 Elliptic Coordinates	66

		2.7.5	Parabolic Coordinates	67

      2.7.5 Parabolic Coordinates 67
      2.7.6 Bipolar Coordinates 68
      2.7.7 Prolate and Oblate Spheroidal Coordinates 68
      2.7.8 Paraboloidal Coordinates 69
  2.8 Exercises 69
  2.9 The Program Operator 70

3 Finite and Infinitesimal Deformations   **77**
  3.1 Deformation Gradient 77
  3.2 Stretch Ratio and Angular Distortion 80
  3.3 Invariants of C and B 83
  3.4 Displacement and Displacement Gradient 84
  3.5 Infinitesimal Deformation Theory 86
  3.6 Transformation Rules for Deformation Tensors 88
  3.7 Some Relevant Formulae 89
  3.8 Compatibility Conditions 92
  3.9 Curvilinear Coordinates 95
  3.10 Exercises 96
  3.11 The Program Deformation 101

4 Kinematics   **109**
  4.1 Velocity and Acceleration 109
  4.2 Velocity Gradient 112
  4.3 Rigid, Irrotational, and Isochoric Motions 113
  4.4 Transformation Rules for a Change of Frame 115
  4.5 Singular Moving Surfaces 116
  4.6 Time Derivative of a Moving Volume 119
  4.7 Worked Exercises 123
  4.8 The Program Velocity 126

5 Balance Equations   **131**
  5.1 General Formulation of a Balance Equation 131
  5.2 Mass Conservation 136
  5.3 Momentum Balance Equation 137
  5.4 Balance of Angular Momentum 140
  5.5 Energy Balance 141
  5.6 Entropy Inequality 143
  5.7 Lagrangian Formulation of Balance Equations 146
  5.8 The Principle of Virtual Displacements 150
  5.9 Exercises 151

6 Constitutive Equations — 155
- 6.1 Constitutive Axioms 155
- 6.2 Thermoviscoelastic Behavior 160
- 6.3 Linear Thermoelasticity 165
- 6.4 Exercises 169

7 Symmetry Groups: Solids and Fluids — 171
- 7.1 Symmetry 171
- 7.2 Isotropic Solids 174
- 7.3 Perfect and Viscous Fluids 177
- 7.4 Anisotropic Solids 181
- 7.5 Exercises 183
- 7.6 The Program LinElasticityTensor 185

8 Wave Propagation — 189
- 8.1 Introduction 189
- 8.2 Cauchy's Problem for Second-Order PDEs 190
- 8.3 Characteristics and Classification of PDEs 194
- 8.4 Examples 196
- 8.5 Cauchy's Problem for a Quasi-Linear First-Order System . 199
- 8.6 Classification of First-Order Systems 201
- 8.7 Examples 202
- 8.8 Second-Order Systems 206
- 8.9 Ordinary Waves 207
- 8.10 Linearized Theory and Waves 211
- 8.11 Shock Waves 215
- 8.12 Exercises 218
- 8.13 The Program PdeEqClass 221
- 8.14 The Program PdeSysClass 227
- 8.15 The Program WavesI 233
- 8.16 The Program WavesII 239

9 Fluid Mechanics — 245
- 9.1 Perfect Fluid 245
- 9.2 Stevino's Law and Archimedes' Principle 246
- 9.3 Fundamental Theorems of Fluid Dynamics 250
- 9.4 Boundary Value Problems for a Perfect Fluid 254
- 9.5 2D Steady Flow of a Perfect Fluid 256
- 9.6 D'Alembert's Paradox and the Kutta–Joukowsky Theorem 264
- 9.7 Lift and Airfoils 267
- 9.8 Newtonian Fluids 272
- 9.9 Applications of the Navier–Stokes Equation 273
- 9.10 Dimensional Analysis and the Navier–Stokes Equation . . 274
- 9.11 Boundary Layer 276

    9.12 Motion of a Viscous Liquid around an Obstacle 281
    9.13 Ordinary Waves in Perfect Fluids 287
    9.14 Shock Waves in Fluids 290
    9.15 Shock Waves in a Perfect Gas 293
    9.16 Exercises 296
    9.17 The Program Potential 298
    9.18 The Program Wing 306
    9.19 The Program Joukowsky 309
    9.20 The Program JoukowskyMap 312

10 Linear Elasticity   **317**
    10.1 Basic Equations of Linear Elasticity 317
    10.2 Uniqueness Theorems 321
    10.3 Existence and Uniqueness of Equilibrium Solutions 323
    10.4 Examples of Deformations 327
    10.5 The Boussinesq–Papkovich–Neuber Solution 329
    10.6 Saint-Venant's Conjecture 330
    10.7 The Fundamental Saint-Venant Solutions 335
    10.8 Ordinary Waves in Elastic Systems 339
    10.9 Plane Waves 345
    10.10 Reflection of Plane Waves in a Half-Space 350
    10.11 Rayleigh Waves 356
    10.12 Reflection and Refraction of SH Waves 359
    10.13 Harmonic Waves in a Layer 362
    10.14 Exercises 365

11 Other Approaches to Thermodynamics   **367**
    11.1 Basic Thermodynamics 367
    11.2 Extended Thermodynamics 370
    11.3 Serrin's Approach 372
    11.4 An Application to Viscous Fluids 375

References   **379**
Index   **383**

Preface

The motion of any body depends both on its characteristics and the forces acting on it. Although taking into account all possible properties makes the equations too complex to solve, sometimes it is possible to consider only the properties that have the greatest influence on the motion. *Models of ideals bodies*, which contain only the most relevant properties, can be studied using the tools of mathematical physics. Adding more properties into a model makes it more realistic, but it also makes the motion problem harder to solve.

In order to highlight the above statements, let us first suppose that a system S of N unconstrained bodies C_i, $i = 1, ..., N$, is sufficiently described by the model of N **material points** whenever the bodies have negligible dimensions with respect to the dimensions of the region containing the trajectories. This means that all the physical properties of C_i that influence the motion are expressed by a positive number, the **mass** m_i, whereas the position of C_i with respect to a frame I is given by the position vector $\mathbf{r}_i(t)$ versus time. To determine the functions $\mathbf{r}_i(t)$, one has to integrate the following system of Newtonian equations:

$$m_i \ddot{\mathbf{r}}_i = \mathbf{F}_i \equiv \mathbf{f}_i(\mathbf{r}_1, ..., \mathbf{r}_N, \dot{\mathbf{r}}_1, ..., \dot{\mathbf{r}}_N, t),$$

$i = 1, ..., N$, where the forces $\mathbf{F}_i$, due both to the external world and to the other points of S, are assigned functions $\mathbf{f}_i$ of the positions and velocities of all the points of S, as well as of time. Under suitable regularity assumptions about the functions $\mathbf{f}_i$, the previous (vector) system of second-order **ordinary** differential equations in the unknowns $\mathbf{r}_i(t)$ has one and only one solution satisfying the given initial conditions

$$\mathbf{r}_i(t_0) = \mathbf{r}_i^0, \quad \dot{\mathbf{r}}_i(t_0) = \dot{\mathbf{r}}_i^0, \quad i = 1, ..., N.$$

A second model that more closely matches physical reality is represented by a system S of **constrained rigid bodies** C_i, $i = 1, ..., N$. In this scheme, the **extension** of C_i and the presence of **constraints** are taken into account. The position of C_i is represented by the three-dimensional region occupied by C_i in the frame I. Owing to the supposed rigidity of both bodies C_i and constraints, the configurations of S are described by $n \leq 6N$ parameters $q_1, ..., q_n$, which are called **Lagrangian coordinates**. Moreover, the mass m_i of C_i is no longer sufficient for describing the physical properties of C_i since we have to know both its density and geometry. To determine the motion of S, that is, the functions $q_1(t), ..., q_n(t)$, the Lagrangian expressions of the kinetic energy $T(q, \dot{q})$ and active forces $Q_h(q, \dot{q})$ are necessary. Then a possible motion of S is a solution of the Lagrange equations

$$\frac{d}{dt}\frac{\partial T}{\partial \dot{q}_h} - \frac{\partial T}{\partial q_h} = Q_h(q, \dot{q}), \quad h = 1, ..., n,$$

satisfying the given initial conditions

$$q_h(t_0) = q_h^0, \quad \dot{q}_h(t_0) = \dot{q}_h^0, \quad h = 1, ..., n,$$

which once again fix the initial configuration and the velocity field of S.

We face a completely different situation when, to improve the description, we adopt the model of **continuum mechanics**. In fact, in this model the bodies are **deformable** and, at the same time, the matter is supposed to be **continuously distributed** over the volume they occupy, so that their molecular structure is completely erased. In this book we will show that the substitution of rigidity with the deformability leads us to determine three scalar functions of three spatial variables and time, in order to find the motion of S. Consequently, the fundamental evolution laws become partial differential equations. This consequence of deformability is the root of the mathematical difficulties of continuum mechanics.

This model must include other characteristics which allow us to describe the different *macroscopic* material behaviors. In fact, bodies undergo different deformations under the influence of the same applied loads. The mathematical description of different materials is the object of the **constitutive equations**. These equations, although they have to describe a wide variety of real bodies, must in any case satisfy some general principles. These principles are called **constitutive axioms** and they reflect general rules of behavior. These rules, although they imply severe restrictions on the form of the constitutive equations, permit us to describe different materials. The constitutive equations can be divided into classes describing the behavior of material categories: elastic bodies, fluids, etc.. The choice of a particular constitutive relation cannot be done a priori but instead relies on experiments, due to the fact that the macroscopic behavior of a body is strictly related to its molecular structure. Since the continuum model erases this structure, the constitutive equation of a particular material has to be determined by experimental procedures. However, the introduction of deformability into the model does not permit us to describe all the phenomena accompanying the motion. In fact, the viscosity of S as well as the friction between S and any external bodies produce heating, which in turn causes heat exchanges among parts of S or between S and its surroundings. Mechanics is not able to describe these phenomena, and we must resort to the ***thermomechanics of continuous systems***. This theory combines the laws of mechanics and thermodynamics, provided that they are suitably generalized to a deformable continuum at a nonuniform temperature.

The situation is much more complex when the continuum carries charges and currents. In such a case, we must take into account Maxwell's equations, which describe the electromagnetic fields accompanying the motion, together with the thermomechanic equations. The coexistence of all these equations gives rise to a compatibility problem: in fact, Maxwell's equations are covariant under Lorentz transformations, whereas thermomechanics laws are covariant under Galileian transformations.

This book is devoted to those readers interested in understanding the basis of continuum mechanics and its fundamental applications: balance laws, constitutive axioms, linear elasticity, fluid dynamics, waves, etc. It is self-contained, as it illustrates all the required mathematical tools, starting from an elementary knowledge of algebra and analysis.

It is divided into 11 chapters. In the first two chapters the elements of linear algebra are presented (vectors, tensors, eigenvalues, eigenvectors, etc.), together with the foundations of vector analysis (curvilinear coordinates, covariant derivative, Gauss and Stokes theorems). In the remaining 9 chapters the foundations of continuum mechanics and some fundamental applications of simple continuous media are introduced. More precisely, the finite deformation theory is discussed in Chapter 3, and the kinetic principles, the singular surfaces, and the general differential formulae for surfaces and volumes are presented in Chapter 4. Chapter 5 contains the general integral balance laws of mechanics, as well as their local Eulerian or Lagrangian forms. In Chapters 6 and 7 the constitutive axioms, the thermo-viscoelastic materials, and their symmetries are discussed. In Chapter 8, starting from the characteristic surfaces, the classification of a quasi-linear partial differential system is discussed, together with ordinary waves and shock waves. The following two chapters cover the application of the general principles presented in the previous chapters to perfect or viscous fluids (Chapter 9) and to linearly elastic systems (Chapter 10). In the last chapter, a comparison of some proposed thermodynamic theories is presented. Finally, in Appendix A the concept of a weak solution is introduced.

This volume has a companion disc containing many programs written with Mathematica®. These programs apply to topics discussed in the book such as the equivalence of applied vector systems, differential operators in curvilinear coordinates, kinematic fields, deformation theory, classification of systems of partial differential equations, motion representation of perfect fluids by complex functions, waves in solids, and so on. This approach has already been adopted by two of the authors in other books (see [5], [6]).[1]

Many other important topics of continuum mechanics are not considered in this volume, which is essentially devoted to the foundations of the theory. In a second volume, for which editing is in progress, nonlinear elasticity, mixtures, phase changes, piezoelastic and magnetoelastic bodies will be discussed. Moreover, as examples of one-dimensional and two-dimensional continua, the elements of bars and plates will be presented.

---

[1] The reader interested in other fundamental books in continuum mechanics can consult, for example, the references [1], [2], [3], [4].

Chapter 1

Elements of Linear Algebra

1.1 Motivation to Study Linear Algebra

In describing reality mathematically, several tools are used. The most familiar is certainly the concept of a function, which expresses the dependence of some quantities $(y_1, \ldots, y_m)$ on others $(x_1, \ldots, x_n)$ by m analytical relations

$$y_i = f_i(x_1, \ldots, x_n), \qquad i = 1, \ldots, m. \tag{1.1}$$

For instance, the coordinate x of a material point moving on a straight line with harmonic motion varies with time t according to the law

$$x = A\sin(\omega t + \varphi),$$

where $\omega = 2\pi\nu$ and $A, \nu,$ and φ are the amplitude, the frequency, and the phase of the harmonic motion, respectively. Similarly, an attractive elastic force **F** depends on the lengthening **s** of the spring, according to the rule

$$F_i = -kx_i, \qquad i = 1, 2, 3.$$

In this formula F_i denotes the ith force component, $k > 0$ is the elastic constant, and x_i is the ith component of **s**. A final example is a dielectric, in which the components D_i of the electric induction **D** depend on the components E_i of the electric field **E**.

Later, it will be shown that the tension in a neighborhood I of a point P of an elastic material S depends on the deformation of I; both the deformation and the tension are described by 9 variables. Therefore to find the relation between deformation and tension, we need 9 functions like (1.1), depending on 9 variables.

The previous examples, and many others, tell us how important it is to study systems of real functions depending on many real variables. In studying mathematical analysis, we learn how difficult this can be, so it is quite natural to start with *linear* relations f_i. In doing this we are making the assumption that small *causes* produce small *effects*. If we interpret the independent variables x_i as *causes* and the dependent variables y_i as *effects*, then the functions (1.1) can be replaced with their Taylor expansions at the origin $(0,\ldots,0)$. Limiting the Taylor expansions to first order gives linear relations, which can be explored using the techniques of *linear algebra*. In this chapter we present some fundamental aspects of this subject.[1]

1.2 Vector Spaces and Bases

Let $\Re$ ($\mathbb{C}$) be the field of real (complex) numbers. A **vector space** on $\Re$ ($\mathbb{C}$) is an arbitrary set $\mathcal{E}$ equipped with two algebraic operations, called **addition** and **multiplication**. The elements of this algebraic structure are called **vectors**.

The addition operation associates to any pair of vectors $\mathbf{u}$, $\mathbf{v}$ of $\mathcal{E}$ their **sum** $\mathbf{u} + \mathbf{v} \in \mathcal{E}$ in a way that satisfies the following formal properties:

$\mathbf{u} + \mathbf{v} = \mathbf{v} + \mathbf{u}$;

only one element $\mathbf{0} \in \mathcal{E}$ exists for which $\mathbf{u} + \mathbf{0} = \mathbf{u}$;

for all $\mathbf{u} \in \mathcal{E}$, only one element $-\mathbf{u} \in \mathcal{E}$ exists such that $\mathbf{u} + (-\mathbf{u}) = \mathbf{0}$. (1.2)

The multiplication operation, which associates the **product** $a\mathbf{u} \in \mathcal{E}$ to any number a belonging to $\Re$ ($\mathbb{C}$) and any vector $\mathbf{u} \in \mathcal{E}$, has to satisfy the formal properties

$$1\mathbf{u} = \mathbf{u},$$
$$(a+b)\mathbf{u} = a\mathbf{u} + b\mathbf{u},$$
$$a(\mathbf{u}+\mathbf{v}) = a\mathbf{u} + a\mathbf{v}. \quad (1.3)$$

Let $W = \{\mathbf{u}_1, \ldots, \mathbf{u}_r\}$ be a set of r distinct vectors belonging to $\mathcal{E}$; r is also called the **order** of W. The vector

$$a_1\mathbf{u}_1 + \cdots + a_r\mathbf{u}_r,$$

where $a_1, \ldots, a_r$ are real (complex) numbers, is called a **linear combination** of $\mathbf{u}_1, \ldots, \mathbf{u}_r$. Moreover, the set $\{\mathbf{u}_1, \ldots, \mathbf{u}_r\}$ is **linearly independent** or **free** if any linear combination

$$a_1\mathbf{u}_1 + \cdots + a_r\mathbf{u}_r = \mathbf{0} \quad (1.4)$$

---
[1] For a more extensive study of the subjects of the first two chapters, see [9], [10], [11], [12].

1.2. Vector Spaces and Bases

implies that $a_1 = \cdots = a_r = 0$. In the opposite case, they are **linearly dependent**.

If it is possible to find linearly independent vector systems of order n, where n is a finite integer, but there is no free system of order $n+1$, then we say that n is the **dimension** of $\mathcal{E}$, which is denoted by $\mathcal{E}_n$. Any linearly independent system W of order n is said to be a **basis** of $\mathcal{E}_n$. When it is possible to find linearly independent vector sets of any order, the dimension of the vector space is said to be infinity.

We state the following theorem without proof.

Theorem 1.1
A set $W = \{\mathbf{e}_1, \ldots, \mathbf{e}_n\}$ of vectors of $\mathcal{E}_n$ is a basis if and only if any vector $\mathbf{u} \in \mathcal{E}_n$ can be expressed as a unique linear combination of the elements of W:
$$\mathbf{u} = u^1 \mathbf{e}_1 + \cdots + u^n \mathbf{e}_n. \tag{1.5}$$

For brevity, from now on we denote a basis by $(\mathbf{e}_i)$. The coefficients u^i are called **contravariant components** of $\mathbf{u}$ with respect to the basis $(\mathbf{e}_i)$. The relations among the contravariant components of the same vector with respect two bases will be analyzed in Section 1.4.

Examples

- The set of the oriented segments starting from a point O of ordinary three-dimensional space constitutes the simplest example of a three-dimensional vector space on $\Re$. Here the addition is defined with the usual parallelogram rule, and the product $a\mathbf{u}$ is defined by the oriented segment having a length $|a|$ times the length of $\mathbf{u}$ and a direction coinciding with the direction of $\mathbf{u}$, if $a > 0$, or with the opposite one, if $a < 0$.

- The set of the matrices
$$\begin{pmatrix} a_{11} & \cdots & a_{1n} \\ \cdots & \cdots & \cdots \\ a_{m1} & \cdots & a_{mn} \end{pmatrix},$$
where the coefficients are real (complex) numbers, equipped with the usual operations of summation of two matrices and the product of a matrix for a real (complex) number, is a vector space. Moreover, it is easy to verify that one basis for this vector space is the set of $m \times n$-matrices that have all the elements equal to zero except for one which is equal to 1. Consequently, its dimension is equal to mn.

- The set $\mathcal{P}$ of all polynomials

$$P(x) = a_0 x^n + a_1 x^{n-1} + \cdots + a_n$$

of the same degree n with real (complex) coefficients is a $(n+1)$-dimensional vector space, since the polynomials

$$P_1(x) = x^n, \quad P_2 = x^{n-1}, \quad \ldots, \quad P_{n+1} = 1 \qquad (1.6)$$

form a basis of $\mathcal{P}$.

- The set F of continuous functions on the interval $[0, 1]$ is a vector space whose dimension is infinity since, for any n, the set of polynomials (1.6) is linearly independent.

From now on we will use Einstein's notation, in which the summation over two repeated indices is understood provided that one is a subscript and the other a superscript. In this notation, relation (1.5) is written as

$$\mathbf{u} = u^i \mathbf{e}_i. \qquad (1.7)$$

Moreover, for the sake of simplicity, all of the following considerations refer to a three-dimensional vector space $\mathcal{E}_3$, although almost all the results are valid for any vector space.

A subset $U \subset \mathcal{E}$ is a ***vector subspace*** of $\mathcal{E}$ if

$$\mathbf{u}, \mathbf{v} \in \mathcal{E} \Rightarrow \mathbf{u} + \mathbf{v} \in \mathcal{E},$$
$$a \in \Re, \mathbf{u} \in \mathcal{E} \Rightarrow a\mathbf{u} \in \mathcal{E}. \qquad (1.8)$$

Let U and V be two subspaces of $\mathcal{E}_3$ having dimensions p and q ($p+q \leq 3$), respectively, and let $U \cap V = \{\mathbf{0}\}$. The ***direct sum*** of U and V, denoted by

$$W = U \oplus V,$$

is the set of all the vectors $\mathbf{w} = \mathbf{u} + \mathbf{v}$, with $\mathbf{u} \in U$ and $\mathbf{v} \in V$.

Theorem 1.2
The direct sum W of the subspaces U and V is a new subspace whose dimension is $p + q$.

PROOF It is easy to verify that W is a vector space. Moreover, let $(\mathbf{u}_1, ..., \mathbf{u}_p)$ and $(\mathbf{v}_1, ..., \mathbf{v}_q)$ be two bases of U and V, respectively. The definition of the direct sum implies that any vector $\mathbf{w} \in W$ can be written as $\mathbf{w} = \mathbf{u} + \mathbf{v}$, where $\mathbf{u} \in U$ and $\mathbf{v} \in V$ are uniquely determined. Therefore,

$$\mathbf{w} = \sum_{i=1}^{p} u^i \mathbf{u}_i + \sum_{i=1}^{q} v^i \mathbf{v}_i,$$

1.3. Euclidean Vector Space

where the coefficients u^i and v^i are again uniquely determined. It follows that the vectors $(\mathbf{u}_1, ..., \mathbf{u}_p, \mathbf{v}_1, ..., \mathbf{v}_q)$ represent a basis of W. ∎

Figure 1.1 illustrates the previous theorem for $p = q = 1$.

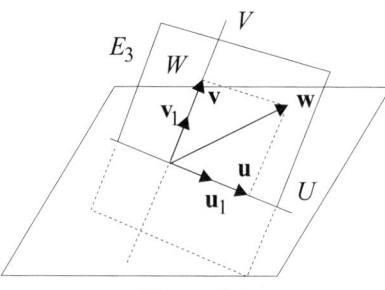

Figure 1.1

1.3 Euclidean Vector Space

Besides the usual operations of vector sum and product of a real number times a vector, the operation of **scalar** or **inner product** of two vectors can be introduced. This operation, which associates a real number $\mathbf{u} \cdot \mathbf{v}$ to any pair of vectors $(\mathbf{u}, \mathbf{v})$, is defined by the following properties:

1. it is distributive with respect to each argument:
$$(\mathbf{u} + \mathbf{v}) \cdot \mathbf{w} = \mathbf{u} \cdot \mathbf{w} + \mathbf{v} \cdot \mathbf{w}, \qquad \mathbf{u} \cdot (\mathbf{v} + \mathbf{w}) = \mathbf{u} \cdot \mathbf{v} + \mathbf{u} \cdot \mathbf{w};$$

2. it is associative:
$$a\mathbf{u} \cdot \mathbf{v} = \mathbf{u} \cdot (a\mathbf{v});$$

3. it is symmetric:
$$\mathbf{u} \cdot \mathbf{v} = \mathbf{v} \cdot \mathbf{u};$$

4. it satisfies the condition
$$\mathbf{u} \cdot \mathbf{u} \geq 0,$$

the equality holding if and only if $\mathbf{u}$ vanishes.

From now on, a space in which a scalar product is defined is referred to as a **Euclidean vector space**.

Property (4) allows us to define the *length* of the vector **u** as the number

$$|\mathbf{u}| = \sqrt{\mathbf{u} \cdot \mathbf{u}};$$

in particular, a vector is a *unit* or *normal* vector if

$$|\mathbf{u}| = 1.$$

Two vectors **u** and **v** are *orthogonal* if

$$\mathbf{u} \cdot \mathbf{v} = \mathbf{0}.$$

The following two inequalities, due to *Minkowski* and *Schwarz*, respectively, can be proved:

$$|\mathbf{u} + \mathbf{v}| \leq |\mathbf{u}| + |\mathbf{v}|, \qquad (1.9)$$
$$|\mathbf{u} \cdot \mathbf{v}| \leq |\mathbf{u}| \, |\mathbf{v}|. \qquad (1.10)$$

The last inequality permits us to define the *angle* $0 \leq \varphi \leq \pi$ between two vectors **u** and **v** by the relation

$$\cos \varphi = \frac{\mathbf{u} \cdot \mathbf{v}}{|\mathbf{u}| \, |\mathbf{v}|},$$

which allows us to give the inner product the following elementary form:

$$\mathbf{u} \cdot \mathbf{v} = |\mathbf{u}| \, |\mathbf{v}| \cos \varphi.$$

A vector system $(\mathbf{u}_1, ..., \mathbf{u}_n)$ is called *orthonormal* if

$$\mathbf{u}_i \cdot \mathbf{u}_j = \delta_{ij},$$

where δ_{ij} is the Kronecker symbol

$$\delta_{ij} = \begin{cases} 1, & i = j, \\ 0, & i \neq j. \end{cases}$$

It is easy to verify that the vectors belonging to an orthonormal system $\{\mathbf{u}_1, \ldots, \mathbf{u}_m\}$ are linearly independent; in fact, it is sufficient to consider the scalar product between any of their vanishing linear combinations and a vector $\mathbf{u}_h$ to verify that the hth coefficient of the combination vanishes. This result implies that the number m of the vectors of an orthogonal system is ≤ 3.

The *Gram–Schmidt orthonormalization procedure* permits us to obtain an orthonormal system starting from any set of three independent

1.3. Euclidean Vector Space

vectors $(\mathbf{u}_1, \mathbf{u}_2, \mathbf{u}_3)$. In fact, if this set is independent, it is always possible to determine the constants λ in the expressions

$$\mathbf{v}_1 = \mathbf{u}_1,$$

$$\mathbf{v}_2 = \lambda_2^1 \mathbf{v}_1 + \mathbf{u}_2,$$

$$\mathbf{v}_3 = \lambda_3^1 \mathbf{v}_1 + \lambda_3^2 \mathbf{v}_2 + \mathbf{u}_3,$$

in such a way that $\mathbf{v}_2$ is orthogonal to $\mathbf{v}_1$ and $\mathbf{v}_3$ is orthogonal to both $\mathbf{v}_1$ and $\mathbf{v}_2$. By a simple calculation, we find that these conditions lead to the following values of the constants λ:

$$\lambda_2^1 = -\frac{\mathbf{u}_2 \cdot \mathbf{v}_1}{|\mathbf{v}_1|^2},$$

$$\lambda_3^2 = -\frac{\mathbf{u}_3 \cdot \mathbf{v}_2}{|\mathbf{v}_2|^2},$$

$$\lambda_3^1 = -\frac{\mathbf{u}_3 \cdot \mathbf{v}_1}{|\mathbf{v}_1|^2}.$$

Then the vectors $\mathbf{v}_h$, $h = 1, 2, 3$, are orthogonal to each other and it is sufficient to consider the vectors $\mathbf{v}_h/|\mathbf{v}_h|$ to obtain the desired orthonormal system.

Let $(\mathbf{e}_i)$ be a basis and let $\mathbf{u} = u^i \mathbf{e}_i$, $\mathbf{v} = v^j \mathbf{e}_j$, be the representations of two vectors in this basis; then the scalar product becomes

$$\mathbf{u} \cdot \mathbf{v} = g_{ij} u^i v^j, \tag{1.11}$$

where

$$g_{ij} = \mathbf{e}_i \cdot \mathbf{e}_j = g_{ji}. \tag{1.12}$$

When $\mathbf{u} = \mathbf{v}$, (1.11) gives the square length of a vector in terms of its components:

$$|\mathbf{u}|^2 = g_{ij} u^i u^j. \tag{1.13}$$

Owing to the property (4) of the scalar product, the quadratic form on the right-hand side of (1.13) is positive definite, so that

$$g \equiv \det(g_{ij}) > 0. \tag{1.14}$$

If the basis is orthonormal, then (1.11) and (1.13) assume the simpler form

$$\mathbf{u} \cdot \mathbf{v} = \sum_{i=1}^{3} u^i v^i, \tag{1.15}$$

$$|\mathbf{u}|^2 = \sum_{i=1}^{3} (u^i)^2. \qquad (1.16)$$

Due to (1.14), the matrix (g_{ij}) has an inverse so that, if g^{ij} are the coefficients of this last matrix, we have

$$g^{ih} g_{hj} = \delta^i_j. \qquad (1.17)$$

If a basis $(\mathbf{e}_i)$ is assigned in the Euclidean space $\mathcal{E}_3$, it is possible to define the **dual** or **reciprocal basis** $(\mathbf{e}^i)$ of $(\mathbf{e}_i)$ with the relations

$$\mathbf{e}^i = g^{ij} \mathbf{e}_j. \qquad (1.18)$$

These new vectors constitute a basis since from any linear combination of them,

$$\lambda_i \mathbf{e}^i = g^{ih} \lambda_i \mathbf{e}_h = \mathbf{0},$$

we derive

$$g^{ih} \lambda_i = 0,$$

when we recall that the vectors $(\mathbf{e}_i)$ are independent. On the other hand, this homogeneous system admits only the trivial solution $\lambda_1 = \lambda_2 = \lambda_3 = 0$, since $\det(g^{ij}) > 0$.

The chain of equalities, see (1.18), (1.12), (1.17),

$$\mathbf{e}^i \cdot \mathbf{e}_j = g^{ih} \mathbf{e}_h \cdot \mathbf{e}_j = g^{ih} g_{hj} = \delta^i_j \qquad (1.19)$$

proves that $\mathbf{e}^i$ is orthogonal to all the vectors $\mathbf{e}_j$, $j \neq i$ (see Figure 1.2 and Exercise 1.2). It is worthwhile to note that, if the basis $(\mathbf{e}_i)$ is orthogonal, then (1.18) implies that

$$\mathbf{e}^i = g^{ii} \mathbf{e}_i = \frac{\mathbf{e}_i}{g_{ii}},$$

so that any vector $\mathbf{e}^i$ of the dual basis is parallel to the vector $\mathbf{e}_i$.

From this relation we have

$$\mathbf{e}_i \cdot \mathbf{e}^i = 1,$$

so that

$$|\mathbf{e}_i||\mathbf{e}^i| \cos 0 = 1,$$

and

$$|\mathbf{e}^i| = \frac{1}{|\mathbf{e}_i|}. \qquad (1.20)$$

In particular, when the basis $(\mathbf{e}_i)$ is orthonormal, $\mathbf{e}^i = \mathbf{e}_i$, $i = 1, \ldots, n$.

1.4. Base Changes

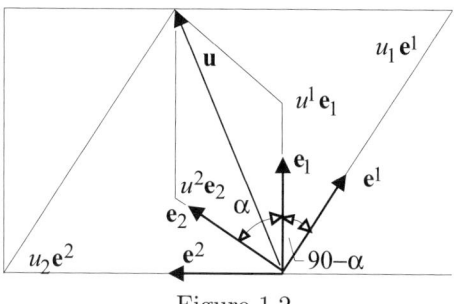

Figure 1.2

With the introduction of the dual basis, any vector **u** admits two representations:

$$\mathbf{u} = u^i \mathbf{e}_i = u_j \mathbf{e}^j. \tag{1.21}$$

The quantities u_i are called **covariant components** of **u**. They can be obtained from the components of **u** with respect to the reciprocal basis or from the projections of **u** onto the axes of the original basis:

$$\mathbf{u} \cdot \mathbf{e}_i = (u_j \mathbf{e}^j) \cdot \mathbf{e}_i = u_j \delta_i^j = u_i. \tag{1.22}$$

In particular, if the basis is orthonormal, then the covariant and contravariant components are equal.

The relation (1.22) shows that the covariant and contravariant components can be defined respectively as

$$u_i = \mathbf{u} \cdot \mathbf{e}_i = g_{ij} u^j, \qquad u^j = \mathbf{u} \cdot \mathbf{e}^j = g^{ji} u_i. \tag{1.23}$$

It follows from (1.23) that the scalar product (1.11) can be written in one of the following equivalent forms:

$$\mathbf{u} \cdot \mathbf{v} = g_{ij} u^i v^j = u_j v^j = g^{ij} u_i v_j. \tag{1.24}$$

1.4 Base Changes

Covariant or contravariant components of a vector are relative to the chosen basis. Consequently, it is interesting to determine the corresponding transformation rules under a change of the basis. Let us consider the base change $(\mathbf{e}_i) \longrightarrow (\mathbf{e}'_i)$; when we note that any vector $\mathbf{e}'_i$ can be represented as a linear combination of the vectors $(\mathbf{e}_i)$, we have[2]

$$\mathbf{e}'_i = A^j_i \mathbf{e}_j, \qquad \mathbf{e}_i = (A^{-1})^j_i \mathbf{e}'_j. \qquad (1.25)$$

In other words, the base change is assigned by giving the nonsingular matrix (A^i_j) of the vector components $(\mathbf{e}'_i)$ with respect to the basis $(\mathbf{e}_i)$.

On the other hand, any vector $\mathbf{u}$ admits the following two representations:

$$\mathbf{u} = u'^i \mathbf{e}'_i = u^j \mathbf{e}_j,$$

from which, taking into account (1.25), there follow the transformation formulae of vector components with respect to base changes:

$$u'^i = (A^{-1})^i_j u^j, \qquad u^i = A^i_j u'^j. \qquad (1.26)$$

In order to determine the corresponding transformation formulae of covariant components, we note that (1.23) and (1.25) imply that

$$u'_i = \mathbf{u} \cdot \mathbf{e}'_i = \mathbf{u} \cdot A^j_i \mathbf{e}_j,$$

and then
$$u'_i = A^j_i u_j \qquad (1.27)$$

or
$$u_i = (A^{-1})^j_i u'_j. \qquad (1.28)$$

A comparison of (1.25), (1.26), and (1.27) justifies the definition of covariant and contravariant components given to u_i and u^i: the transformation of covariant components involves the coefficient of the base transformation (1.25), whereas the transformation of the contravariant components involves the coefficients of the inverse base transformation.

It is also interesting to determine the transformation formulae of the quantities g_{ij} and g^{ij} associated with the scalar product. To this end, we note that (1.12) and (1.25) imply that

$$g'_{ij} = \mathbf{e}'_i \cdot \mathbf{e}'_j = A^h_i A^k_j \mathbf{e}_h \cdot \mathbf{e}_k;$$

that is,
$$g'_{ij} = A^h_i A^k_j g_{hk}. \qquad (1.29)$$

Similarly, from the invariance of the scalar product with respect to a base change and from (1.24), we have

$$g'^{ij} u'_i v'_j = g^{hk} u_h v_k,$$

---

[2]It is easy to verify, by using considerations similar to those used at the end of the previous section, that the independence of the vectors $(\mathbf{e}'_i)$ is equivalent to requiring that $\det(A^i_j) \neq 0$.

1.5. Vector Product

and taking into account (1.28), we obtain

$$g'^{ij} = (A^{-1})^i_h (A^{-1})^j_k g^{hk}. \tag{1.30}$$

From (1.18), (1.30), and (1.25), it also follows that

$$\mathbf{e}'^i = (A^{-1})^i_h \mathbf{e}^h. \tag{1.31}$$

Finally, let $(\mathbf{e}_i)$ and $(\mathbf{e}'_i)$ be two bases of $\mathcal{E}_3$ related to each other by the transformation formulae (1.25). When

$$\det(A^i_j) > 0, \tag{1.32}$$

we say that the two bases have the same **orientation**. It is easy to recognize that this connection between the two bases represents an equivalence relation R in the set B of the bases of $\mathcal{E}_3$. Moreover, R divides B in two equivalence classes; any two bases, each of them belonging to one class, define the two possible orientations of $\mathcal{E}_3$.

1.5 Vector Product

Let $(\mathbf{e}_i)$ be a basis of $\mathcal{E}_3$. The **vector product** or **cross product** is an internal operation which associates, to any pair of vectors $\mathbf{u}$ and $\mathbf{v}$ of $\mathcal{E}_3$, a new vector $\mathbf{u} \times \mathbf{v}$ whose *covariant* components are

$$(\mathbf{u} \times \mathbf{v})_i = \sqrt{g}\, e_{ijh} u^j v^h, \tag{1.33}$$

where u^i and v^j denote the contravariant components of $\mathbf{u}$ and $\mathbf{v}$ and e_{ijh} is the **Levi–Civita symbol**, defined as follows:

$$e_{ijh} = \begin{cases} 0, & \text{if two indices are equal;} \\ 1, & \text{if the permutation } ijh \text{ is even;} \\ -1, & \text{if the permutation } ijh \text{ is odd.} \end{cases} \tag{1.34}$$

The attribute even or odd of the index sequence ijh depends on the number of index exchanges we have to do in order to reproduce the fundamental sequence 123. For example, 231 is even, since two exchanges are needed: the first one produces 132 and the second, 123. On the other hand, 213 is odd, since just one exchange of indices is sufficient to come back to 123.

From the definition (1.34), we see that the exchange of two indices in the permutation symbol implies a sign change; that is,

$$e_{ijk} = -e_{kji} = e_{kij} = -e_{ikj}.$$

Moreover, we remark that the product of two permutation symbols can be expressed by the Kronecker symbol δ using the following identity

$$e_{miq}e_{jkq} = \delta_{mj}\delta_{ik} - \delta_{mk}\delta_{ij} = \begin{vmatrix} \delta_{mj} & \delta_{mk} \\ \delta_{ij} & \delta_{ik} \end{vmatrix}. \tag{1.35}$$

This formula can be memorized as follows: the first two indices of the first e appear in the determinant as row indices, whereas the first two indices of the second e are column indices.

It is easy to verify that the composition law (1.33) is distributive, associative, and skew-symmetric. In order to recognize that the present definition of the cross product $\mathbf{u} \times \mathbf{v}$ coincides with the elementary one, we consider an orthonormal basis $(\mathbf{e}_i)$ such that $|\mathbf{u}|\mathbf{e}_1 = \mathbf{u}$ and the plane $(\mathbf{e}_1, \mathbf{e}_2)$ contains $\mathbf{v}$. In this basis, $\mathbf{u}$ and $\mathbf{v}$ have components $(|\mathbf{u}|, 0, 0)$ and $(|\mathbf{v}| \cos \varphi, |\mathbf{v}| \sin \varphi, 0)$. Here φ is the angle between $\mathbf{v}$ and $\mathbf{e}_1$; that is, the angle between $\mathbf{v}$ and $\mathbf{u}$. Finally, due to (1.33), we easily derive that $\mathbf{u} \times \mathbf{v}$ is orthogonal to the plane defined by $\mathbf{u}, \mathbf{v}$ and its component along $\mathbf{e}_3$ is $|\mathbf{u}||\mathbf{v}| \sin \varphi$.

It remains to verify that the right-hand side of (1.33) is independent of the basis or, equivalently, that the quantities on the left-hand side of (1.33) transform with the law (1.27). First, from (1.29) we derive

$$\sqrt{g'} = \pm \det(A^i_j) \sqrt{g}, \tag{1.36}$$

where the $+$ sign holds if $\det(A^i_j) > 0$ and the $-$ sign holds in the opposite case. Moreover, the determinant definition implies that

$$e_{ijh} = \frac{1}{\det(A^q_p)} A^l_i A^m_j A^n_h \, e_{lmn}, \tag{1.37}$$

so that, taking into account (1.33) and (1.36), we write

$$(\mathbf{u} \times \mathbf{v})'_i = \sqrt{g'} \, e_{ijh} u'^j v'^h = \pm A^l_i \, e_{lmn} \sqrt{g} \, u^m v^n.$$

Then it is possible to conclude that the cross product defines a vector only for base changes which preserve the orientation (Section 1.4). Vectors having this characteristic are called **axial vectors** or **pseudovectors**, whereas the vectors transforming with the law (1.27) are called **polar vectors**.

Examples of axial vectors are given by the moment $\mathbf{M}_O$ of a force $\mathbf{F}$ with respect to a pole O or by the angular velocity ω, whose direction changes with the basis orientation. This is not true for the force, velocity, etc., which are examples of polar vectors (see Exercise 10).

At a less formal but more intuitive level, the nature of the vector under consideration can be recognized by looking at its behavior under a reflection (see Figure 1.3): if the reflection preserves the direction, we are in the presence of an axial vector, while if the reflection induces a change of its direction, the vector is polar.

1.6. Mixed Product

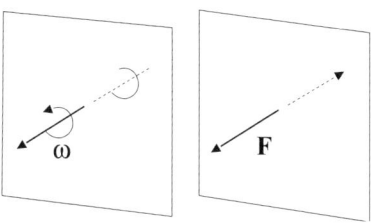

Figure 1.3

Axial vector ω (angular velocity)
Polar vector **F** (force)

1.6 Mixed Product

The **mixed product** of three vectors **u**, **v**, **w**, is the scalar quantity $(\mathbf{u} \times \mathbf{v}) \cdot \mathbf{w}$. When (1.24) and (1.33) are taken into account, we obtain the following coordinate form of the mixed product:

$$(\mathbf{u} \times \mathbf{v}) \cdot \mathbf{w} = \sqrt{g}\, e_{ijh} w^i u^j v^h. \tag{1.38}$$

The skew-symmetry properties of the Levi–Civita symbol (see Section 1.5) allow us to verify the following cyclic property of the mixed product:

$$(\mathbf{u} \times \mathbf{v}) \cdot \mathbf{w} = (\mathbf{w} \times \mathbf{u}) \cdot \mathbf{v} = (\mathbf{v} \times \mathbf{w}) \cdot \mathbf{u}.$$

It is easy to recognize the geometric meaning of the mixed product. In fact, it has been already said (Section 1.5) that the vector product is a vector orthogonal to the plane defined by **u** and **v**, having the norm equal to the area of the parallelogram σ with sides $|\mathbf{u}|$ and $|\mathbf{v}|$. The scalar product $(\mathbf{u} \times \mathbf{v}) \cdot \mathbf{w}$ is equal to the component **w** along $(\mathbf{u} \times \mathbf{v})$ times the length of $(\mathbf{u} \times \mathbf{v})$. Finally, the mixed product represents the volume of the parallelepiped with edges **u**, **v**, and **w**.

1.7 Elements of Tensor Algebra

Let $\mathcal{E}_3$ be a Euclidean vector space. The mapping

$$\mathbf{T} : \mathcal{E}_3 \to \mathcal{E}_3 \tag{1.39}$$

is *linear* if, using the notation $\mathbf{T}(\mathbf{u}) = \mathbf{Tu}$, we have

$$\mathbf{T}(a\mathbf{u} + b\mathbf{v}) = a\mathbf{Tu} + b\mathbf{Tv}, \tag{1.40}$$

for all $a, b \in \Re$ and for all $\mathbf{u}, \mathbf{v} \in \mathcal{E}_3$.

We will let $\mathrm{Lin}(\mathcal{E}_3)$ denote the set of linear mappings of $\mathcal{E}_3$ into $\mathcal{E}_3$. These applications are also called *endomorphisms*, **Euclidean double tensors**, or **2-*tensors*** of $\mathcal{E}_3$.

The *product* $\mathbf{ST}$ of two tensors $\mathbf{S}$ and $\mathbf{T}$ is the composition of $\mathbf{S}$ and $\mathbf{T}$; that is, for all $\mathbf{u} \in \mathcal{E}_3$ we have

$$(\mathbf{ST})\mathbf{u} = \mathbf{S}(\mathbf{Tu}). \tag{1.41}$$

Generally, $\mathbf{ST} \neq \mathbf{TS}$; if $\mathbf{ST} = \mathbf{TS}$, then the two tensors are said to *commute*. The *transpose* $\mathbf{T}^T$ of $\mathbf{T}$ is the new tensor defined by the condition

$$\mathbf{u} \cdot \mathbf{T}^T \mathbf{v} = \mathbf{v} \cdot (\mathbf{Tu}), \quad \forall \mathbf{u}, \mathbf{v} \in \mathcal{E}_3. \tag{1.42}$$

The following properties can be easily established:

$$\begin{aligned}(\mathbf{T} + \mathbf{S})^T &= \mathbf{T}^T + \mathbf{S}^T, \\ (\mathbf{TS})^T &= \mathbf{S}^T \mathbf{T}^T, \\ (\mathbf{T}^T)^T &= \mathbf{T}.\end{aligned}$$

A tensor $\mathbf{T}$ is *symmetric* if

$$\mathbf{T} = \mathbf{T}^T, \tag{1.43}$$

and *skew-symmetric* if

$$\mathbf{T} = -\mathbf{T}^T. \tag{1.44}$$

The *tensor product* of vectors $\mathbf{u}$ and $\mathbf{v}$ is the tensor $\mathbf{u} \otimes \mathbf{v}$ such that

$$(\mathbf{u} \otimes \mathbf{v})\mathbf{w} = \mathbf{u}\, \mathbf{v} \cdot \mathbf{w}, \quad \forall \mathbf{w} \in \mathcal{E}_3. \tag{1.45}$$

The set $\mathrm{Lin}(\mathcal{E}_3)$ of tensors defined on $\mathcal{E}_3$ is itself a vector space with respect to the ordinary operations of addition and the product of a mapping times a real number. For this space the following theorem holds:

1.7. Elements of Tensor Algebra

Theorem 1.3
The vector space $\text{Lin}(\mathcal{E}_3)$ of Euclidean double tensors is 9-dimensional. The tensor systems $(\mathbf{e}_i \otimes \mathbf{e}_j), (\mathbf{e}_i \otimes \mathbf{e}^j), (\mathbf{e}^i \otimes \mathbf{e}^j)$ are bases of $\text{Lin}(\mathcal{E}_3)$ so that for all $\mathbf{u} \in \mathcal{E}_3$, the following representations of Euclidean double tensors hold:

$$\mathbf{T} = T^{ij}\mathbf{e}_i \otimes \mathbf{e}_j = T^i_j \mathbf{e}_i \otimes \mathbf{e}^j = T_{ij}\mathbf{e}^i \otimes \mathbf{e}^j. \tag{1.46}$$

The matrices $(T^{ij}), (T^i_j)$, and (T_{ij}) are called the contravariant, mixed, and covariant components of $\mathbf{T}$, respectively. Finally, the relations among the different components are given by the equations

$$T_{ij} = g_{jh}T^h_i = g_{ih}g_{jk}T^{hk}. \tag{1.47}$$

PROOF To verify that the system $(\mathbf{e}_i \otimes \mathbf{e}_j)$ is a basis of $\text{Lin}(\mathcal{E}_3)$, it is sufficient to prove that it is a linearly independent system and that any tensor $\mathbf{T}$ can be expressed as a linear combination of the elements of $(\mathbf{e}_i \otimes \mathbf{e}_j)$. First, from

$$\lambda^{ij}\mathbf{e}_i \otimes \mathbf{e}_j = \mathbf{0},$$

and from (1.45), (1.12), we derive

$$\lambda^{ij}\mathbf{e}_i \otimes \mathbf{e}_j(\mathbf{e}_h) = \lambda^{ij}\mathbf{e}_i \, \mathbf{e}_j \cdot \mathbf{e}_h = \lambda^{ij}g_{jh}\mathbf{e}_i = \mathbf{0}$$

for any $h = 1, 2, 3$. But the vectors $(\mathbf{e}_i)$ form a basis of $\mathcal{E}_3$, so that the previous relation implies, for any index i, the homogeneous system

$$\lambda^{ij}g_{jh} = 0$$

of 3 equations in the three unknowns λ^{ij}, where i is fixed. Since the determinant $\det(g_{ij})$ of this system is different from zero, see (1.14), all the unknowns λ^{ij} with fixed i vanish. From the arbitrariness of i, the theorem follows.

Similarly, from the linear combination

$$\lambda^i_j \mathbf{e}_i \otimes \mathbf{e}^j = \mathbf{0},$$

when (1.19) is taken into account, we have

$$\lambda^i_j \mathbf{e}_i \otimes \mathbf{e}^j(\mathbf{e}_h) = \lambda^i_j \mathbf{e}_i \, \mathbf{e}^j \cdot \mathbf{e}_h = \lambda^i_j \delta^j_h \mathbf{e}_i = \lambda^i_h \mathbf{e}_i = \mathbf{0},$$

so that $\lambda^i_h = 0$, for any choice of i and h. In the same way, the independence of tensors $(\mathbf{e}^i \otimes \mathbf{e}^j)$ can be proved.

To show that any tensor $\mathbf{T}$ can be written as a linear combination of any one of these systems, we start by noting that, from (1.21) and the linearity of $\mathbf{T}$, we have

$$\mathbf{T}\mathbf{u} = u_j \mathbf{T}\mathbf{e}^j.$$

On the other hand, $\mathbf{T}\mathbf{e}^j$ is a vector of $\mathcal{E}_3$ and therefore can be represented in the following form, see (1.21):

$$\mathbf{T}\mathbf{e}^j = T^{ij}\mathbf{e}_i.$$

If the definition of covariant components (1.23) is recalled together with (1.45), we have

$$\mathbf{T}\mathbf{u} = T^{ij}u_j\mathbf{e}_i = T^{ij}\mathbf{e}_i \otimes \mathbf{e}_j(\mathbf{u}).$$

Finally, owing to the arbitrariness of $\mathbf{u}$, we obtain

$$\mathbf{T} = T^{ij}\mathbf{e}_i \otimes \mathbf{e}_j.$$

To show $(1.46)_2$, it is sufficient to remark that

$$\mathbf{T}\mathbf{u} = u^j\mathbf{T}\mathbf{e}_j,$$

so that, introducing the notation $\mathbf{T}\mathbf{e}_j = T^i_j\mathbf{e}_i$, we derive

$$\mathbf{T}\mathbf{u} = T^i_j u^j \mathbf{e}_i.$$

Noting that (1.23) and (1.22) hold, we have $u^j = g^{ji}u_i = g^{ji}\mathbf{e}_i \cdot \mathbf{u} = \mathbf{e}^j \cdot \mathbf{u}$, and the previous relation gives

$$\mathbf{T}\mathbf{u} = T^i_j \mathbf{e}_i \otimes \mathbf{e}^j(\mathbf{u}).$$

From this expression, due to the arbitrariness of $\mathbf{u}$, $(1.46)_2$ is derived. In a similar way, $(1.46)_3$ can be proved.

Relation (1.47) is easily verified using (1.46) and the definition of the reciprocal basis (1.18), so that the theorem is proved. ∎

Example 1.1
Verify that the matrices of contravariant or covariant components of a symmetric (skew-symmetric) tensor are symmetric (skew-symmetric), using the definition of transpose and symmetry.

The symmetry of $\mathbf{T}$ ($\mathbf{T} = \mathbf{T}^T$) and (1.42) imply that

$$\mathbf{v} \cdot \mathbf{T}\mathbf{u} = \mathbf{u} \cdot \mathbf{T}\mathbf{v},$$

so that

$$v_i T^{ij} u_j = u_i T^{ij} v_j = u_j T^{ji} v_i.$$

The arbitrariness of $\mathbf{u}$ and $\mathbf{v}$ leads to $T^{ij} = T^{ji}$. In a similar way, it can be proved that $T_{ij} = -T_{ji}$ when $\mathbf{T}$ is skew-symmetric.

1.7. Elements of Tensor Algebra

Example 1.2
Using the different representations (1.46), verify that the contravariant, mixed and covariant components of the unit or identity tensor **I**

$$\mathbf{I}\mathbf{u} = \mathbf{u}, \quad \forall \mathbf{u} \in \mathcal{E}_3,$$

are given by the following matrices, respectively:

$$(g^{ij}), \quad (\delta^i_j), \quad (g_{ij}).$$

Starting from the condition defining **I**, written in terms of components,

$$(I^{ij}\mathbf{e}_i \otimes \mathbf{e}_j) \cdot (u_h \mathbf{e}^h) = u_i \mathbf{e}^i,$$

we derive the equation

$$I^{ih} u_h \mathbf{e}_i = g^{hi} u_h \mathbf{e}_i,$$

which implies that $I^{ih} = g^{ih}$. In a similar way the others results follow.

The *image* of **T**, denoted by Im(**T**), is the subset of $\mathcal{E}_3$ such that

$$\text{Im}(\mathbf{T}) = \{\mathbf{v} \in \mathcal{E}_3 \mid \exists \mathbf{u} \in \mathcal{E}_3 : \mathbf{v} = \mathbf{T}\mathbf{u}\}, \tag{1.48}$$

whereas the *kernel* of **T**, denoted by Ker(**T**), is the subset of $\mathcal{E}_3$ such that

$$\text{Ker}(\mathbf{T}) = \{\mathbf{u} \in \mathcal{E}_3 \mid \mathbf{T}\mathbf{u} = \mathbf{0}\}. \tag{1.49}$$

Theorem 1.4
Im(**T**) *and* Ker(**T**) *are vector subspaces of* $\mathcal{E}_3$. *Moreover,* **T** *has an inverse if and only if* Ker(**T**) = $\{\mathbf{0}\}$.

PROOF   In fact, if $\mathbf{v}_1 = \mathbf{T}\mathbf{u}_1 \in \text{Im}(\mathbf{T})$ and $\mathbf{v}_2 = \mathbf{T}\mathbf{u}_2 \in \text{Im}(\mathbf{T})$, we have

$$a\mathbf{v}_1 + b\mathbf{v}_2 = a\mathbf{T}\mathbf{u}_1 + b\mathbf{T}\mathbf{u}_2 = \mathbf{T}(a\mathbf{u}_1 + b\mathbf{u}_2),$$

so that $a\mathbf{v}_1 + b\mathbf{v}_2 \in \text{Im}(\mathbf{T})$. Moreover, if $\mathbf{u}_1, \mathbf{u}_2 \in \text{Ker}(\mathbf{T})$, we have $\mathbf{T}\mathbf{u}_1 = \mathbf{T}\mathbf{u}_2 = \mathbf{0}$ and therefore

$$\mathbf{0} = a\mathbf{T}\mathbf{u}_1 + b\mathbf{T}\mathbf{u}_2 = \mathbf{T}(a\mathbf{u}_1 + b\mathbf{u}_2),$$

so that $a\mathbf{u}_1 + b\mathbf{u}_2 \in \text{Ker}(\mathbf{T})$. Finally, if **T** has an inverse, the condition $\mathbf{T}(\mathbf{0}) = \mathbf{0}$ implies that the inverse image of the zero vector of $\mathcal{E}_3$ contains only the zero vector, i.e., Ker(**T**) = $\{\mathbf{0}\}$. Conversely, if this condition implied the existence of two vectors $\mathbf{u}', \mathbf{u}'' \in \mathcal{E}_3$ such that $\mathbf{T}\mathbf{u}' = \mathbf{T}\mathbf{u}''$, we would have $\mathbf{T}(\mathbf{u}' - \mathbf{u}'') = \mathbf{0}$. From this relation it would follow that $\mathbf{u}' - \mathbf{u}'' \in \text{Ker}(\mathbf{T})$ and therefore $\mathbf{u}' - \mathbf{u}'' = \mathbf{0}$. ∎

As a consequence of the previous results, we can say that **T** is an isomorphism if and only if

$$\text{Im}(\mathbf{T}) = \mathcal{E}_3, \qquad \text{Ker}(\mathbf{T}) = \{\mathbf{0}\}. \tag{1.50}$$

Theorem 1.5
Let us consider a tensor $\mathbf{T} \in \text{Lin}(\mathcal{E}_3)$ *and a basis* $(\mathbf{e}_i)$ *of* $\mathcal{E}_3$. *Then the following conditions are equivalent:*

1. **T** *is an isomorphism;*

2. *the vectors* $\mathbf{T}(\mathbf{e}_i)$ *represent a basis* $\mathcal{E}_3$;

3. *the representative matrix* (T^i_j) *of* **T** *with respect to the basis* $(\mathbf{e}_i)$ *is not singular.*

PROOF

1. $\Longrightarrow$ 2.

    In fact, from the conditions $\lambda^i \mathbf{T}\mathbf{e}_i = \mathbf{T}(\lambda^i \mathbf{e}_i) = \mathbf{0}$ and $(1.50)_2$, it follows that $\lambda^i \mathbf{e}_i = \mathbf{0}$; consequently, $\lambda^i = 0$, since $(\mathbf{e}_i)$ is a basis and the vectors $\mathbf{T}\mathbf{e}_i$ are independent. Moreover, if **T** is an isomorphism, then for any $\mathbf{v} \in \mathcal{E}_3$ there exists a unique $\mathbf{u} \in \mathcal{E}_3$ such that $\mathbf{v} = \mathbf{T}\mathbf{u} = u^i \mathbf{T}\mathbf{e}_i$, so that the vectors $\mathbf{T}\mathbf{e}_i$ represent a basis of $\mathcal{E}_3$.

2. $\Longrightarrow$ 3.

    In fact, the condition $\lambda^i \mathbf{T}\mathbf{e}_i = \mathbf{0}$ can be written $\lambda^i T^h_i \mathbf{e}_h = \mathbf{0}$ so that, due to (2), the homogeneous system $\lambda^i T^h_i = 0$ must admit only the vanishing solution and therefore the matrix (T^i_j) is not singular.

3. $\Longrightarrow$ 1.

    In terms of components, $\mathbf{v} = \mathbf{T}\mathbf{u}$ is expressed by the system

    $$v^i = T^i_h u^h. \tag{1.51}$$

    For any choice of the vector $\mathbf{v} \in \mathcal{E}_3$, this system can be considered as a linear system of n equations with n unknowns u^h, which admits one and only one solution, when (T^i_j) is not singular. Consequently, **T** is an isomorphism. ∎

To conclude this section, the transformation rules under a base change $(\mathbf{e}_i) \longrightarrow (\mathbf{e}'_i)$ of the components of a Euclidean second-order tensor will be derived. First, from $(1.46)_1$, we have

$$\mathbf{T} = T^{ij} \mathbf{e}_i \otimes \mathbf{e}_j = T'^{hk} \mathbf{e}'_h \otimes \mathbf{e}'_k,$$

1.8. Eigenvalues and Eigenvectors

so that, recalling $(1.25)_2$, we find

$$(A^{-1})^h_i (A^{-1})^k_j T^{ij} \mathbf{e}'_h \otimes \mathbf{e}'_k = T'^{hk} \mathbf{e}'_h \otimes \mathbf{e}'_k.$$

Since the tensors $(\mathbf{e}'_h \otimes \mathbf{e}'_k)$ form a basis of $\text{Lin}(\mathcal{E}_3)$, we have

$$T'^{hk} = (A^{-1})^h_i (A^{-1})^k_j T^{ij}. \tag{1.52}$$

Similarly, starting from $(1.46)_{2,3}$ and taking into account (1.25), we can derive the following transformation formulae for mixed and covariant components:

$$T'^h_k = (A^{-1})^h_i A^j_k T^i_j, \tag{1.53}$$

$$T'_{hk} = A^i_h A^j_k T_{ij}. \tag{1.54}$$

Remark In the following sections we will often use the notation $\mathbf{u} \cdot \mathbf{T}$. It denotes the linear mapping $\mathcal{E}_3 \to \mathcal{E}_3$ that, in terms of components, is written

$$v^i = u^j T^i_j. \tag{1.55}$$

When the tensor $\mathbf{T}$ is symmetric, we have

$$\mathbf{u} \cdot \mathbf{T} = \mathbf{T}\mathbf{u}.$$

Remark It is worthwhile noting that the relations (1.51) to (1.54) can be written adopting matrix notation. For instance, in agreement with the convention that the product of two matrices is calculated by rows times columns, the following form can be given to (1.54):

$$T' = A^T T A, \tag{1.56}$$

where T and T' are the matrices formed with the components of $\mathbf{T}$ and $\mathbf{T}'$ with respect to the bases $(\mathbf{e}_i)$ and $(\mathbf{e}'_i)$, respectively, and A denotes the matrix of the base change $\mathbf{e}_i \to \mathbf{e}'_i$, see (1.25). It is also important to note that in the literature, usually, the same symbol denotes both the tensor $\mathbf{T}$ and its representative matrix T in a fixed basis. Consequently, relation (1.56) is also written as

$$\mathbf{T}' = \mathbf{A}^T \mathbf{T} \mathbf{A}.$$

1.8 Eigenvalues and Eigenvectors of a Euclidean Second-Order Tensor

A real number λ is an *eigenvalue* of the tensor $\mathbf{T}$ if there exists a nonvanishing vector $\mathbf{u}$ called an *eigenvector* belonging to the eigenvalue λ, satisfying the *eigenvalue equation*

$$\mathbf{Tu} = \lambda \mathbf{u}. \tag{1.57}$$

The equation (1.57) reflects the following problem, which has many applications: the tensor $\mathbf{T}$ associates to any vector $\mathbf{u}$ the vector $\mathbf{v}$, generally different from $\mathbf{u}$, both in norm and direction. The vectors which are not rotated by the linear application $\mathbf{T}$ are just the solutions of (1.57).

We easily recognize that the eigenvectors belonging to the same eigenvalue λ form a vector subspace of $\mathcal{E}_3$, called the *characteristic space* $V(\lambda)$ associated with the eigenvalue λ. The *geometric multiplicity* of λ is the dimension m of this subspace. The *spectrum* of $\mathbf{T}$ is the list $\lambda_1 \leq \lambda_2 \leq \ldots$ of its eigenvalues.

Theorem 1.6
The following properties hold:

1. *the eigenvalues of a positive definite tensor*

$$\mathbf{v} \cdot \mathbf{Tv} > 0 \quad \forall \mathbf{v} \in \mathcal{E}_3,$$

   are positive;

2. *the characteristic spaces of symmetric tensors are orthogonal to each other.*

PROOF If $\mathbf{T}$ is a symmetric positive definite tensor, λ its eigenvalue, and $\mathbf{u}$ one of the eigenvectors corresponding to λ, we have

$$\mathbf{u} \cdot \mathbf{Tu} = \lambda \, |\mathbf{u}|^2.$$

Since $|\mathbf{u}|^2 > 0$, the previous equation implies $\lambda > 0$. Moreover, if λ_1 and λ_2 are distinct eigenvalues of $\mathbf{T}$, and $\mathbf{u}_1$, $\mathbf{u}_2$ are the two corresponding eigenvectors, we can write

$$\mathbf{Tu}_1 = \lambda_1 \mathbf{u}_1, \quad \mathbf{Tu}_2 = \lambda_2 \mathbf{u}_2,$$

so that

$$\mathbf{u}_2 \cdot \mathbf{Tu}_1 = \lambda_1 \mathbf{u}_1 \cdot \mathbf{u}_2, \quad \mathbf{u}_1 \cdot \mathbf{Tu}_2 = \lambda_2 \mathbf{u}_1 \cdot \mathbf{u}_2.$$

1.8. Eigenvalues and Eigenvectors

From the symmetry ($\mathbf{T} = \mathbf{T}^T$) of $\mathbf{T}$, we derive $\mathbf{u}_2 \cdot \mathbf{T}\mathbf{u}_1 = \mathbf{u}_1 \cdot \mathbf{T}\mathbf{u}_2$. Consequently, we find
$$(\lambda_2 - \lambda_1)\mathbf{u}_1 \cdot \mathbf{u}_2 = 0,$$
and the theorem is proved, since $\lambda_2 - \lambda_1 \neq 0$. ∎

Let $(\mathbf{e}_i)$ be a fixed basis; when the representation of $\mathbf{T}$ in $(1.46)_1$ is used, the eigenvalue equation becomes
$$T^{ij}\mathbf{e}_i \otimes \mathbf{e}_j(\mathbf{u}) = T^{ij}\mathbf{e}_i\, \mathbf{e}_j \cdot \mathbf{u} = \lambda u^i \mathbf{e}_i,$$
or equivalently,
$$T^{ij} u_j \mathbf{e}_i = \lambda g^{ij} u_j \mathbf{e}_i.$$
The unknowns of this homogeneous system are the *covariant* components of the eigenvector $\mathbf{u}$:
$$(T^{ij} - \lambda g^{ij})u_j = 0. \tag{1.58}$$
This system has nonvanishing solutions if and only if the eigenvalue λ is a solution of the **characteristic equation**
$$\det(T^{ij} - \lambda g^{ij}) = 0, \tag{1.59}$$
which is a 3rd-degree equation in the unknown λ. The multiplicity of λ, as a root of (1.59), is called the **algebraic multiplicity** of the eigenvalue.

If the representation of $\mathbf{T}$ in $(1.46)_2$ is used, we have
$$T^i_j u^j \mathbf{e}_i = \lambda u^i \mathbf{e}_i = \lambda \delta^i_j u^j \mathbf{e}_i,$$
so that the eigenvalue equation, written in mixed components, becomes
$$(T^i_j - \lambda \delta^i_j)u^j = 0, \tag{1.60}$$
and the characteristic equation is
$$\det(T^i_j - \lambda \delta^i_j) = 0. \tag{1.61}$$

Finally, using the representation $(1.46)_3$, we have
$$(T_{ij} - \lambda g_{ij})u^j = 0, \tag{1.62}$$
$$\det(T_{ij} - \lambda g_{ij}) = 0. \tag{1.63}$$

Remark Developing (1.61), we obtain a 3rd-degree equation having the following form:
$$-\lambda^3 + I_1 \lambda^2 - I_2 \lambda + I_3 = 0, \tag{1.64}$$

where the coefficients I_1, I_2, and I_3 are called the **first, second**, and **third invariant**, respectively, and are defined as follows:

$$I_1 = T_i^i, \qquad I_2 = \frac{1}{2}(T_i^i T_j^j - T_j^i T_i^j), \qquad I_3 = \det(T_j^i). \tag{1.65}$$

It is clear that the roots of equation (1.64) are independent of the basis; that is, the eigenvalues are characteristic of the tensor **T** if and only if the coefficients of the characteristic equation (1.64) are invariant with respect to base changes. We leave to reader to prove this property (see hint in Exercise 1). If another kind of components is used, the expressions for the invariant coefficients change. For example, for covariant components, the expressions become

$$I_1 = g^{ij} T_{ij}, \qquad I_2 = \frac{1}{2}(g^{ij} T_{ij} g^{hk} T_{hk} - g^{ih} T_{hj} g^{jk} T_{ik}), \tag{1.66}$$

$$I_3 = \det(g^{ih} T_{hj}) = \det(g^{ih}) \det(T_{hj}). \tag{1.67}$$

Theorem 1.7
*If **T** is a symmetric tensor, then*

1. *the geometric and algebraic multiplicities of an eigenvalue λ coincide;*

2. *it is always possible to determine at least one orthonormal basis $(\mathbf{u}_i)$ whose elements are eigenvectors of **T**;*

3. *in these bases, one of the following representation of **T** is possible:*

   i) if all the eigenvalues are distinct, then

$$\mathbf{T} = \sum_{i=1}^{3} \lambda_i \mathbf{u}_i \otimes \mathbf{u}_i; \tag{1.68}$$

ii) if $\lambda_1 = \lambda_2 \neq \lambda_3$, then

$$\mathbf{T} = \lambda_1 (\mathbf{I} - \mathbf{u}_3 \otimes \mathbf{u}_3) + \lambda_3 \mathbf{u}_3 \otimes \mathbf{u}_3; \tag{1.69}$$

iii) if $\lambda_1 = \lambda_2 = \lambda_3 \equiv \lambda$, then

$$\mathbf{T} = \lambda \mathbf{I}, \tag{1.70}$$

*where **I** is the identity tensor of $\mathcal{E}_3$.*

PROOF We start by assuming that the property (1) has already been proved. Consequently, three cases are possible: the three eigenvalues are distinct, two are equal but different from the third one, or all of them are

1.8. Eigenvalues and Eigenvectors

equal. In the first case, the eigenvector subspace $V(\lambda_i)$ belonging to the eigenvalue λ_i is one-dimensional and orthogonal to the others (see Theorem 1.5). Therefore, if a unit eigenvector $\mathbf{u}_i$ is fixed in each of these subspaces, an orthonormal basis is obtained.

In the second case, the eigenvalue $\lambda_1 = \lambda_2$ is associated with a two-dimensional eigenvector subspace $V(\lambda_1)$, so that it is always possible to find an orthonormal eigenvector pair $(\mathbf{u}_1, \mathbf{u}_2)$ in $V(\lambda_1)$. If a third unit vector $\mathbf{u}_3$ is chosen in the one-dimensional eigenvector subspace $V(\lambda_3)$, orthogonal to the previous one, an orthonormal eigenvector basis $(\mathbf{u}_i)$ is obtained.

Finally, if all the eigenvalues are equal to λ, the subspace $V(\lambda)$ is three-dimensional that is, it coincides with the whole space $\mathcal{E}_3$ and it is certainly possible to determine an orthonormal eigenvector basis.

In conclusion, an orthonormal eigenvector basis $(\mathbf{u}_1, \mathbf{u}_2, \mathbf{u}_3)$ can always be found. Since this basis is orthonormal, the contravariant, mixed, and covariant components of any tensor $\mathbf{T}$ coincide; moreover, the reciprocal basis is identical to $(\mathbf{u}_1, \mathbf{u}_2, \mathbf{u}_3)$ and the tensor g_{ij} has components δ_{ij}.

Finally, all types of components of the tensor $\mathbf{T}$ in the basis $(\mathbf{u}_1, \mathbf{u}_2, \mathbf{u}_3)$ are given by the diagonal matrix

$$(T^{ij}) = \begin{pmatrix} \lambda_1 & 0 & 0 \\ 0 & \lambda_2 & 0 \\ 0 & 0 & \lambda_3 \end{pmatrix}. \tag{1.71}$$

In fact, using any one of the representation formulae (1.46), the eigenvalue equation $\mathbf{T}\mathbf{u}_h = \lambda_h \mathbf{u}_h$ can be written in the form

$$T^{ij}\delta_{hj}\mathbf{u}_i = \lambda_h \delta_h^i \mathbf{u}_i,$$

from which we derive

$$T^{ih} = \lambda_h \delta_h^i,$$

and (1.68) is proved. In the case (ii) (1.68) becomes

$$\mathbf{T} = \lambda_1(\mathbf{u}_1 \otimes \mathbf{u}_1 + \mathbf{u}_2 \otimes \mathbf{u}_2) + \lambda_3 \mathbf{u}_3 \otimes \mathbf{u}_3.$$

But the unit matrix can be written as

$$\mathbf{I} = \mathbf{u}_1 \otimes \mathbf{u}_1 + \mathbf{u}_2 \otimes \mathbf{u}_2 + \mathbf{u}_3 \otimes \mathbf{u}_3 \tag{1.72}$$

and (1.69) is also verified. Finally, (1.70) is a trivial consequence of (1.68) and (1.72). ∎

The relation (1.68) is called a ***spectral decomposition*** of $\mathbf{T}$.

The following square-root theorem is important in finite deformation theory (see Chapter 3):

Theorem 1.8
If $\mathbf{T}$ is a positive definite symmetric tensor, then one and only one positive definite symmetric tensor $\mathbf{U}$ exists such that

$$\mathbf{U}^2 = \mathbf{T}. \tag{1.73}$$

PROOF Since $\mathbf{T}$ is symmetric and positive definite from the previous theorem it follows that

$$\mathbf{T} = \sum_{i=1}^{3} \lambda_i \mathbf{u}_i \otimes \mathbf{u}_i,$$

where $\lambda_i > 0$ and $(\mathbf{u}_i)$ is an orthonormal basis. Consequently, we can define the symmetric positive definite tensor

$$\mathbf{U} = \sum_{i=1}^{3} \sqrt{\lambda_i} \mathbf{u}_i \otimes \mathbf{u}_i. \tag{1.74}$$

In order to verify that $\mathbf{U}^2 = \mathbf{T}$, it will be sufficient to prove that they coincide when applied to eigenvectors. In fact,

$$\mathbf{U}^2 \mathbf{u}_h = \left(\sum_{i=1}^{3} \sqrt{\lambda_i} \mathbf{u}_i \otimes \mathbf{u}_i \right) \left(\sum_{i=1}^{3} \sqrt{\lambda_i} \mathbf{u}_i \otimes \mathbf{u}_i \right) \mathbf{u}_h$$

$$= \left(\sum_{i=1}^{3} \sqrt{\lambda_i} \mathbf{u}_i \otimes \mathbf{u}_i \right) \sqrt{\lambda_h} \mathbf{u}_h = \lambda_h \mathbf{u}_h = \mathbf{T} \mathbf{u}_h.$$

To verify that $\mathbf{U}$ is uniquely defined, suppose that there is another symmetric positive definite tensor $\mathbf{V}$ such that $\mathbf{V}^2 = \mathbf{T}$. Then we have

$$(\mathbf{U}^2 - \mathbf{V}^2)(\mathbf{u}_i) = 0 \Rightarrow \mathbf{U}^2 = \mathbf{V}^2.$$

Recalling (1.74) and the spectral representation (1.68), we conclude that the eigenvalues of $\mathbf{V}$ are $\pm\sqrt{\lambda_i}$. But $\mathbf{V}$ is also supposed to be symmetric and positive definite so that it coincides with $\mathbf{U}$. ∎

1.9 Orthogonal Tensors

A 2-tensor $\mathbf{Q}$ is *orthogonal* if, for any pair of vectors in $\mathcal{E}_3$,

$$\mathbf{Qu} \cdot \mathbf{Qv} = \mathbf{u} \cdot \mathbf{v}. \tag{1.75}$$

1.9. Orthogonal Tensors

In other words, a 2-tensor is orthogonal if, for any pair of vectors **u**, **v**,

- the length of **Qu** is equal to the length of **u**; and
- the angle between **Qu** and **Qv** is equal to the angle between **u** and **v**.

In particular, an orthogonal tensor transforms an orthonormal basis $(\mathbf{e}_i)$ into a system of orthonormal vectors $(\mathbf{Q}\mathbf{e}_i)$. In Section 1.3 it was noted that such a system is linearly independent, so that $(\mathbf{Q}\mathbf{e}_i)$ is again a basis. Moreover, from Theorem 1.4 we can say that an orthogonal tensor is an isomorphism and its representative matrix $Q = (Q_i^j)$ in a given basis $(\mathbf{e}_i)$ is not singular:

$$\det Q \neq 0. \tag{1.76}$$

In terms of this matrix, the condition (1.75) becomes

$$g_{hk} Q_i^h Q_j^k = g_{ij}, \tag{1.77}$$

which implies the following result: In any orthonormal basis $(\mathbf{e}_i)$, an orthogonal tensor is represented by an orthogonal matrix:

$$Q^T Q = I \Leftrightarrow Q^T = Q^{-1}$$
$$\det Q = \pm 1. \tag{1.78}$$

It is clear that the composition of two orthogonal tensors leads again to an orthogonal tensor, that the identity is an orthogonal tensor, and that the inverse of an orthogonal tensor is again orthogonal. Therefore, the set of all the orthogonal tensors is a group. We note that these tensors are also called **rotations**. In particular those rotations for which $\det Q = 1$ are called **proper rotations**. Usually, the group of rotations is denoted by $O(3)$, and the subgroup of proper rotations is denoted by $SO(3)$.

The following is very important:

Theorem 1.9
A nonidentical rotation **Q** *always has the eigenvalue* $\lambda = 1$ *with geometric multiplicity equal to 1; i.e., a one-dimensional subspace* A *exists which is invariant with respect to* **Q**. *Moreover, if* $A_\perp$ *is the subspace of all the vectors which are orthogonal to* A *and* $\mathbf{Q}_\perp \neq \mathbf{I}$ *is the restriction of* **Q** *to* $A_\perp$, *then* $\lambda = 1$ *is the only eigenvalue.*

PROOF The eigenvalue equation for **Q** in an orthonormal basis $(\mathbf{e}_i)$ is:

$$(Q_j^i - \lambda \delta_j^i) u^j = 0, \tag{1.79}$$

where (Q_j^i) is orthogonal, and the characteristic equation is

$$P_3(\lambda) = -\lambda^3 + I_1 \lambda^2 - I_2 \lambda + I_3 = 0, \tag{1.80}$$

with the invariants given by the following expressions:

$$I_1 = Q_1^1 + Q_2^2 + Q_3^3,$$
$$I_2 = \det\begin{pmatrix} Q_2^2 & Q_2^3 \\ Q_3^2 & Q_3^3 \end{pmatrix} + \det\begin{pmatrix} Q_1^1 & Q_1^3 \\ Q_3^1 & Q_3^3 \end{pmatrix} + \det\begin{pmatrix} Q_1^1 & Q_1^2 \\ Q_2^1 & Q_2^2 \end{pmatrix},$$
$$I_3 = \det(Q_i^j) = 1. \tag{1.81}$$

We note that the determinants appearing in (1.81) are the cofactors A_j^i of Q_1^1, Q_2^2, and Q_3^3, respectively. On the other hand, the condition $\det(Q_i^j) = 1$ implies that

$$A_j^i = (Q^{-1})_j^i = (Q^T)_j^i$$

that is,

$$A_1^1 = Q_1^1, \quad A_2^2 = Q_2^2, \quad A_3^3 = Q_3^3. \tag{1.82}$$

Therefore, the characteristic equation becomes

$$P_3(\lambda) = -\lambda^3 + I_1 \lambda^2 - I_1 \lambda + 1 = 0 \tag{1.83}$$

and $\lambda = 1$ is a root.

Next, we have to verify that the eigenspace A belonging to $\lambda = 1$ is one-dimensional. Since the dimension of A is equal to $3 - p$, where $p \leq 2$ is the rank of the matrix $(Q_j^i - \delta_j^i)$, we have to prove that $p = 2$. The value $p = 0$ is not admissible since in such a case Q would be the unit matrix. If p were equal to 1, all the 2×2–minors of the above matrix should vanish:

$$\det\begin{pmatrix} Q_1^1 - 1 & Q_1^2 \\ Q_2^1 & Q_2^2 - 1 \end{pmatrix} = A_3^3 - Q_1^1 - Q_2^2 + 1 = 0,$$

$$\det\begin{pmatrix} Q_1^1 - 1 & Q_1^3 \\ Q_3^1 & Q_3^3 - 1 \end{pmatrix} = A_2^2 - Q_1^1 - Q_3^3 + 1 = 0,$$

$$\det\begin{pmatrix} Q_2^2 - 1 & Q_2^3 \\ Q_3^2 & Q_3^3 - 1 \end{pmatrix} = A_1^1 - Q_2^2 - Q_3^3 + 1 = 0.$$

These relations and (1.82) would imply $Q_1^1 = Q_2^2 = Q_3^3 = 1$. From this result and the orthogonality conditions

$$(Q_1^1)^2 + (Q_1^2)^2 + (Q_1^3)^2 = 1,$$
$$(Q_2^1)^2 + (Q_2^2)^2 + (Q_2^3)^2 = 1,$$
$$(Q_3^1)^2 + (Q_3^2)^2 + (Q_3^3)^2 = 1,$$

it should follow that $Q_j^i = \delta_j^i$, which violates the hypothesis that Q is not the unit matrix.

1.9. Orthogonal Tensors

It remains to prove that $\lambda = 1$ is the only eigenvalue of $\mathbf{Q}$. First, we note that $A_\perp$ is a 2-dimensional subspace. If $(\mathbf{u}_1, \mathbf{u}_2)$ is an orthonormal basis of $A_\perp$ and $\mathbf{u}_3$ is a unit vector of A, it is clear that $(\mathbf{u}_1, \mathbf{u}_2, \mathbf{u}_3)$ is an orthonormal basis of the whole vector space $\mathcal{E}_3$. Since the rotation $\mathbf{Q}$ does not modify the scalar product of two vectors, we have

$$\mathbf{Qu}_1 = \overline{Q}_1^1 \mathbf{u}_1 + \overline{Q}_1^2 \mathbf{u}_2,$$
$$\mathbf{Qu}_2 = \overline{Q}_2^1 \mathbf{u}_1 + \overline{Q}_2^2 \mathbf{u}_2,$$
$$\mathbf{Qu}_3 = \mathbf{u}_3.$$

Therefore, the representative matrix $\overline{Q}$ of $\mathbf{Q}$ in the basis $(\mathbf{u}_1, \mathbf{u}_2, \mathbf{u}_3)$ is

$$\overline{Q} = \begin{pmatrix} \overline{Q}_1^1 & \overline{Q}_1^2 & 0 \\ \overline{Q}_2^1 & \overline{Q}_2^2 & 0 \\ 0 & 0 & 1 \end{pmatrix},$$

and the orthogonality conditions are

$$(\overline{Q}_1^1)^2 + (\overline{Q}_1^2)^2 = 1,$$
$$(\overline{Q}_2^1)^2 + (\overline{Q}_2^2)^2 = 1,$$
$$\overline{Q}_1^1 \overline{Q}_2^1 + \overline{Q}_1^2 \overline{Q}_2^2 = 0.$$

These relations imply the existence of an angle $\varphi \in (0, 2\pi)$ such that

$$\overline{Q} = \begin{pmatrix} \cos\varphi & -\sin\varphi & 0 \\ \sin\varphi & \cos\varphi & 0 \\ 0 & 0 & 1 \end{pmatrix},$$

(note that $\varphi = 0$ and $\varphi = 2\pi$ correspond to the unit matrix). Therefore, in the basis $(\mathbf{u}_1, \mathbf{u}_2, \mathbf{u}_3)$ the characteristic equation becomes

$$P_3(\lambda) = (1 - \lambda)(\lambda^2 - 2\lambda \cos\varphi + 1) = 0,$$

so that, if $\mathbf{Q}_\perp \neq \mathbf{I}$ (i.e. $\varphi \neq \pi$), then the only real eigenvalue is $\lambda = 1$. ∎

The one-dimensional eigenspace belonging to the eigenvalue $\lambda = 1$ is called the *rotation axis*.

1.10 Cauchy's Polar Decomposition Theorem

In this section we prove a very important theorem, called *Cauchy's polar decomposition theorem*.

Theorem 1.10

Let $\mathbf{F}$ be a 2-tensor for which $\det \mathbf{F} > 0$ in a Euclidean vector space $\mathcal{E}_3$. Then there exist positive definite, symmetric tensors $\mathbf{U}, \mathbf{V}$ and a proper rotation $\mathbf{R}$ such that

$$\mathbf{F} = \mathbf{RU} = \mathbf{VR}, \qquad (1.84)$$

where $\mathbf{U}, \mathbf{V},$ and $\mathbf{R}$ are uniquely defined by the relations

$$\mathbf{U} = \sqrt{\mathbf{F}^T \mathbf{F}}, \qquad \mathbf{V} = \sqrt{\mathbf{F}\mathbf{F}^T},$$
$$\mathbf{R} = \mathbf{F}\mathbf{U}^{-1} = \mathbf{V}^{-1}\mathbf{F}. \qquad (1.85)$$

PROOF To simplify the notation, an orthonormal basis $(\mathbf{e}_1, \mathbf{e}_2, \mathbf{e}_3)$ is chosen in $\mathcal{E}_3$. Moreover, we denote by T the representative matrix of any 2-tensor $\mathbf{T}$ with respect to the basis $(\mathbf{e}_1, \mathbf{e}_2, \mathbf{e}_3)$.

First, we will prove that the decompositions are unique. In fact, using the matrix notation, if $F = R'U' = V'R'$ are any other two decompositions, we obtain

$$F^T F = U' R'^T R' U' = U'^2,$$

so that, from Theorem 1.8, U' is uniquely determined and coincides with $(1.85)_1$. Consequently, $R' = F(U')^{-1} = FU^{-1} = R$ and then $V' = R'U'R'^T = RUR^T = V$.

To prove the existence, we note that the tensor $C = F^T F$ is symmetric and positive definite. Therefore, Theorem 1.8 implies that $\mathbf{U}$ exists and that it is the only tensor satisfying the equation $U^2 = C$. Choosing $R = FU^{-1}$, we obtain $F = RU$. The following relations prove that R is a proper rotation:

$$R^T R = U^{-1} F^T F U^{-1} = U^{-1} U^2 U^{-1} = I,$$
$$\det R = \det F \det U^{-1} > 0.$$

Finally, the second decomposition is obtained by taking $V = RUR^T$. ∎

1.11 Higher Order Tensors

The definition of a Euclidean second-order tensor introduced in the previous section is generalized here to Euclidean higher order tensors. A **3rd-order Euclidean tensor** or, in short, a **3-tensor**, is a mapping

$$\mathbf{T} : \mathcal{E}_3 \times \mathcal{E}_3 \longrightarrow \mathcal{E}_3,$$

which is linear with respect to each variable. The Levi–Civita symbol introduced by (1.34) is an example of a 3rd-order tensor.

If a basis $(\mathbf{e}_i)$ is introduced in $\mathcal{E}_3$, then for all $\mathbf{u}, \mathbf{v} \in \mathcal{E}_3$, we have the relation

$$\mathbf{T}(\mathbf{u}, \mathbf{v}) = \mathbf{T}(u^i \mathbf{e}_i, v^j \mathbf{e}_j) = u^i v^j \mathbf{T}(\mathbf{e}_i, \mathbf{e}_j),$$

from which, introducing the notation

$$\mathbf{T}(\mathbf{e}_i, \mathbf{e}_j) = T_{ij}^k \mathbf{e}_k,$$

we also have

$$\mathbf{T}(\mathbf{u}, \mathbf{v}) = T_{ij}^k u^i v^j \mathbf{e}_k. \tag{1.86}$$

The **tensor product** of the vectors $\mathbf{u}, \mathbf{v}, \mathbf{w}$ is the following 3rd-order tensor:

$$\mathbf{u} \otimes \mathbf{v} \otimes \mathbf{w}(\mathbf{x}, \mathbf{y}) = \mathbf{u}(\mathbf{v} \cdot \mathbf{x})(\mathbf{w} \cdot \mathbf{y}) \qquad \forall \mathbf{x}, \mathbf{y} \in \mathcal{E}_3. \tag{1.87}$$

From (1.23), equation (1.86) becomes

$$\mathbf{T}(\mathbf{u}, \mathbf{v}) = T_{ij}^k (\mathbf{e}^i \cdot \mathbf{u})(\mathbf{e}^j \cdot \mathbf{v}) \mathbf{e}_k = T_{ij}^k \mathbf{e}_k \otimes \mathbf{e}^i \otimes \mathbf{e}^j (\mathbf{u}, \mathbf{v}),$$

so that, from the arbitrariness of vectors $\mathbf{u}, \mathbf{v}$, we obtain

$$\mathbf{T} = T_{ij}^k \mathbf{e}_k \otimes \mathbf{e}^i \otimes \mathbf{e}^j. \tag{1.88}$$

Relation (1.86) can be written using different types of components of the tensor $\mathbf{T}$. For example, recalling the definition of reciprocal basis, we derive

$$\mathbf{T} = T_j^{lk} \mathbf{e}_l \otimes \mathbf{e}^j \otimes \mathbf{e}_k = T^{lmk} \mathbf{e}_l \otimes \mathbf{e}_m \otimes \mathbf{e}_k = T_{pqr} \mathbf{e}^p \otimes \mathbf{e}^q \otimes \mathbf{e}^r,$$

where

$$T_j^{lk} = g^{li} T_{ij}^k, \qquad T^{lmk} = g^{li} g^{mj} T_{ij}^k, \qquad T_{pqr} = g_{pk} T_{qr}^k,$$

denote the different components of the tensor.

A 3rd-order tensor can be also defined as a linear mapping

$$\mathbf{T} : \mathcal{E}_3 \longrightarrow \mathrm{Lin}(\mathcal{E}_3), \tag{1.89}$$

by again introducing the tensor product (1.87) as follows:

$$\mathbf{u} \otimes \mathbf{v} \otimes \mathbf{w}(\mathbf{x}) = (\mathbf{w} \cdot \mathbf{x})\mathbf{u} \otimes \mathbf{v}.$$

In fact, from the equation

$$\mathbf{T}\mathbf{u} = u^i \mathbf{T}\mathbf{e}_i$$

and the representation

$$\mathbf{T}\mathbf{e}_i = T_i^{jk} \mathbf{e}_j \otimes \mathbf{e}_k,$$

we derive

$$\mathbf{T} = T_i^{jk} \mathbf{e}_j \otimes \mathbf{e}_k \otimes \mathbf{e}^i.$$

Finally, a 4th-order tensor is a linear mapping $\mathbf{T} : \mathcal{E}_3 \times \mathcal{E}_3 \times \mathcal{E}_3 \longrightarrow \mathcal{E}_3$ or, equivalently, a linear mapping $\mathbf{T} : \text{Lin}(\mathcal{E}_3) \longrightarrow \text{Lin}(\mathcal{E}_3)$. Later, we will prove that, in the presence of small deformations of an elastic system, the deformation state is described by a symmetric 2-tensor E_{hk}, while the corresponding stress state is expressed by another symmetric tensor T_{hk}. Moreover, the relation between these two tensors is linear:

$$T_{hk} = \mathbb{C}_{hk}^{lm} E_{lm},$$

where the 4th-order tensor $\mathbb{C}_{hk}^{lm}$ is called the linear elasticity tensor.

1.12 Euclidean Point Space

In this section a mathematical model of the ordinary three-dimensional Euclidean space will be given. A set E_n, whose elements will be denoted by capital letters $A, B, \ldots$, will be called an *n-dimensional affine space* if a mapping

$$f : (A, B) \in E_n \times E_n \to f(A, B) \equiv \overrightarrow{AB} \in \mathcal{E}_n, \tag{1.90}$$

exists between the pairs of elements of E_n and the vectors of a real n-dimensional vector space $\mathcal{E}_n$, such that

1.
$$\overrightarrow{AB} = -\overrightarrow{BA},$$
$$\overrightarrow{AB} = \overrightarrow{AC} + \overrightarrow{CB}; \tag{1.91}$$

2. for any $O \in E_n$ and any vector $\mathbf{u} \in \mathcal{E}_n$, one and only one element $P \in E_n$ exists such that

$$\overrightarrow{OP} = \mathbf{u}. \tag{1.92}$$

1.12. Euclidean Point Space

Then the elements of E_n will be called **points**; moreover, the points A, B of E_n that correspond to $\overrightarrow{AB}$, are called the initial and final point of $\overrightarrow{AB}$, respectively.

A **frame of reference** $(O, \mathbf{e}_i)$ in an n-dimensional affine space E_n is the set of a point $O \in E_n$ and a basis $(\mathbf{e}_i)$ of $\mathcal{E}_n$. The point O is called the **origin** of the frame. For a fixed frame, the relation (1.92) can be written as

$$\overrightarrow{OP} = u^i \mathbf{e}_i, \qquad (1.93)$$

where the real numbers u^i denote the contravariant components of $\mathbf{u}$ in the basis $(\mathbf{e}_i)$. These components are also called the **rectilinear coordinates** of P in the frame $(O, \mathbf{e}_i)$. In order to derive the transformation formulae of the rectilinear coordinates (Section 1.1) of a point $P \in E_3$ for the frame change $(O, \mathbf{e}_i) \longrightarrow (O', \mathbf{e}'_i)$, we note that

$$\overrightarrow{OP} = \overrightarrow{OO'} + \overrightarrow{O'P}. \qquad (1.94)$$

Representing the vectors of (1.94) in both the frames, we have

$$u^i \mathbf{e}_i = u^i_{O'} \mathbf{e}_i + u'^j \mathbf{e}'_j,$$

where $u^i_{O'}$ denote the coordinates of O' with respect to $(O, \mathbf{e}_i)$. Recalling (1.25), from the previous relation we derive

$$u^i = u^i_{O'} + A^i_j u'^j. \qquad (1.95)$$

Finally, if $\mathcal{E}_n$ is a Euclidean vector space, then E_n is called a **Euclidean point space**. In this case, the distance between two points A and B can be defined as the length of the unique vector $\mathbf{u}$ corresponding to the pair (A, B). To determine the expression for the distance in terms of the rectilinear coordinates of the above points, we note that, in a given frame of reference $(O, \mathbf{e}_i)$, we have:

$$\overrightarrow{AB} = \overrightarrow{AO} + \overrightarrow{OB} = \overrightarrow{OB} - \overrightarrow{OA} = (u^i_B - u^i_A)\mathbf{e}_i, \qquad (1.96)$$

where u^i_B, u^i_A denote the rectilinear coordinates of B and A, respectively. Therefore, the distance $|AB|$ can be written as

$$|AB| = \sqrt{g_{ij}(u^i_B - u^i_A)(u^j_B - u^j_A)}, \qquad (1.97)$$

and it assumes the Pitagoric form

$$|AB| = \sqrt{\sum_{i=1}^{n}(u^i_B - u^i_A)^2} \qquad (1.98)$$

when the basis $(\mathbf{e}_i)$ is orthonormal.

1.13 Exercises

1. Prove the invariance under base changes of the coefficients of the characteristic equation (1.64). Use the transformation formulae (1.53) of the mixed components of a tensor.

2. Using the properties of the cross product, determine the reciprocal basis $(\mathbf{e}^i)$ of $(\mathbf{e}_i), i = 1, 2, 3$.

   The vectors of the reciprocal basis $(\mathbf{e}^i)$ satisfy the condition (1.19)

   $$\mathbf{e}^i \cdot \mathbf{e}_j = \delta^i_j,$$

   which constitutes a linear system of n^2 equations in the n^2 unknowns represented by the components of the vectors $\mathbf{e}^i$. For $n = 3$, the reciprocal vectors are obtained by noting that, since $\mathbf{e}^1$ is orthogonal to both $\mathbf{e}_2$ and $\mathbf{e}_3$, we can write

   $$\mathbf{e}^1 = k(\mathbf{e}_2 \times \mathbf{e}_3).$$

   On the other hand, we also have

   $$\mathbf{e}^1 \cdot \mathbf{e}_1 = k(\mathbf{e}_2 \times \mathbf{e}_3) \cdot \mathbf{e}_1 = 1,$$

   so that

   $$\frac{1}{k} = \mathbf{e}_1 \cdot \mathbf{e}_2 \times \mathbf{e}_3.$$

   Finally,

   $$\mathbf{e}^1 = k\mathbf{e}_2 \times \mathbf{e}_3, \quad \mathbf{e}^2 = k\mathbf{e}_3 \times \mathbf{e}_1, \quad \mathbf{e}^3 = k\mathbf{e}_1 \times \mathbf{e}_2.$$

   These relations also show that $k = 1$ and the reciprocal basis coincides with $(\mathbf{e}_i)$, when this is orthonormal.

3. Evaluate the eigenvalues and eigenvectors of the tensor $\mathbf{T}$ whose components in the orthonormal basis $(\mathbf{e}_1, \mathbf{e}_2, \mathbf{e}_3)$ are

   $$\begin{pmatrix} 4 & 2 & -1 \\ 2 & 4 & 1 \\ -1 & 1 & 3 \end{pmatrix}. \tag{1.99}$$

   The eigenvalue equation is

   $$\mathbf{T}\mathbf{u} = \lambda \mathbf{u}.$$

1.13. Exercises

This has nonvanishing solutions if and only if the eigenvalue λ is a solution of the characteristic equation

$$\det(\mathbf{T} - \lambda \mathbf{I}) = \mathbf{0}. \tag{1.100}$$

In our case, the relation (1.100) requires that

$$\det \begin{pmatrix} 4-\lambda & 2 & -1 \\ 2 & 4-\lambda & 1 \\ -1 & 1 & 3-\lambda \end{pmatrix} = 0,$$

corresponding to the following 3rd-degree algebraic equation in the unknown λ:

$$-\lambda^3 + 11\lambda^2 - 34\lambda + 24 = 0,$$

whose solutions are

$$\lambda_1 = 1, \quad \lambda_2 = 4, \quad \lambda_3 = 6.$$

The components of the corresponding eigenvectors can be obtained by solving the following homogeneous system:

$$T^{ij} u_j = \lambda \delta^{ij} u_j. \tag{1.101}$$

For $\lambda = 1$, equations (1.101) become

$$\begin{aligned} 3u_1 + 2u_2 - u_3 &= 0, \\ 2u_1 + 3u_2 + u_3 &= 0, \\ -u_1 + u_2 + 2u_3 &= 0. \end{aligned}$$

Imposing the normalization condition

$$u_i u_i = 1$$

to these equations gives

$$u_1 = \frac{1}{\sqrt{3}}, \quad u_2 = -\frac{1}{\sqrt{3}}, \quad u_3 = \frac{1}{\sqrt{3}}.$$

Proceeding in the same way for $\lambda = 4$ and $\lambda = 6$, we obtain the components of the other two eigenvectors:

$$\left(-\frac{1}{\sqrt{6}}, \frac{1}{\sqrt{6}}, \frac{2}{\sqrt{6}}\right), \quad \left(\frac{1}{\sqrt{2}}, \frac{1}{\sqrt{2}}, 0\right).$$

From the symmetry of $\mathbf{T}$, the three eigenvectors are orthogonal (verify).

4. Let $\mathbf{u}$ and λ be the eigenvectors and eigenvalues of the tensor $\mathbf{T}$. Determine the eigenvectors and eigenvalues of $\mathbf{T}^{-1}$.

   If $\mathbf{u}$ is an eigenvector of $\mathbf{T}$, then

   $$\mathbf{T}\mathbf{u} = \lambda \mathbf{u}.$$

   Multiplying by $\mathbf{T}^{-1}$, we obtain the condition

   $$\mathbf{T}^{-1}\mathbf{u} = \frac{1}{\lambda}\mathbf{u},$$

   which shows that $\mathbf{T}^{-1}$ and $\mathbf{T}$ have the same eigenvectors, while the eigenvalues of $\mathbf{T}^{-1}$ are the reciprocal of the eigenvalues of $\mathbf{T}$.

5. Let

   $$g_{11} = g_{22} = 1, g_{12} = 0,$$

   be the coefficients of the scalar product $\mathbf{g}$ with respect to the basis $(\mathbf{e}_1, \mathbf{e}_2)$; determine the components of $\mathbf{g}$ in the new basis defined by the transformation matrix (1.99) and check whether or not this new basis is orthogonal.

6. Given the basis $(\mathbf{e}_1, \mathbf{e}_2, \mathbf{e}_3)$ in which the coefficients of the scalar product are

   $$g_{11} = g_{22} = g_{33} = 2, g_{12} = g_{13} = g_{23} = 1,$$

   determine the reciprocal vectors and the scalar product of the two vectors $\mathbf{u} = (1, 0, 0)$ and $\mathbf{v} = (0, 1, 0)$.

7. Given the tensor $\mathbf{T}$ whose components in the orthogonal basis $(\mathbf{e}_1, \mathbf{e}_2, \mathbf{e}_3)$ are

   $$\begin{pmatrix} 1 & 0 & 0 \\ 0 & 1 & 1 \\ 0 & 2 & 3 \end{pmatrix},$$

   determine its eigenvalues and eigenvectors.

8. Evaluate the eigenvalues and eigenvectors of a tensor $\mathbf{T}$ whose components are expressed by the matrix (1.99), but use the basis $(\mathbf{e}_1, \mathbf{e}_2, \mathbf{e}_3)$ of the Exercise 6.

9. Evaluate the component matrix of the most general symmetric tensor $\mathbf{T}$ in the basis $(\mathbf{e}_1, \mathbf{e}_2, \mathbf{e}_3)$, supposing that its eigenvectors $\mathbf{u}$ are directed along the vector $\mathbf{n} = (0, -1, 1)$ or are orthogonal to it; that is, verify the condition

   $$\mathbf{n} \cdot \mathbf{u} = -u_2 + u_3 = 0.$$

   Hint: Take the basis formed by the eigenvectors $\mathbf{e}'_1 = (1, 0, 0)$, $\mathbf{e}'_2 = (0, 1, 1)$ and $\mathbf{n}$.

1.13. Exercises

10. An axial vector **u** transforms according to the law

$$u_i = \pm A_i^j u_j.$$

Verify that the vector product $\mathbf{u} \times \mathbf{v}$ is an axial vector when both u and v are polar and a polar vector when one of them is axial.

1.14 The Program VectorSys

Aim of the Program VectorSys

The program `VectorSys` determines, for any system Σ of applied vectors, an equivalent vector system Σ' and, when the scalar invariant vanishes, its central axis. Moreover, it plots in the space both the system Σ and Σ', as well as the central axis.

Description of the Problem and Relative Algorithm

Two systems $\Sigma = \{(A_i, \mathbf{v}_i)\}_{i=1,\cdots,n}$ and $\Sigma' = \{(B_j, \mathbf{w}_j)\}_{j=1,\cdots,m}$ of applied vectors are **equivalent** if they have the same resultants and moments with respect to *any* pole O. It can be proved that Σ and Σ' are equivalent if and only if

$$\mathbf{R} = \mathbf{R}', \mathbf{M}_P = \mathbf{M}'_P, \tag{1.102}$$

where $\mathbf{R}$, $\mathbf{R}'$ denote their respective resultants and $\mathbf{M}_P$, $\mathbf{M}'_P$ their moments with respect to a *fixed* pole P.

The **scalar invariant** of a system of applied vectors Σ is the scalar product

$$I = \mathbf{M}_P \cdot \mathbf{R}, \tag{1.103}$$

which is *independent of the pole* P. If $\mathbf{R} \neq \mathbf{0}$, then the locus of points Q satisfying the condition

$$\mathbf{M}_Q \times \mathbf{R} = \mathbf{0} \tag{1.104}$$

is a straight line parallel to $\mathbf{R}$, which is called the **central axis** of Σ. Its parametric equations are the components of the vector [3]

$$\mathbf{u} = \frac{\mathbf{M}_P \times \mathbf{R}}{R^2} + t\mathbf{R}, \tag{1.105}$$

where P is any point of $\mathcal{E}_3$ and R is the norm of $\mathbf{R}$.

It is possible to show that any system Σ of applied vectors is equivalent to its resultant $\mathbf{R}$, applied at an arbitrary point P, and a torque equal to the moment $\mathbf{M}_P$ of Σ with respect to P. Moreover, there are systems of applied vectors which are equivalent to either a vector or a torque (see [8], [9]).

---

[3] Equation (1.105) is a direct consequence of (1.104) and the formula

$$\mathbf{M}_O = \mathbf{M}_P + (P - O) \times \mathbf{R} \qquad \forall O, P \in \Re^3.$$

1.11. The Program VectorSys

More precisely, the following results hold:

1. If $I = 0$

   (a) and $\mathbf{R} \neq \mathbf{0}$, then the system Σ is equivalent to its resultant $\mathbf{R}$ applied at any point A of the central axis;

   (b) and $\mathbf{R} = \mathbf{0}$, then the system Σ is equivalent to any torque having the moment $\mathbf{M}_P$ of Σ with respect to P.

2. If $I \neq 0$, then the system Σ is equivalent to its resultant $\mathbf{R}$ applied at any point P and a torque with moment $\mathbf{M}_P$.

A system of applied vectors is **equivalent to zero** if for all $P \in \Re^3$ we have $\mathbf{R} = \mathbf{M}_P = \mathbf{0}$.

Command Line of the Program VectorSys

VectorSys[A_,V_,P_]

Parameter List

Input Data

A = list of points of application of the vectors of Σ;

V = list of components of the vectors of Σ;

P = pole with respect to which the moment of Σ is evaluated.

Output Data

equivalent system Σ';

central axis of Σ;

plot of Σ;

plot of Σ';

plot of the central axis of Σ.

Worked Examples

1. Let $\Sigma = \{(A_i, \mathbf{v}_i)\}_{i=1,\cdots,3}$ be the following applied vectors system:

$$A_1 \equiv (0,1,0), \mathbf{v}_1 \equiv (1,0,1),$$
$$A_2 \equiv (1,0,0), \mathbf{v}_2 \equiv (2,1,0),$$
$$A_3 \equiv (0,1,2), \mathbf{v}_3 \equiv (3,0,0).$$

and let $P \equiv (0,0,0)$ be the pole with respect to which the moment of Σ is evaluated.

To apply the program VectorSys to Σ, we input the following data:

```
A = {{0, 1, 0}, {1, 0, 0}, {0, 1, 2}};
V = {{1, 0, 1}, {2, 1, 0}, {3, 0, 0}};
P = {0, 0, 0};
VectorSys[A, V, P]
```

The corresponding output is [4]

```
The vector system is equivalent to the resultant R = {6,
1, 1} applied at P and a torque Mp = {1, 6, -3}.
```

2. Consider the system $\Sigma = \{(A_i, \mathbf{v}_i)\}_{i=1,2}$, where

$$A_1 \equiv (0,1,0), \quad \mathbf{v}_1 \equiv (1,0,0),$$
$$A_2 \equiv (0,-1,0), \mathbf{v}_2 \equiv (-1,0,0)$$

and a pole $P \equiv (0,0,0)$.

To apply the program VectorSys, we input

```
A = {{0, 1, 0}, {0, -1, 0}};
V = {{1, 0, 0}, {-1, 0, 0}};
P = {0, 0, 0};
VectorSys[A, V, P]
```

The corresponding output is

```
The vector system is equivalent to the torque Mp = {0, 0,
-2}.
```

3. Let $P \equiv (0,0,0)$ be the pole and let $\Sigma = \{(A_i, \mathbf{v}_i)\}_{i=1,2}$ be a vector system, where

$$A_1 \equiv (1,0,0), \mathbf{v}_1 \equiv (0,1,0),$$
$$A_2 \equiv (2,0,0), \mathbf{v}_2 \equiv (0,-1/2,0).$$

---
[4] Due to space limitations, the graphic output is not displayed in the text.

1.11. The Program VectorSys

Input:
```
A = {{1, 0, 0}, {2, 0, 0}};
V = {{0, 1, 0}, {0, -1/2, 0}};
P = {0, 0, 0};
VectorSys[A, V, P]
```

Output:

The central axis is identified by

x(t) = 0,

y(t) = $\dfrac{t}{2}$, $\forall t \in \Re$

x(t) = 0,

and the system is equivalent to the resultant R = $\left\{0, \dfrac{t}{2}, 0\right\}$ applied at any point of the central axis.

4. Let $\Sigma = \{(A_i, \mathbf{v}_i)\}_{i=1,2}$ be a vector system with

$$A_1 \equiv (1,0,0), \mathbf{v}_1 \equiv (0,1,0),$$
$$A_2 \equiv (2,0,0), \mathbf{v}_2 \equiv (0,1,0),$$

and the pole at the origin.

Input:
```
A = {{1, 0, 0}, {2, 0, 0}};
V = {{0, 1, 0}, {0, 1, 0}};
P = {0, 0, 0};
VectorSys[A, V, P]
```

Output:

The central axis is identified by

x(t) = $-\dfrac{3}{2}$,

y(t) = 2t, $\forall t \in \Re$

x(t) = 0,

and the system is equivalent to the resultant R = {0, 2, 0} applied at any point of the central axis.

Exercises

Apply the program `VectorSys` to the following vector systems:

1. $\Sigma = \{(A_i, \mathbf{v}_i)\}_{i=1,\cdots,3}$, where

$$A_1 \equiv (0,1,0), \mathbf{v}_1 \equiv (1,0,1),$$
$$A_2 \equiv (1,0,0), \mathbf{v}_2 \equiv (2,1,0),$$
$$A_3 \equiv (2,0,3), \mathbf{v}_3 \equiv (3,0,0).$$

2. $\Sigma = \{(A_i, \mathbf{v}_i)\}_{i=1,\cdots,4}$, where

$$A_1 \equiv (0,1,0), \mathbf{v}_1 \equiv (1,0,1),$$
$$A_2 \equiv (1,0,0), \mathbf{v}_2 \equiv (-2,1,3),$$
$$A_3 \equiv (2,1,0), \mathbf{v}_3 \equiv (1,3,-1),$$
$$A_4 \equiv (1,3,1), \mathbf{v}_4 \equiv (-1,-2,4).$$

3. $\Sigma = \{(A_i, \mathbf{v}_i)\}_{i=1,2}$, where

$$A_1 \equiv (0,0,1), \mathbf{v}_1 \equiv (1,0,0),$$
$$A_2 \equiv (0,0,0), \mathbf{v}_2 \equiv (-1,0,0).$$

4. $\Sigma = \{(A_i, \mathbf{v}_i)\}_{i=1,\cdots,3}$, where

$$A_1 \equiv (1,0,0), \quad \mathbf{v}_1 \equiv (1,0,0),$$
$$A_2 \equiv (1/2,0,1), \mathbf{v}_2 \equiv (0,1,0),$$
$$A_3 \equiv (0,1,0), \quad \mathbf{v}_3 \equiv (0,0,1).$$

5. $\Sigma = \{(A_i, \mathbf{v}_i)\}_{i=1,2}$, where

$$A_1 \equiv (0,0,0), \mathbf{v}_1 \equiv (1,0,0),$$
$$A_2 \equiv (1,0,0), \mathbf{v}_2 \equiv (-1,0,0).$$

1.15 The Program EigenSystemAG

Aim of the Program EigenSystemAG

The program `EigenSystemAG`, evaluates the eigenvalues of a square matrix and determines the algebraic and geometric multiplicity of each of them as well as the corresponding eigenspace.

Description of the Algorithm

Given the mixed components T_i^j of a 2-tensor **T**, the program determines the spectrum and the characteristic space of each eigenvalue.

Command Line of the Program EigenSystemAG

`EigenSystemAG[matrix]`

Parameter List

Input Data

`matrix` = mixed components of a 2-tensor.

Output Data

the eigenvalues and their algebraic and geometric multiplicities;

the eigenspaces relative to each eigenvector;

an eigenvector basis when the tensor can be put in a diagonal form.

Worked Examples

1. Consider the square matrix

$$A = \begin{pmatrix} 1 & 1 & 0 & 1 \\ 1 & 0 & 1 & 0 \\ 0 & 0 & 1 & 0 \\ 0 & 1 & 0 & 1 \end{pmatrix}.$$

To apply the program, it is sufficient to input the following data:

`matrix = {{1, 1, 0, 1}, {1, 0, 1, 0}, {0, 0, 1, 0}, {0, 1, 0, 1}};`

`EigenSystemAG[matrix]`

The output is

Algebraic and geometric multiplicity of distinct eigenvalues:

$\lambda_1 = 0$: AlgMult $= 2$,   GeoMult $= 1$
Eigenspace relative to λ_1: $V_1 = \{\{0, -1, 0, 1\}\}$.

$\lambda_2 = 1$: AlgMult $= 1$,   GeoMult $= 1$
Eigenspace relative to λ_2: $V_2 = \{\{-1, 0, 1, 0\}\}$

$\lambda_3 = 2$: AlgMult $= 1$,   GeoMult $= 1$
Eigenspace relative to λ_3: $V_1 = \{\{2, 1, 0, 1\}\}$.

The matrix cannot be reduced to a diagonal form!

2. Consider the symmetric 2-tensor

$$\mathbf{T} = \begin{pmatrix} 1 & 1 \\ 1 & 2 \end{pmatrix}.$$

To find an eigenvector basis with respect to which $\mathbf{T}$ is diagonal, the program EigenSystemAg can be used as follows:

matrix $= \{\{1, 1\}, \{1, 2\}\}$;

EigenSystemAG[matrix]

The output is

Algebraic and geometric multiplicity of distinct eigenvalues:

$\lambda_1 = \dfrac{1}{2}(3 - \sqrt{5})$: AlgMult $= 1$,   GeoMult $= 1$

Eigenspace relative to λ_1: $V_1 = \left\{ \left\{ \dfrac{1}{2}(-1-\sqrt{5}), 1 \right\} \right\}$.

$\lambda_2 = \dfrac{1}{2}(3 + \sqrt{5})$: AlgMult $= 1$,   GeoMult $= 1$

Eigenspace relative to λ_2: $V_2 = \left\{ \left\{ \dfrac{1}{2}(-1+\sqrt{5}), 1 \right\} \right\}$

The matrix is diagonal in the basis of its eigenvectors:

$\mathcal{B}_1 = \left\{ \left\{ \dfrac{1}{2}(-1-\sqrt{5}), 1 \right\}, \left\{ \dfrac{1}{2}(-1+\sqrt{5}), 1 \right\} \right\}.$

1.11. The Program EigenSystemAG

43

Exercises

Apply the program `EigenSystemAG` to the following 2-tensors:

1. $\mathbf{T} = \begin{pmatrix} 0 & 0 & 0 & 0 \\ 1 & 0 & 1 & 0 \\ 0 & 0 & 1 & 0 \\ 0 & 1 & 0 & 1 \end{pmatrix}$;

2. $\mathbf{T} = \begin{pmatrix} 1 & 1 & 0 \\ 0 & 1 & 0 \\ 0 & 1 & 0 \end{pmatrix}$;

3. $\mathbf{T} = \begin{pmatrix} 1 & 0 & 0 & 0 \\ 0 & 1 & 0 & 0 \\ 0 & 0 & 1 & 0 \\ 0 & 0 & 0 & 1 \end{pmatrix}$;

4. $\mathbf{T} = \begin{pmatrix} 1 & 0 & 0 & 0 \\ 0 & 0 & 0 & 0 \\ 0 & 1 & 0 & 0 \\ 0 & 0 & 0 & 0 \end{pmatrix}$.

Chapter 2

Vector Analysis

2.1 Curvilinear Coordinates

Let $(O, \mathbf{u}_i)$ be a frame in the Euclidean three-dimensional space E_3. At the beginning of Chapter 1, we stated that any point $P \in E_3$ is defined by its linear coordinates (u^i) with respect to the frame $(O, \mathbf{u}_i)$; moreover, these coordinates coincide with the contravariant components of the position vector $\mathbf{u} = \overrightarrow{OP}$ in the basis $(\mathbf{u}_i)$.

Let
$$u^i = f^i(y^1, y^2, y^3), \quad i = 1, 2, 3, \tag{2.1}$$
be three functions of class $C^1(D)$ in the domain $D \subset \Re^3$, and let $D' \subset \Re^3$ be the image of D under the functions f^i. Suppose that

1. these functions generate a bijective correspondence between D and D';

2. the inverse functions
$$y^i = g^i(u^1, u^2, u^3), \quad i = 1, 2, 3,$$
are also of class C^1 in D'.

In particular, these conditions imply that the Jacobian determinant of the functions f^i does not vanish at any point of D:
$$\det\left(\frac{\partial u^i}{\partial y^j}\right) \neq 0. \tag{2.2}$$

If the functions f^i are linear,
$$u^i = u^i_{O'} + A^i_j y^j,$$
and $\det(A^i_j) > 0$, then the relations (2.1) define a linear coordinate change in E_3, see (1.95), relative to the frame change $(O, \mathbf{u}_i) \to (O', \mathbf{u}'_i)$; in this case, D and D' coincide with $\Re^3$.

More generally, when the functions f^i are *nonlinear* and the variables y^i are used to define the points of E_3, we say that **curvilinear coordinates** (y^i) have been introduced in the region $\Omega \subseteq E_3$.

The **coordinate curves** are those curves of E_3 whose parametric equations are obtained from the functions (2.1) by fixing two of the variables y^i and varying the third one. From our hypotheses, it follows that three coordinate curves cross at any point $P \in \Omega$; moreover, the vectors

$$\mathbf{e}_i = \left(\frac{\partial u^j}{\partial y^i}\right)_P \mathbf{u}_j \tag{2.3}$$

are linearly independent, since from the condition

$$\lambda^i \mathbf{e}_i = \mathbf{0},$$

we have

$$\lambda^i \left(\frac{\partial u^j}{\partial y^i}\right)_P \mathbf{u}_j = \mathbf{0}.$$

The independence of the vectors $\mathbf{u}_i$ leads to the following homogeneous system:

$$\lambda^i \left(\frac{\partial u^j}{\partial y^i}\right)_P = 0, \qquad j = 1, 2, 3,$$

which, from (2.2), admits only the trivial solution $\lambda^1 = \lambda^2 = \lambda^3 = 0$. The three vectors $(\mathbf{e}_i)$ that are tangent to the coordinate curves at P form the **natural** or **holonomic frame** $(P, \mathbf{e}_i)$ at P for the space E_3.

We note that in Cartesian coordinates the base vectors are unit vectors, tangent to coordinate curves. In nonorthogonal coordinates the natural base vectors are tangent to coordinate curves, but in general are neither unit vectors nor orthogonal to each other.

In addition to transformations from linear coordinates to curvilinear coordinates, it is useful to consider the transformations from curvilinear coordinates (y^i) to other curvilinear coordinates (y'^i), which are defined by regular bijective functions

$$y'^i = y'^i(y^1, y^2, y^3), \qquad i = 1, 2, 3. \tag{2.4}$$

Once again, the hypothesis that the functions (2.4) are bijective implies

$$\det\left(\frac{\partial y'^i}{\partial y^j}\right) > 0, \tag{2.5}$$

at any regular point. If $u^i = f^i(y^j)$, $u^i = f'^i(y'^j)$, and $y^j = y^j(y'^h)$ denote, respectively, the coordinate changes $(y^i) \longrightarrow (u^i)$, $(y'^i) \longrightarrow (u^i)$, $(y'^i) \longrightarrow (y^i)$, the chain rule gives

$$\frac{\partial u^i}{\partial y'^j} = \frac{\partial u^i}{\partial y^h} \frac{\partial y^h}{\partial y'^j}.$$

2.1. Curvilinear Coordinates

When the condition (2.3) is taken into account, this relation becomes

$$\mathbf{e}'_j = \frac{\partial y^h}{\partial y'^j}\mathbf{e}_h. \qquad (2.6)$$

Then we can conclude that *as a consequence of a coordinate transformation* $y'^j = y'^j(y^h)$, *the natural basis at P transforms according to the law* (2.6); *that is, the matrix of the base change coincides with the Jacobian matrix of the inverse functions* $y^i = y^i(y'^j)$.

From the results of Chapter 1, it follows that the relation between the components of any vector $\mathbf{v}$ at P, with respect to the bases $\mathbf{e}_i$ and $\mathbf{e}'_i$, is expressed by the following formulae:

$$v'^i = \frac{\partial y'^i}{\partial y^h}v^h. \qquad (2.7)$$

All the considerations of Chapter 1 can be repeated at any point P of E_3, when curvilinear coordinates are adopted, provided that the natural basis at P is chosen. For instance, the inner product of the vectors $\mathbf{v} = v^i \mathbf{e}_i$ and $\mathbf{w} = w^i \mathbf{e}_i$ applied at $P(y^i)$ is written as

$$\mathbf{v} \cdot \mathbf{w} = g_{ij}(y^h)v^i w^j, \qquad (2.8)$$

where the *functions*

$$g_{ij}(y^h) = \mathbf{e}_i \cdot \mathbf{e}_j = \sum_{h=1}^{3} \frac{\partial u^h}{\partial y^i}\frac{\partial u^h}{\partial y^j} \qquad (2.9)$$

denote the inner products of the vectors of the natural basis at P. It should be noted that, in curvilinear coordinates, the inner products (2.9) depend on the point; that is, the quantities g_{ij} are functions of the coordinates (y^i). Also in this case, the (local) reciprocal bases can be defined by the positions

$$\mathbf{e}^i = g^{ij}\mathbf{e}_j, \qquad (2.10)$$

where (g^{ij}) is the inverse matrix of (g_{ij}).

The square of the distance between two points with coordinates (u^i) and $(u^i + du^i)$ in the *orthonormal* frame $(O, \mathbf{u}_i)$ is given by

$$ds^2 = \sum_{i=1}^{3}(du^i)^2.$$

Differentiating (2.1) and taking into account the relations (2.9), we obtain

$$ds^2 = g_{ij}(y^h)dy^i dy^j. \qquad (2.11)$$

The functions $g_{ij}(y^h)$ are called **metric coefficients** in the curvilinear coordinates (y^i). From (2.6), in the coordinate change $y'^j = y'^j(y^h)$, the metric coefficients transform according to the rule

$$g'_{ij} = \frac{\partial y^h}{\partial y'^i} \frac{\partial y^k}{\partial y'^j} g_{hk}. \tag{2.12}$$

2.2 Examples of Curvilinear Coordinates

Let
$$\begin{cases} u^1 = r\cos\varphi, \\ u^2 = r\sin\varphi, \end{cases} \tag{2.13}$$

be the transformation from polar coordinates (r, φ) to Cartesian coordinates (u^1, u^2) in the plane E_2. These functions are regular and bijective for all $(r, \varphi) \in D = (0, \infty) \times [0, 2\pi)$ and for all $(u^1, u^2) \in D'$, where D' denotes E_2 without the origin. At any point of D, the Jacobian matrix J is written as

$$J = \begin{pmatrix} \cos\varphi & -r\sin\varphi \\ \sin\varphi & r\cos\varphi \end{pmatrix},$$

and its determinant is given by $r > 0$. The coordinate curves, obtained by fixing either φ or r in (2.13), are the half-lines going out from the origin, without the origin itself, as well as the circles with their centers at the origin and radius r (see Figure 2.1).

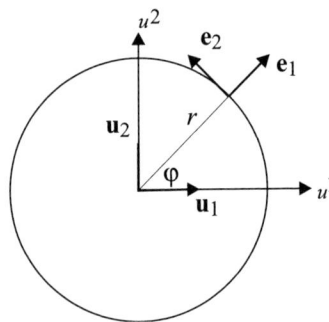

Figure 2.1

In the frame $(O, \mathbf{u}_1, \mathbf{u}_2)$, where $\mathbf{u}_1$, and $\mathbf{u}_2$ are unit vectors along the axes Ou^1 and Ou^2 respectively, the tangent vectors in polar coordinates are

$$\begin{cases} \mathbf{e}_1 = \cos\varphi\,\mathbf{u}_1 + \sin\varphi\,\mathbf{u}_2, \\ \mathbf{e}_2 = -r\sin\varphi\,\mathbf{u}_1 + r\cos\varphi\,\mathbf{u}_2, \end{cases}$$

2.2. Examples of Curvilinear Coordinates

so that the metric coefficients are

$$g_{11} = \mathbf{e}_1 \cdot \mathbf{e}_1 = 1, \qquad g_{12} = \mathbf{e}_1 \cdot \mathbf{e}_2 = 0, \qquad g_{22} = \mathbf{e}_2 \cdot \mathbf{e}_2 = r^2,$$

and the metric becomes

$$ds^2 = dr^2 + r^2 d\varphi^2.$$

More generally, let

$$\begin{cases} u^1 = r\cos\varphi\sin\theta, \\ u^2 = r\sin\varphi\sin\theta, \\ u^3 = r\cos\theta, \end{cases} \qquad (2.14)$$

be the transformation from spherical coordinates (r, φ, θ) to Cartesian coordinates (u^1, u^2, u^3). These functions are regular and bijective for all $(r, \varphi, \theta) \in D = (0, \infty) \times [0, 2\pi) \times [0, \pi)$ and for all $(u^1, u^2, u^3) \in D'$, where D' is E_3 without the origin. At any point of D, the Jacobian matrix is written as

$$J = \begin{pmatrix} \cos\varphi\sin\theta & -r\sin\varphi\sin\theta & r\cos\varphi\cos\theta \\ \sin\varphi\sin\theta & r\cos\varphi\sin\theta & r\sin\varphi\cos\theta \\ \cos\theta & 0 & -r\sin\theta \end{pmatrix}$$

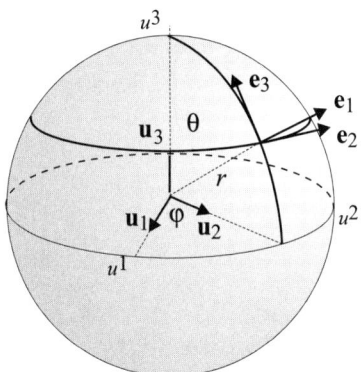

Figure 2.2

and its determinant is $-r^2 \sin\theta \neq 0$. The coordinate curves are respectively the half-lines going out from the origin and different from Ou^3, the circles with center at the origin and radius r lying in the plane $\varphi = const.$ (the meridians), and finally, the circles with center $(0, 0, r\cos\theta)$ and radius r (the parallels). In the frame $(O, \mathbf{u}_1, \mathbf{u}_2, \mathbf{u}_3)$, where $\mathbf{u}_1, \mathbf{u}_2$, and $\mathbf{u}_3$ are unit vectors along the axes Ou^1, Ou^2, and Ou^3, the components of the tangent vectors to these coordinate curves are given by the columns of the Jacobian matrix

$$\begin{cases} \mathbf{e}_1 = \cos\varphi\sin\theta\,\mathbf{u}_1 + \sin\varphi\sin\theta\,\mathbf{u}_2 + \cos\theta\,\mathbf{u}_3, \\ \mathbf{e}_2 = -r\sin\varphi\sin\theta\,\mathbf{u}_1 + r\cos\varphi\sin\theta\,\mathbf{u}_2, \\ \mathbf{e}_3 = r\cos\varphi\cos\theta\,\mathbf{u}_1 + r\sin\varphi\cos\theta\,\mathbf{u}_2 - r\sin\theta\,\mathbf{u}_3, \end{cases}$$

the metric coefficients are

$$\begin{aligned} g_{11} &= \mathbf{e}_1\cdot\mathbf{e}_1 = 1, \quad g_{12} = \mathbf{e}_1\cdot\mathbf{e}_2 = g_{13} = \mathbf{e}_1\cdot\mathbf{e}_3 = 0, \\ g_{23} &= \mathbf{e}_2\cdot\mathbf{e}_3 = 0, \quad g_{22} = \mathbf{e}_2\cdot\mathbf{e}_2 = r^2\sin^2\theta, \\ g_{33} &= \mathbf{e}_3\cdot\mathbf{e}_3 = r^2, \end{aligned} \qquad (2.15)$$

and the square of the distance between two points becomes

$$ds^2 = dr^2 + r^2\sin^2\theta\,d\varphi^2 + r^2 d\theta^2.$$

We leave the reader to verify that in the transformation from cylindrical coordinates to Cartesian coordinates,

$$\begin{cases} u^1 = r\cos\varphi, \\ u^2 = r\sin\varphi, \\ u^3 = u^3, \end{cases} \qquad (2.16)$$

we have

$$J = \begin{pmatrix} \cos\varphi & -r\sin\varphi & 0 \\ \sin\varphi & r\cos\varphi & 0 \\ 0 & 0 & 1 \end{pmatrix}, \ \det J = r,$$

$$\begin{cases} \mathbf{e}_1 = \cos\varphi\,\mathbf{u}_1 + \sin\varphi\,\mathbf{u}_2, \\ \mathbf{e}_2 = -r\sin\varphi\,\mathbf{u}_1 + r\cos\varphi\,\mathbf{u}_2, \\ \mathbf{e}_3 = \mathbf{u}_3, \end{cases}$$

$$g_{11} = \mathbf{e}_1\cdot\mathbf{e}_1 = 1,\ g_{12} = \mathbf{e}_1\cdot\mathbf{e}_2 = g_{13} = \mathbf{e}_1\cdot\mathbf{e}_3 = g_{23} = 0,$$
$$g_{22} = \mathbf{e}_2\cdot\mathbf{e}_2 = r^2,\ g_{33} = \mathbf{e}_3\cdot\mathbf{e}_3 = 1,$$
$$ds^2 = dr^2 + r^2 d\varphi^2 + (du^3)^2.$$

2.3   Differentiation of Vector Fields

A *vector field* is a mapping $\mathbf{v}: P \in U \subset E_3 \longrightarrow \mathbf{v}(P) \in \mathcal{E}_3$ which associates a vector $\mathbf{v}(P)$ to any point P belonging to an open domain U of E_3. If (y^i) denotes a curvilinear coordinate system in U and $(P, \mathbf{e}_i)$ is the

2.3. Differentiation of Vector Fields

associated natural frame (see previous section), we have $\mathbf{v}(P) = v^i(y^h)\mathbf{e}_i$ in U. The vector field $\mathbf{v}$ is said to be ***differentiable of class*** C^k ($k \geq 0$) in U if its components $v^i(y^h)$ are of class C^k. For the sake of simplicity, from now on the class of a vector field will not be stated because it will be plain from the context.

If $f : U \longrightarrow \Re$ is a differentiable function of the coordinates y^i, then the ***gradient*** of f is the vector field

$$\nabla f \equiv f_{,i}\,\mathbf{e}^i = g^{ij} f_{,j}\,\mathbf{e}_i, \tag{2.17}$$

where the short notation $\partial f/\partial y^i = f_{,i}$ has been used. We can easily verify that the definition (2.17) is independent of the coordinates, since the quantities appearing as covariant or contravariant components in (2.17) have the right transformation properties.

It has already been noted at the beginning of this chapter that the natural frame $(P, \mathbf{e}_i)$ depends on the point $P \in U$. In view of the next applications, it is interesting to evaluate the variation of the fields $\mathbf{e}_i$ along the coordinate curves; that is, the derivatives $\mathbf{e}_{i,j}$. From (2.3), we have

$$\mathbf{e}_{i,h} = u^k{}_{,ih}\,\mathbf{u}_k, \tag{2.18}$$

so that, if the ***Christoffel symbols*** Γ^k_{ih} are introduced by the relations

$$\mathbf{e}_{i,h} = \Gamma^k_{ih}\mathbf{e}_k, \tag{2.19}$$

then the derivatives on the left-hand side will be known with respect to the basis $(\mathbf{e}_i)$, provided that the expressions of Christoffel symbols have been determined. In order to find these expressions, we note that (2.18) and (2.19) imply the symmetry of these symbols with respect to the lower indices. Moreover, the relations $\mathbf{e}_i \cdot \mathbf{e}_j = g_{ij}$ give the equations

$$g_{ij,h} = \mathbf{e}_{i,h} \cdot \mathbf{e}_j + \mathbf{e}_i \cdot \mathbf{e}_{j,h}$$

which, from (2.19), can be written as

$$g_{ij,h} = g_{jk}\Gamma^k_{ih} + g_{ik}\Gamma^k_{jh}. \tag{2.20}$$

Cyclically permuting the indices i, j, h the other relations are obtained:

$$\begin{aligned}g_{hi,j} &= g_{ik}\Gamma^k_{hj} + g_{hk}\Gamma^k_{ij},\\ g_{jh,i} &= g_{hk}\Gamma^k_{ji} + g_{jk}\Gamma^k_{hi}.\end{aligned} \tag{2.21}$$

Adding the equation $(2.21)_1$ to (2.20) and subtracting $(2.21)_2$, we derive the following formula:

$$\Gamma^l_{jh} = \frac{1}{2}g^{li}(g_{ij,h} + g_{hi,j} - g_{jh,i}). \tag{2.22}$$

Finally, the condition $\mathbf{e}^j \cdot \mathbf{e}_i = \delta_i^j$ implies that

$$\mathbf{e}^j{}_{,h} \cdot \mathbf{e}_i + \mathbf{e}^j \cdot \mathbf{e}_{i,h} = 0$$

or

$$\mathbf{e}^j{}_{,h} \cdot \mathbf{e}_i = -\Gamma^j_{ih}.$$

This relation shows that the opposites of the Christoffel symbols coincide with the covariant components of the vector field $\mathbf{e}^j{}_{,h}$; consequently,

$$\mathbf{e}^j{}_{,h} = -\Gamma^j_{ih}\mathbf{e}^i. \tag{2.23}$$

Consider any vector field $\mathbf{v} = v^k \mathbf{e}_k$ expressed in the curvilinear coordinates (y^i). Then a simple calculation shows that relation (2.19) implies

$$\mathbf{v}_{,h} = (v^k{}_{,h} + \Gamma^k_{lh} v^l)\mathbf{e}_k. \tag{2.24}$$

If the field $\mathbf{v}$ is written in the form $\mathbf{v} = v_k \mathbf{e}^k$ and (2.23) is taken into account, then instead of (2.24) we obtain

$$\mathbf{v}_{,h} = (v_{k,h} - \Gamma^l_{kh} v_l)\mathbf{e}^k. \tag{2.25}$$

To extend (2.24), (2.25) to any Euclidean second-order tensor field $\mathbf{T}$, it is sufficient to use one of the possible representation formulae

$$\mathbf{T} = T^{kl}\mathbf{e}_k \otimes \mathbf{e}_l = T^k_l \mathbf{e}_k \otimes \mathbf{e}^l = T_{kl}\mathbf{e}^k \otimes \mathbf{e}^l$$

to derive

$$\begin{aligned}
\mathbf{T}_{,h} &= (T^{kl}{}_{,h} + \Gamma^k_{ph} T^{pl} + \Gamma^l_{ph} T^{kp})\mathbf{e}_k \otimes \mathbf{e}_l, \\
\mathbf{T}_{,h} &= (T^k_l{}_{,h} + \Gamma^k_{ph} T^p_l - \Gamma^p_{lh} T^k_p)\mathbf{e}_k \otimes \mathbf{e}^l, \\
\mathbf{T}_{,h} &= (T_{kl,h} - \Gamma^p_{kh} T_{pl} - \Gamma^p_{lh} T_{kp})\mathbf{e}^k \otimes \mathbf{e}^l.
\end{aligned} \tag{2.26}$$

All of these derivative formulae reduce to the ordinary ones when Cartesian coordinates are used, since in these coordinates the Christoffel symbols vanish.

The **_covariant derivative_** of the vector field $\mathbf{v}$ is the second-order tensor field

$$\begin{aligned}
\nabla \mathbf{v} = \mathbf{v}_{,h} \otimes \mathbf{e}^h &= (v^k{}_{,h} + \Gamma^k_{lh} v^l)\mathbf{e}_k \otimes \mathbf{e}^h \\
&= (v_{k,h} - \Gamma^l_{kh} v_l)\mathbf{e}^k \otimes \mathbf{e}^h.
\end{aligned} \tag{2.27}$$

Similarly, the **_covariant derivative_** of a second-order tensor field $\mathbf{T}$ is the third-order tensor field defined as follows:

$$\nabla \mathbf{T} = \mathbf{T}_{,h} \otimes \mathbf{e}^h. \tag{2.28}$$

We note that relations (2.22) and (2.26)$_3$ imply that

$$\mathbf{g}_{,h} = \mathbf{0}, \qquad \nabla \mathbf{g} = \mathbf{0}. \tag{2.29}$$

2.3. Differentiation of Vector Fields

To generalize these elementary concepts, we introduce the ***divergence*** of the vector field $\mathbf{v}$ in the following way, see (2.24):

$$\nabla \cdot \mathbf{v} = \mathbf{v}_{,h} \cdot \mathbf{e}^h = v^h_{,h} + \Gamma^h_{lh} v^l. \tag{2.30}$$

More generally, the ***divergence*** of a second-order tensor field $\mathbf{T}$ is given by the vector field

$$\nabla \cdot \mathbf{T} = \mathbf{T}_{,h} \cdot \mathbf{e}^h = (T^{hl}_{,h} + \Gamma^h_{ph} T^{pl} + \Gamma^l_{ph} T^{hp}) \mathbf{e}_l. \tag{2.31}$$

Now, the identity

$$\Gamma^i_{ih} = \frac{1}{\sqrt{g}} (\sqrt{g})_{,h}, \tag{2.32}$$

where

$$g = \det(g_{ij}),$$

can be proved starting from (2.20) and the relation

$$\frac{\partial g}{\partial g_{ij}} = g g^{ij},$$

which, in turn, will be shown in Section 3.7; in fact,

$$\frac{1}{\sqrt{g}} (\sqrt{g})_{,h} = \frac{1}{\sqrt{g}} \left(\frac{1}{2\sqrt{g}} \frac{\partial g}{\partial g_{ij}} g_{ij,h} \right) = \frac{1}{2} g^{ij} g_{ij,h}$$

$$= \frac{1}{2} g^{ij} \left(g_{jk} \Gamma^k_{ih} + g_{ik} \Gamma^k_{jh} \right)$$

$$= \frac{1}{2} \left(\Gamma^i_{ih} + \Gamma^j_{jh} \right) = \Gamma^i_{ih}.$$

By using the relation (2.32), we can write the divergence of a vector or tensor field in the following equivalent forms:

$$\begin{aligned} \nabla \cdot \mathbf{v} &= \frac{1}{\sqrt{g}} \left(\sqrt{g} v^h \right)_{,h}, \\ \nabla \cdot \mathbf{T} &= \left(\frac{1}{\sqrt{g}} \left(\sqrt{g} T^{hl} \right)_{,h} + \Gamma^l_{ph} T^{hp} \right) \mathbf{e}_l. \end{aligned} \tag{2.33}$$

Let us introduce the skew-symmetric tensor field

$$\nabla \mathbf{v} - (\nabla \mathbf{v})^T = \mathbf{v}_{,h} \otimes \mathbf{e}^h - (\mathbf{v}_{,h} \otimes \mathbf{e}^h)^T.$$

Recalling $(2.27)_3$ and the definition of transpose tensor (1.42), we have

$$\nabla \mathbf{v} - (\nabla \mathbf{v})^T = (v_{k,h} - \Gamma^l_{kh} v_l) \mathbf{e}^k \otimes \mathbf{e}^h - (v_{k,h} - \Gamma^l_{kh} v_l) \mathbf{e}^h \otimes \mathbf{e}^k$$

$$= (v_{k,h} - v_{h,k}) \mathbf{e}^k \otimes \mathbf{e}^h.$$

The axial vector[1]

$$\nabla \times \mathbf{v} = \frac{1}{2\sqrt{g}} \epsilon^{ihk} (v_{k,h} - v_{h,k}) \mathbf{e}_i \tag{2.34}$$

is called the **curl** of **v**. A simple mnemonic device for deducing the components of $\nabla \times \mathbf{v}$ is obtained by noting that the components coincide with the algebraic complements of $\mathbf{e}_1, \mathbf{e}_2, \mathbf{e}_3$ in the matrix

$$\begin{pmatrix} \mathbf{e}_1 & \mathbf{e}_2 & \mathbf{e}_3 \\ \dfrac{\partial}{\partial y^1} & \dfrac{\partial}{\partial y^2} & \dfrac{\partial}{\partial y^3} \\ v_1 & v_2 & v_3 \end{pmatrix}.$$

It is well known that the curves $x^i(t)$ that are solutions of the system

$$\frac{dx^i}{dt} = v^i(x^j)$$

are called **integral curves** of the vector field **v**. Moreover, in analysis it is proved that the directional derivative of a function f along the integral curves of the vector field **u** is expressed by the formula

$$\frac{df}{dt} = \mathbf{u} \cdot \nabla f = u^h f_{,h}. \tag{2.35}$$

Similarly, the **directional derivative** of the vector field **v** along the integral curves of **u** is defined as the new vector field

$$\frac{d\mathbf{v}}{dt} = \mathbf{u} \cdot \nabla \mathbf{v}. \tag{2.36}$$

Applying (2.27) to **v**, we find that the previous definition assumes the following form:

$$\frac{d\mathbf{v}}{dt} = u^h \left(v^i{}_{,h} + \Gamma^i_{jh} v^j \right) \mathbf{e}_i. \tag{2.37}$$

Finally, the **Laplacian** of f is the function

$$\Delta f = \nabla \cdot \nabla f. \tag{2.38}$$

It is easy to verify from (2.17), (2.30), and (2.32) that the Laplacian of f can be written as follows:

$$\Delta f = \frac{1}{\sqrt{g}} \left(\sqrt{g} g^{hj} f_{,j} \right)_{,h}, \tag{2.39}$$

where, as usual, $g = \det(g_{ij})$.

---

[1] It is worth noting that all the previous definitions are independent of the dimension of the space, whereas the definition (2.34) of curl refers to a three-dimensional space.

2.4 The Stokes and Gauss Theorems

In this section the following notations are introduced: $f : U \longrightarrow \Re$ is a function of class C^1 in an open region U of $\Re^2$ or $\Re^3$, $C \subset U$ is any domain having a regular boundary ∂C, and (u^i) are Cartesian coordinates relative to the frame $(O, \mathbf{u}_i)$. In analysis the famous **Gauss's theorem** is proved:

$$\int_{\partial C} f \cos nu^i \, d\sigma = \int_C f_{,i} \, dc, \tag{2.40}$$

where $\cos nu^i$ is the cosine of the angle between the external unit vector $\mathbf{n}$ orthogonal to ∂C and the unit vector $\mathbf{u}_i$ along the axis u^i. Applying the previous formula to any component with respect to $(O, \mathbf{u}_i)$ of a vector field $\mathbf{v}$ and adding the relations obtained, we derive the **vector form of Gauss's theorem**:

$$\int_{\partial C} \mathbf{v} \cdot \mathbf{n} \, d\sigma = \int_C \nabla \cdot \mathbf{v} \, dc. \tag{2.41}$$

Similarly, applying the formula (2.40) to the components T^{i1}, T^{i2}, T^{i3} of a tensor field and adding the results, we have

$$\int_{\partial C} \mathbf{n} \cdot \mathbf{T} \, d\sigma = \int_C \nabla \cdot \mathbf{T} \, dc. \tag{2.42}$$

Another remarkable result is expressed by **Stokes's theorem**:

$$\int_{\partial \sigma} \mathbf{v} \cdot \mathbf{t} \, ds = \int_\sigma \nabla \times \mathbf{v} \cdot \mathbf{n} \, d\sigma, \tag{2.43}$$

where σ is a regular surface having a regular boundary $\partial \sigma$, $\mathbf{t}$ is the unit vector tangent to the curve $\partial \sigma$, and $\mathbf{n}$ is the external unit vector orthogonal to σ. The orientation on $\partial \sigma$ (that is of $\mathbf{t}$) is counter-clockwise with respect to $\mathbf{n}$ (see Figure 2.3).

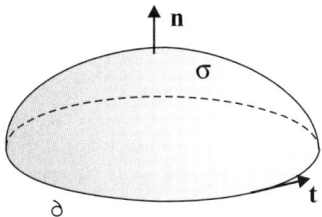

Figure 2.3

2.5 Singular Surfaces

Let C be a compact domain of E_3 having a regular boundary ∂C, let σ be a regular oriented surface, and let $\mathbf{n}$ be the unit vector normal to σ. We suppose that σ divides C into two regions C^- and C^+, where C^+ is the part of C containing $\mathbf{n}$.

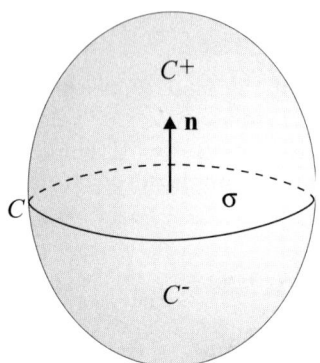

Figure 2.4

Let $\psi(P)$ be a function of class $C^1(C-\sigma)$ with the following finite limits:

$$\lim_{P \longrightarrow \mathbf{r},\, P \in C^+} \psi(P) \equiv \psi^+(\mathbf{r}), \quad \lim_{P \longrightarrow \mathbf{r},\, P \in C^-} \psi(P) \equiv \psi^-(\mathbf{r})$$

for all $\mathbf{r} \in \sigma$. The functions $\psi^+(\mathbf{r}), \psi^-(\mathbf{r})$ are supposed to be continuous on σ. If the **jump** of $\psi(P)$ in crossing σ,

$$[[\psi(\mathbf{r})]] = \psi^+(\mathbf{r}) - \psi^-(\mathbf{r}),$$

does not vanish at some points $\mathbf{r} \in \sigma$, then the surface σ is called a ***0-order singular surface*** for ψ.

Similarly, when $[[\psi]] = 0$ on σ, and the limits

$$\lim_{P \longrightarrow \mathbf{r},\, P \in C^+} \frac{\partial \psi}{\partial x_i}(P) \equiv \left(\frac{\partial \psi}{\partial x_i}\right)^+(\mathbf{r}), \quad \lim_{P \longrightarrow \mathbf{r},\, P \in C^-} \frac{\partial \psi}{\partial x_i}(P) \equiv \left(\frac{\partial \psi}{\partial x_i}\right)^-(\mathbf{r}),$$

exist and are continuous functions of the point $\mathbf{r} \in \sigma$, the surface σ is called a ***first-order singular surface*** for ψ if, at some points of σ, the jump below does not vanish:

$$\left[\left[\frac{\partial \psi}{\partial x_i}(P)\right]\right] = \left(\frac{\partial \psi}{\partial x_i}\right)^+(\mathbf{r}) - \left(\frac{\partial \psi}{\partial x_i}\right)^-(\mathbf{r}).$$

2.5. Singular Surfaces

At this point, the meaning of a **k-order singular surface** for ψ should be clear. Moreover, in the previous definitions ψ can be regarded as a component of a vector or tensor field.

Theorem 2.1 (Generalized Gauss's theorem)
If σ is a 0-order singular surface for the vector field $\mathbf{v}(P)$ of class $C^1(C - \sigma)$, then the following result holds:

$$\int_{\partial C} \mathbf{v} \cdot \mathbf{N}\, d\sigma = \int_C \nabla \cdot \mathbf{v}\, dC + \int_\sigma [[\mathbf{v}(P)]] \cdot \mathbf{n}\, d\sigma, \qquad (2.44)$$

where $\mathbf{N}$ and $\mathbf{n}$ denote the vectors normal to ∂C and σ, respectively; moreover, $\mathbf{n}$ lies in the region C^+.

PROOF In our hypotheses Gauss's theorem is applicable to both the domains C^- and C^+:

$$\int_{\partial C^- - \sigma} \mathbf{v} \cdot \mathbf{N}\, d\sigma + \int_\sigma \mathbf{v}^-(\mathbf{r}) \cdot \mathbf{n}\, d\sigma = \int_{C^-} \nabla \cdot \mathbf{v}\, dC,$$

$$\int_{\partial C^+ - \sigma} \mathbf{v} \cdot \mathbf{N}\, d\sigma - \int_\sigma \mathbf{v}^+(\mathbf{r}) \cdot \mathbf{n}\, d\sigma = \int_{C^+} \nabla \cdot \mathbf{v}\, dC,$$

where we note that at any point of $\partial C^- \cap \sigma$ we have $\mathbf{N} = \mathbf{n}$, while on $\partial C^+ \cap \sigma$ we have $\mathbf{N} = -\mathbf{n}$. Formula (2.44) is obtained by adding the two previous relations. ∎

A similar formula can be proved for a tensor, starting from (2.42):

$$\int_{\partial C} \mathbf{n} \cdot \mathbf{T}\, d\sigma = \int_C \nabla \cdot \mathbf{T}\, dC + \int_\sigma \mathbf{n} \cdot [[\mathbf{T}(P)]]\, d\sigma. \qquad (2.45)$$

To generalize Stokes's theorem, we consider a vector field $\mathbf{v}(P)$ for which σ is a 0-order singular surface, together with a regular surface S contained in C and intersecting σ along the curve γ.
The following theorem holds (see Figure 2.5):

Theorem 2.2 (Generalized Stokes's theorem)
If $\sigma \subset C$ is a 0-order singular surface for the vector field $\mathbf{v}(P)$, then for any surface $S \subset C, S \cap \sigma \neq \emptyset$, we have

$$\int_{\partial S} \mathbf{v} \cdot \mathbf{t}\, ds = \int_S \nabla \times \mathbf{v} \cdot \mathbf{N}\, dS + \int_\gamma [[\mathbf{v}]] \cdot \boldsymbol{\tau}\, ds, \qquad (2.46)$$

where $\mathbf{N}$ is the unit vector normal to S and τ is the unit vector tangent to the curve γ, which is oriented in such a way that its orientation appears counter-clockwise with respect to $\mathbf{N}$, applied at a point of $\partial S \cap C^+$.

PROOF Formula (2.46) is proved by applying Stokes's theorem (2.43) to both the surfaces $S^- = S \cap C^-$ and $S^+ = S \cap C^+$ and subtracting the results obtained. ∎

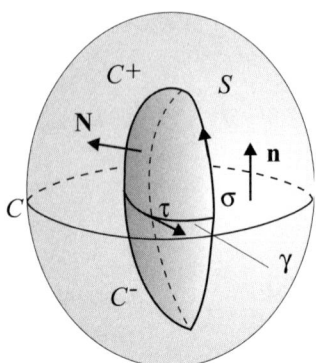

Figure 2.5

The following theorem is due to Hadamard:

Theorem 2.3 (Hadamard)
If ψ is a function of class $C^1(C - \sigma)$ and σ is a first-order singular surface for ψ, then on any regular curve γ of σ we have

$$\tau \cdot [[\nabla \psi]] = 0, \tag{2.47}$$

where τ is the unit vector tangent to γ.

PROOF Let $\mathbf{r}(s)$ be the equation of γ and let $\mathbf{f}(\lambda, s)$ be a one-parameter family of curves contained in C^- such that

$$\lim_{\lambda \longrightarrow 0} \mathbf{f}(\lambda, s) = \mathbf{f}(0, s) = \mathbf{r}(s).$$

From the properties of ψ, on any curve $\mathbf{f}(\lambda, s)$, where λ is fixed, we obtain

$$\frac{d\psi}{ds}(\mathbf{f}(\lambda, s)) = \frac{\partial \mathbf{f}}{\partial s}(\lambda, s) \cdot \nabla \psi(\mathbf{f}(\lambda, s)).$$

The following result follows from the continuity of all the functions involved, when $\lambda \longrightarrow 0$:

$$\frac{d\psi^-}{ds}(\mathbf{r}(s)) = \tau \cdot \nabla \psi(\mathbf{r}(s))^-.$$

2.5. Singular Surfaces

Similar reasoning, starting from C^+, leads to the corresponding relation

$$\frac{d\psi^+}{ds}(\mathbf{r}(s)) = \boldsymbol{\tau} \cdot \nabla \psi(\mathbf{r}(s))^+,$$

and the theorem is proved by taking into account the condition $\psi^+ = \psi^-$ on σ. ∎

Relation (2.47) is usually referred as the **geometric compatibility condition** for a weak discontinuity of the field ψ. This condition states that if $[[\psi]] = 0$, then only the normal derivative can be discontinuous, and the term *geometric* emphasizes the fact that this condition does not depend on the actual motion of the discontinuity. This is formally expressed by the following theorem.

Theorem 2.4
In the hypotheses of the previous theorem, we have

$$[[\nabla \psi]]_\mathbf{r} = a(\mathbf{r})\mathbf{n} \quad \forall \mathbf{r} \in \sigma. \tag{2.48}$$

PROOF In fact, if γ is any regular curve on σ crossing the point $\mathbf{r}$, from (2.47) it follows that $[[\nabla \psi]]_\mathbf{r}$ is orthogonal to any tangent vector to σ at the point $\mathbf{r}$. Consequently, either $[[\nabla \psi]]_\mathbf{r}$ vanishes or it is orthogonal to σ. ∎

If ψ is a component of any r-tensor field $\mathbf{T}$ at r, and σ is a first-order singular surface of $\mathbf{T}$, then from (2.48) we have

$$[[\nabla \mathbf{T}]]_\mathbf{r} = \mathbf{n} \otimes \mathbf{a}(\mathbf{r}) \quad \forall \mathbf{r} \in \sigma, \tag{2.49}$$

where $\mathbf{a}(\mathbf{r})$ is an r-tensor.

In particular, for a vector field $\mathbf{v}$ the previous formula gives

$$\begin{aligned}{} [[\nabla \mathbf{v}]]_\mathbf{r} &= \mathbf{n} \otimes \mathbf{a}(\mathbf{r}), \\ [[\nabla \cdot \mathbf{v}]]_\mathbf{r} &= \mathbf{n} \cdot \mathbf{a}(\mathbf{r}), \\ [[\nabla \times \mathbf{v}]]_\mathbf{r} &= \mathbf{n} \times \mathbf{a}(\mathbf{r}), \end{aligned} \tag{2.50}$$

where $\mathbf{a}(\mathbf{r})$ is a vector field on σ.

If σ is a second-order singular surface for the function ψ of class $C^2(C-\sigma)$ and formula $(2.50)_1$ is applied to $\nabla \psi$, we derive

$$\left[\left[\frac{\partial^2 \psi}{\partial x_j \partial x_i}\right]\right] = n_j a_i. \tag{2.51}$$

But the derivation order can be inverted, and we also have

$$a_i n_j = a_j n_i,$$

so that

$$\frac{a_i}{n_i} = \frac{a_j}{n_j} \equiv \lambda$$

and $a_i = \lambda n_i$. Consequently, the relation (2.51) becomes

$$[[\nabla\nabla\psi]]_{\mathbf{r}} = \mathbf{n} \otimes \mathbf{n}(\lambda(\mathbf{r})). \tag{2.52}$$

More generally, for any r-tensor $\mathbf{T}$, we have

$$[[\nabla\nabla\mathbf{T}]]_{\mathbf{r}} = \mathbf{n} \otimes \mathbf{n} \otimes \mathbf{A}(\mathbf{r}), \tag{2.53}$$

where $\mathbf{A}(\mathbf{r})$ is an r-tensor field.

2.6 Useful Formulae

It often happens that a practical problem of continuum mechanics exhibits symmetry properties which suggest the use of particular coordinate systems. For instance, if the problem exhibits cylindrical symmetry, it is convenient to adopt cylindrical coordinates; on the other hand, if the symmetry is spherical, it is convenient to adopt spherical coordinates, and so on. Therefore, it is useful to determine the expressions for divergence, curl, and the Laplacian in these coordinate systems.

Here we limit our discussion to *orthogonal* curvilinear coordinates, so that the metric is written as

$$ds^2 = g_{11}(dy^1)^2 + g_{22}(dy^2)^2 + g_{33}(dy^3)^2. \tag{2.54}$$

Noting that the inverse matrix of (g_{ij}) has the form

$$(g^{ij}) = \begin{pmatrix} \dfrac{1}{g_{11}} & 0 & 0 \\ 0 & \dfrac{1}{g_{22}} & 0 \\ 0 & 0 & \dfrac{1}{g_{33}} \end{pmatrix},$$

from (2.17) we obtain

$$\nabla f = \frac{1}{g_{ii}} \frac{\partial f}{\partial y^i} \mathbf{e}_i. \tag{2.55}$$

2.6. Useful Formulae

Moreover, if we introduce the notation

$$h = g_{11}g_{22}g_{33},$$

then the relation $(2.33)_1$ becomes

$$\nabla \cdot \mathbf{v} = \frac{1}{\sqrt{h}} \frac{\partial}{\partial y^i} \left(\sqrt{h}\, v^i \right), \qquad (2.56)$$

and (2.34) gives

$$\nabla \times \mathbf{v} = \frac{1}{\sqrt{h}} \left[\left(\frac{\partial v_3}{\partial y^2} - \frac{\partial v_2}{\partial y^3} \right) \mathbf{e}_1 + \left(\frac{\partial v_1}{\partial y^3} - \frac{\partial v_3}{\partial y^1} \right) \mathbf{e}_2 \right. \\ \left. + \left(\frac{\partial v_2}{\partial y^1} - \frac{\partial v_1}{\partial y^2} \right) \mathbf{e}_3 \right]. \qquad (2.57)$$

Finally, from (2.39) we derive

$$\Delta f = \frac{1}{\sqrt{h}} \frac{\partial}{\partial y^i} \left(\sqrt{h}\, \frac{1}{g_{ii}} \frac{\partial f}{\partial y^i} \right). \qquad (2.58)$$

The curvilinear components of a vector or tensor do not in general have all the same dimensions as they do in rectangular Cartesian coordinates. For example, the cylindrical contravariant components of a differential vector are $(dr, d\theta, dz)$, which do not have the same dimensions. In physical applications, it is usually desirable to represent each vector component in the same terms as the vector itself, and for this reason we refer to *physical components*. In a general curvilinear coordinate system, the physical components of a vector at a point are defined as the vector components along the natural base vectors, so that we introduce at any point a local basis $\mathbf{a}_i$ formed by *unit* vectors which are parallel to the vectors $\mathbf{e}_i$ of the natural basis, that is,

$$\mathbf{a}_i = \frac{1}{\sqrt{g_{ii}}} \mathbf{e}_i \text{ (no sum over } i\text{);} \qquad (2.59)$$

similarly, we introduce vectors $\mathbf{a}^i$ parallel to the vectors $\mathbf{e}^i$ of the dual basis:

$$\mathbf{a}^i = \sqrt{g_{ii}}\, \mathbf{e}^i.$$

Since these bases are formed by unit vectors, the components of any tensor field with respect to them have a direct physical meaning. We remark that generally these new bases are not natural; that is, there exist no curvilinear coordinates (x'^i) of which $(\mathbf{a}_i)$ are the natural bases. In other words, we are not searching for the expressions for divergence, curl, and the Laplacian in new curvilinear coordinates, but rather we wish to write (2.56), (2.57), and (2.58) in terms of the components with respect to the basis $(\mathbf{a}_i)$.

First, since $(\mathbf{a}_i)$ is an orthonormal basis, it coincides with the reciprocal basis and there is no difference between covariant and contravariant components of any tensor with respect to it:

$$\mathbf{v} = \bar{v}^i \mathbf{a}_i = \bar{v}_i \mathbf{a}_i.$$

From the identity

$$\mathbf{v} = \bar{v}^i \mathbf{a}_i = v^i \mathbf{e}_i$$

and (2.59), we obtain (no sum over i)

$$\bar{v}^i = \sqrt{g_{ii}} v^i = \sqrt{g_{ii}} \, g^{ii} v_i = \frac{v_i}{\sqrt{g_{ii}}}. \qquad (2.60)$$

It is easy to verify, starting from (2.59) and (2.60), that in the new basis $(\mathbf{a}_i)$, we can write (2.27), (2.55), (2.56), and (2.57) as

$$\nabla f = \frac{1}{\sqrt{g_{ii}}} \frac{\partial f}{\partial y^i} \mathbf{a}_i, \qquad (2.61)$$

$$\nabla \cdot \mathbf{v} = \frac{1}{\sqrt{h}} \left[\frac{\partial}{\partial y^1} \left(\sqrt{g_{22} g_{33}} \, \bar{v}^1 \right) + \frac{\partial}{\partial y^2} \left(\sqrt{g_{11} g_{33}} \, \bar{v}^2 \right) \right.$$
$$\left. + \frac{\partial}{\partial y^3} \left(\sqrt{g_{11} g_{22}} \, \bar{v}^3 \right) \right], \qquad (2.62)$$

$$\nabla \times \mathbf{v} = \frac{1}{\sqrt{h}} \left(\frac{\partial(\sqrt{g_{33}} \, \bar{v}_3)}{\partial y^2} - \frac{\partial(\sqrt{g_{22}} \, \bar{v}_2)}{\partial y^3} \right) \sqrt{g_{11}} \mathbf{a}_1$$
$$+ \frac{1}{\sqrt{h}} \left(\frac{\partial(\sqrt{g_{11}} \, \bar{v}_1)}{\partial y^3} - \frac{\partial(\sqrt{g_{33}} \, \bar{v}_3)}{\partial y^1} \right) \sqrt{g_{22}} \mathbf{a}_2 \qquad (2.63)$$
$$+ \frac{1}{\sqrt{h}} \left(\frac{\partial(\sqrt{g_{22}} \, \bar{v}_2)}{\partial y^1} - \frac{\partial(\sqrt{g_{11}} \, \bar{v}_1)}{\partial y^2} \right) \sqrt{g_{33}} \mathbf{a}_3,$$

$$\nabla \mathbf{v} = \frac{1}{\sqrt{g_{hh}} \sqrt{g_{ll}}} \left((\sqrt{g_{ll}} \, \bar{v}_l)_{,h} - \Gamma^k_{lh} \sqrt{g_{kk}} \, \bar{v}_k \right) \mathbf{a}^h \otimes \mathbf{a}^l, \qquad (2.64)$$

while (2.58) remains the same.

2.7 Some Curvilinear Coordinates

This section is devoted to the description of some curvilinear coordinate systems, usually used to solve some problems characterized by specific symmetry properties. In Section 2.2, we described polar, spherical, and cylindrical coordinates; in this section, we evaluate some differential formulae of functions, vectors, and 2-tensors with respect to the basis $(\mathbf{a}_i)$.

2.7. Some Curvilinear Coordinates

For the other curvilinear coordinates described here,[2] we will present only the formulae that transform them to Cartesian coordinates as well as the plots of coordinate curves.[3] However, all the other metric characteristics and the differential formulae in these coordinates can be easily obtained by using the program Operator in the package Mechanics, which will be described in the last section of this chapter.

2.7.1 Generalized Polar Coordinates

The curvilinear coordinates (r, φ), related to Cartesian coordinates (u^1, u^2) by the functions

$$\begin{cases} u^1 = ra\cos\varphi, \\ u^2 = rb\sin\varphi, \end{cases} \quad a, b \in \Re^+,$$

are called **generalized polar coordinates** (see Section 2.2). The coordinate curves are the half-lines coming out from the origin (see Figure 2.6)

$$\tan\varphi = \frac{au^2}{bu^1}$$

as well as the ellipses

$$\frac{(u^1)^2}{a^2} + \frac{(u^2)^2}{b^2} = r^2.$$

The vectors tangent to the coordinate curves are

$$\begin{cases} \mathbf{e}_1 = a\cos\varphi\,\mathbf{u}_1 + b\sin\varphi\,\mathbf{u}_2, \\ \mathbf{e}_2 = -ra\sin\varphi\,\mathbf{u}_1 + rb\cos\varphi\,\mathbf{u}_2, \end{cases}$$

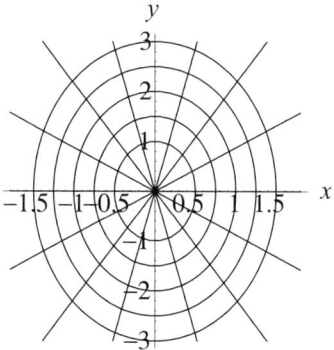

Figure 2.6

---

[2] Except for the generalized polar coordinates, all the other ones are orthogonal.
[3] These plots have been obtained by the built-in function ContourPlot of Mathematica®.

where $(\mathbf{u}_1, \mathbf{u}_2)$ represent the unit vectors along the axes Ou^1 and Ou^2, respectively. Consequently, the metric matrix becomes

$$\begin{pmatrix} a^2 \cos^2 \varphi + b^2 \sin^2 \varphi & r \sin \varphi \cos \varphi \left(b^2 - a^2\right) \\ r \sin \varphi \cos \varphi \left(b^2 - a^2\right) & r^2 \left(b^2 \cos^2 \varphi + a^2 \sin^2 \varphi\right) \end{pmatrix}.$$

Since $g_{12} = g_{21} = r \sin \varphi \cos \varphi \left(b^2 - a^2\right)$, the natural basis and the angle between $\mathbf{e}_1$ and $\mathbf{e}_2$ vary with the point; moreover, this angle is generally not a right angle. Finally, the nonvanishing Christoffel symbols are

$$\Gamma^1_{22} = -r, \ \Gamma^2_{12} = \frac{1}{r}.$$

We note that, when $a = b = 1$, the transformation above reduces to the polar transformation (2.13).

Let F be any function, let $\mathbf{v}$ be a vector field, and let $\mathbf{T}$ be a 2-tensor. If the components are all evaluated in the unit basis of a polar coordinate system, then the following formulae hold:

$$\nabla F = \frac{\partial F}{\partial r} \mathbf{a}_1 + \frac{1}{r} \frac{\partial F}{\partial \varphi} \mathbf{a}_2,$$

$$\Delta F = \frac{1}{r} \frac{\partial}{\partial r} \left(r \frac{\partial F}{\partial r} \right) + \frac{1}{r^2} \frac{\partial^2 F}{\partial \varphi^2},$$

$$\nabla \cdot \mathbf{v} = \frac{1}{r} \frac{\partial (r v_r)}{\partial r} + \frac{1}{r} \frac{\partial v_\varphi}{\partial \varphi},$$

$$\nabla \cdot \mathbf{T} = \left[\frac{1}{r} \frac{\partial}{\partial r} (r T_{rr}) + \frac{1}{r} \frac{\partial T_{r\varphi}}{\partial \varphi} - \frac{1}{r} T_{\varphi\varphi} \right] \mathbf{a}_1$$
$$+ \left[\frac{1}{r} \frac{\partial (r T_{r\varphi})}{\partial r} + \frac{1}{r} \frac{\partial}{\partial \varphi} (T_{\varphi\varphi}) + \frac{1}{r} T_{\varphi r} \right] \mathbf{a}_2.$$

2.7.2 Cylindrical Coordinates

Let P' be the orthogonal projection of a point $P(u^1, u^2, u^3)$ onto the plane $u^3 = 0$, and let (r, θ) be the polar coordinates of P' in this plane (see Figure 2.1). The variables (r, θ, z) are called **cylindrical coordinates** and are related to Cartesian coordinates (u^1, u^2, u^3) by the equations

$$\begin{cases} u^1 = r \cos \theta, \\ u^2 = r \sin \theta, \\ u^3 = z. \end{cases}$$

By using the results of the previous sections, it is possible to evaluate the following expressions for the differential operators in the unit basis

2.7. Some Curvilinear Coordinates

$\{\mathbf{a}_i\}_{i=1,2,3}$ that is associated to the natural basis $\{\mathbf{e}_i\}_{i=1,2,3}$ (see Section 2.2):

$$\nabla F = \frac{\partial F}{\partial r}\mathbf{a}_1 + \frac{1}{r}\frac{\partial F}{\partial \theta}\mathbf{a}_2 + \frac{\partial F}{\partial z}\mathbf{a}_3,$$

$$\Delta F = \frac{1}{r}\frac{\partial}{\partial r}\left(r\frac{\partial F}{\partial r}\right) + \frac{1}{r^2}\frac{\partial^2 F}{\partial \theta^2} + \frac{\partial^2 F}{\partial z^2},$$

$$\nabla \cdot \mathbf{v} = \frac{1}{r}\frac{\partial(rv_r)}{\partial r} + \frac{1}{r}\frac{\partial v_\theta}{\partial \theta} + \frac{\partial v_z}{\partial z},$$

$$\nabla \times \mathbf{v} = \left(\frac{1}{r}\frac{\partial v_z}{\partial \theta} - \frac{\partial v_\theta}{\partial z}\right)\mathbf{a}_1 + \left(\frac{\partial v_r}{\partial z} - \frac{\partial v_z}{\partial r}\right)\mathbf{a}_2$$
$$+ \left(\frac{1}{r}\frac{\partial(rv_\theta)}{\partial r} - \frac{1}{r}\frac{\partial v_r}{\partial \theta}\right)\mathbf{a}_3.$$

$$\nabla \cdot \mathbf{T} = \left[\frac{1}{r}\frac{\partial}{\partial r}(rT_{rr}) + \frac{1}{r}\frac{\partial T_{\theta r}}{\partial \theta} + \frac{\partial T_{zr}}{\partial z} - \frac{1}{r}T_{\theta\theta}\right]\mathbf{a}_1$$
$$+ \left[\frac{1}{r}\frac{\partial T_{\theta\theta}}{\partial \theta} + \frac{\partial T_{z\theta}}{\partial z} + \frac{1}{r}\frac{\partial}{\partial r}(rT_{r\theta}) + \frac{1}{r}T_{\theta r}\right]\mathbf{a}_2$$
$$+ \left[\frac{\partial T_{zz}}{\partial z} + \frac{1}{r}\frac{\partial}{\partial r}(rT_{rz}) + \frac{1}{r}\frac{\partial T_{\theta z}}{\partial \theta}\right]\mathbf{a}_3,$$

where $F, \mathbf{v}$, and $\mathbf{T}$ denote a scalar, vector, and 2-tensor whose components are evaluated in $(\mathbf{a}_i)$.

2.7.3 Spherical Coordinates

The **spherical coordinates** (see Section 2.2) (r, θ, φ) are connected to Cartesian coordinates by the relations

$$\begin{cases} u^1 = r\sin\theta\cos\varphi, \\ u^2 = r\sin\theta\sin\varphi, \\ u^3 = r\cos\theta, \end{cases}$$

and the differential operators in the unit basis are

$$\nabla F = \frac{\partial F}{\partial r}\mathbf{a}_1 + \frac{1}{r}\frac{\partial F}{\partial \theta}\mathbf{a}_2 + \frac{1}{r\sin\theta}\frac{\partial F}{\partial \varphi}\mathbf{a}_3,$$

$$\Delta F = \frac{1}{r^2}\frac{\partial}{\partial r}\left(r^2\frac{\partial F}{\partial r}\right) + \frac{1}{r^2\sin\theta}\frac{\partial}{\partial \theta}\left(\sin\theta\frac{\partial F}{\partial \theta}\right) + \frac{1}{r^2\sin^2\theta}\frac{\partial^2 F}{\partial \varphi^2},$$

$$\nabla \cdot \mathbf{v} = \frac{1}{r^2}\frac{\partial(r^2 v_r)}{\partial r} + \frac{1}{r\sin\theta}\frac{\partial}{\partial \theta}(\sin\theta v_\theta) + \frac{1}{r\sin\theta}\frac{\partial v_\varphi}{\partial \varphi},$$

$$\nabla \times \mathbf{v} = \frac{1}{r\sin\theta}\left(\frac{\partial}{\partial\theta}(v_\varphi \sin\theta) - \frac{\partial v_\theta}{\partial\varphi}\right)\mathbf{a}_1$$
$$+ \frac{1}{r}\left(\frac{1}{\sin\theta}\frac{\partial v_r}{\partial\varphi} - \frac{\partial(rv_\varphi)}{\partial r}\right)\mathbf{a}_2 + \frac{1}{r}\left(\frac{\partial(rv_\theta)}{\partial r} - \frac{\partial v_r}{\partial\theta}\right)\mathbf{a}_3.$$

We omit the very long expression of $\nabla \cdot \mathbf{T}$, which can be obtained by the program Operator.

2.7.4 Elliptic Coordinates

Let P_1 and P_2 denote two fixed points $(-c, 0)$ and $(c, 0)$ on the u^1-axis. Let r_1 and r_2 be the distances from P_1 and P_2 of a variable point P in the plane Ou^1u^2. The **elliptic coordinates** are the variables

$$\left(\xi \equiv \frac{r_1 + r_2}{2c}, \eta \equiv \frac{r_1 - r_2}{2c}, z\right),$$

where $\xi \leq 1$ and $-1 \leq \eta \leq 1$. These variables are related to Cartesian coordinates by the following relations (see Figure 2.7):

$$\begin{cases} u^1 = c\xi\eta, \\ u^2 = c\sqrt{(\xi^2 - 1)(1 - \eta^2)}, \quad c \in \Re \\ u^3 = z. \end{cases}$$

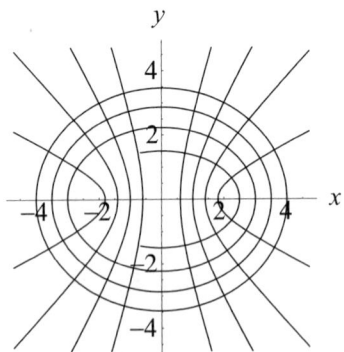

Figure 2.7 Elliptic coordinates

2.7. Some Curvilinear Coordinates

2.7.5 Parabolic Coordinates

The **parabolic coordinates** (ξ, η, z) are related to Cartesian coordinates (u^1, u^2, u^3) by the relations (see Figure 2.8)

$$\begin{cases} u^1 = \frac{1}{2}\left(\eta^2 - \xi^2\right), \\ u^2 = \xi\eta, \\ u^3 = -z. \end{cases}$$

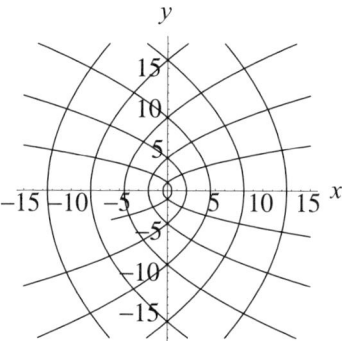

Figure 2.8 Parabolic coordinates

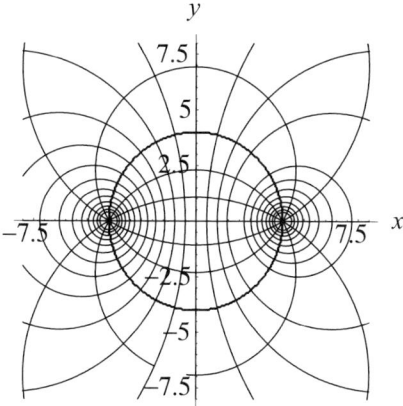

Figure 2.9 Bipolar coordinates

2.7.6 Bipolar Coordinates

The transformations

$$\begin{cases} u^1 = \dfrac{a \sinh \xi}{(\cosh \xi - \cos \eta)}, \\ u^2 = \dfrac{a \sin \eta}{(\cosh \xi - \cos \eta)}, \quad a \in \Re \\ u^3 = z, \end{cases}$$

define the relation between the Cartesian coordinates (u^1, u^2, u^3) and the **bipolar coordinates** (ξ, η, z) (see Figure 2.9).

2.7.7 Prolate and Oblate Spheroidal Coordinates

The transformation equations

$$\begin{cases} u^1 = c\xi\eta, \\ u^2 = c\sqrt{(\xi^2 - 1)(1 - \eta^2)} \cos \varphi, \quad c \in \Re \\ u^3 = c\sqrt{(\xi^2 - 1)(1 - \eta^2)} \sin \varphi, \end{cases}$$

where

$$\xi \geq 1, \quad -1 \leq \eta \leq 1, \quad 0 \leq \varphi \leq 2\pi,$$

express the relation between **prolate spheroidal coordinates** (ξ, η, φ) and Cartesian coordinates (u^1, u^2, u^3).

Similarly, the **oblate spheroidal coordinates** (ξ, η, φ) are defined by the following transformation formulae:

$$\begin{cases} u^1 = c\xi\eta \sin \varphi, \\ u^2 = c\sqrt{(\xi^2 - 1)(1 - \eta^2)}, \quad c \in \Re \\ u^3 = c\xi\eta \cos \varphi, \end{cases}$$

with the same conditions on ξ, η, and φ.

The spheroidal coordinates in the space are obtained by rotating the ellipses of Figure 2.7 around one of its symmetry axes. More specifically, if the rotation is around the major axis x, then the spheroidal prolate coordinates are generated; if the rotation is around the minor axis y, then the coordinates are called oblate. To conclude, we note that the spheroidal coordinates reduce to the spherical ones when the focal distance $2c$ and the eccentricity go to zero.

2.7.8 Paraboloidal Coordinates

The variables (ξ, η, φ), defined by the functions

$$\begin{cases} u^1 = \xi\eta \cos\varphi, \\ u^2 = \xi\eta \sin\varphi, \\ u^3 = \dfrac{1}{2}(\xi^2 - \eta^2), \end{cases}$$

are called **paraboloidal coordinates**. These coordinates are obtained by rotating the parabolas of Figure 2.8 around their symmetry axis x.

2.8 Exercises

1. Write formulae (2.54)–(2.63) in cylindrical and spherical coordinates by using the results of Section 2.2.

2. Find the restrictions on the components of a vector field $\mathbf{u}(r, \varphi)$ in order for it to be uniform ($\nabla \mathbf{u} = \mathbf{0}$) in polar coordinates (r, φ).

3. Verify that $\nabla \times \nabla f = \mathbf{0}$ for any function f and that $\nabla \cdot \nabla \times \mathbf{u} = 0$ for any vector field $\mathbf{u}$.

4. Determine the directional derivative of the function $x^2 - xy + y^2$ along the vector $\mathbf{n} = \mathbf{i} + 2\mathbf{j}$, where $\mathbf{i}$ and $\mathbf{j}$ are unit vectors along the Cartesian axes Ox and Oy.

5. Using the components, prove that

$$\nabla \times (\mathbf{u} \times \mathbf{v}) = \mathbf{v} \cdot \nabla \mathbf{u} - \mathbf{v}\nabla \cdot \mathbf{u} + \mathbf{u}\nabla \cdot \mathbf{v} - \mathbf{u} \cdot \nabla \mathbf{v}.$$

6. Verify that the volume V enclosed by the surface σ can be written as

$$V = \frac{1}{6}\int_\sigma \nabla(r^2) \cdot \mathbf{n}\, d\sigma,$$

where $r^2 = x_i x_i$ and $\mathbf{n}$ is the external unit vector normal to σ.

7. With the notation of the previous exercise, prove the formula

$$V\mathbf{I} = \int_\sigma \mathbf{r} \otimes \mathbf{n}\, d\sigma,$$

where $\mathbf{I}$ is the identity matrix.

2.9 The Program Operator

Aim of the Program

When a transformation from *curvilinear* to *Cartesian coordinates* is given, the program `Operator` evaluates the Jacobian matrix of this transformation, the vectors of the natural basis, the metric coefficients, and all the significant nonvanishing Christoffel symbols. Moreover, if the curvilinear coordinates are *orthogonal*, then all the differential operators are evaluated with respect to the unit basis for functions, vector fields, and 2-tensor fields. In particular, if all these fields are assigned in Cartesian coordinates, but we wish to find the operators in curvilinear coordinates, then the program first supplies the components in these coordinates and then evaluates the expressions for the operators. Finally, fixing for the input parameter `characteristic` the attribute `symbolic`, the program gives all the differential operators in symbolic form, without making explicit the functional dependence of the components on the curvilinear coordinates. In all other cases, the parameter `characteristic` has the attribute `numeric`.

Description of the Algorithm

The program uses the theory presented in the previous sections to determine both the geometric characteristics and the differential operator expressions of functions, vector fields, and 2-tensor fields.

Command Line of the Program Operator

`Operator[tensor, var, transform, characteristic, option, simplifyoption]`

Parameter List

Input Data

`tensor` = {}, {function}, or components of a vector or 2-tensor field in the unit basis associated with the curvilinear coordinates `var` or in the Cartesian coordinates $\{x, y, z\}$;

`var` = curvilinear or Cartesian coordinates;

`transform` = right-hand side of coordinate transformation from curvilinear coordinates `var` to Cartesian coordinates, or of the identity transformation;

2.9. The Program Operator

- **characteristic** = **symbolic** or **numeric**. The first choice is adopted when the functional dependence of the fields on the variables **var** is not given, and in this case the output is symbolic. The second choice refers to other cases.

- **option** = **metric**, **operator**, or **all**. The program gives outputs relative to metric characteristics of the curvilinear coordinates, to differential operators of **tensor**, or to both of them.

- **simplifyoption** = **true** or **false**. In the first case, the program gives many simplifications of symbolic expressions.

Output Data

Jacobian transformation matrix;

natural basis;

metric matrix;

inverse of the metric matrix;

Christoffel symbols;

unit basis associated with the natural one;

gradient and Laplacian of a scalar field;

gradient, divergence, and curl of a vector field;

gradient and divergence of a 2-tensor field;

Use Instructions

We have already said that the input datum **transform** represents the right-hand sides of the transformations from curvilinear coordinates **var** to Cartesian coordinates. In the program, the latter are expressed by the letters x, y, z. Consequently, the program fails if the input parameter **option** is equal to **operator**, **all**, or if the introduced coordinates invert the orientation of one or more Cartesian axes. To overcome this difficulty, we can choose variables in **transform** which are different from the corresponding Cartesian ones. For example, if we wish to evaluate the differential operators of the function F in parabolic coordinates, which have the transformation equations

$$\begin{cases} x = \frac{1}{2}(\eta^2 - \xi^2), \\ y = \xi\eta, \\ z = -z, \end{cases}$$

in the case where the orientation of z-axis is inverted, the previous transformation can be written as

$$\begin{cases} x = \dfrac{1}{2}(\eta^2 - \xi^2), \\ y = \xi\eta, \\ z = -k. \end{cases}$$

Moreover, we recall that the program does not evaluate the differential operators when the curvilinear coordinates are not orthogonal. In this case, it gives only the geometric characteristic.

Worked Examples

1. In order to determine the metric characteristics of the elliptic coordinates $\{\xi, \eta, z\}$ (see Section 2.7), it is sufficient to input the following data:

   tensor = {};

   var = {ξ, η, z};

   transform = {cξη, c√((ξ² − 1)(1 − η²)), z};

   characteristic = symbolic;

   option = metric;

   simplifyoption = true;

   Operator[tensor, var, transform, characteristic, option, simplifyoption]

   The program gives the following output:

   Jacobian matrix

   $$\begin{pmatrix} c\eta & c\xi & 0 \\ c\xi\sqrt{\dfrac{1-\eta^2}{-1+\xi^2}} & -c\eta\sqrt{\dfrac{1-\xi^2}{-1+\eta^2}} & 0 \\ 0 & 0 & 1 \end{pmatrix}$$

   Natural basis

   $$\mathbf{e}_1 = c\eta \mathbf{u}_1 - c\xi\sqrt{\dfrac{1-\eta^2}{-1+\xi^2}}\mathbf{u}_2$$

   $$\mathbf{e}_2 = c\xi \mathbf{u}_1 - c\eta\sqrt{\dfrac{1-\xi^2}{-1+\eta^2}}\mathbf{u}_2$$

   $$\mathbf{e}_3 = \mathbf{u}_3$$

2.9. The Program Operator

Metric matrix

$$\begin{pmatrix} \dfrac{c^2\left(-\eta^2+\xi^2\right)}{-1+\xi^2} & 0 & 0 \\ 0 & \dfrac{c^2\left(\eta^2-\xi^2\right)}{-1+\eta^2} & 0 \\ 0 & 0 & 1 \end{pmatrix}$$

Inverse metric matrix

$$\begin{pmatrix} \dfrac{-1+\xi^2}{c^2\left(-\eta^2+\xi^2\right)} & 0 & 0 \\ 0 & \dfrac{-1+\eta^2}{c^2\left(\eta^2-\xi^2\right)} & 0 \\ 0 & 0 & 1 \end{pmatrix}$$

Christoffel symbols

$$\Gamma^1_{11} = \dfrac{\xi\left(\eta^2-1\right)}{\left(\xi^2-1\right)\left(\xi^2-\eta^2\right)}$$

$$\Gamma^1_{12} = \dfrac{\eta}{\eta^2-\xi^2}$$

$$\Gamma^1_{22} = \dfrac{\xi\left(\xi^2-1\right)}{\left(\eta^2-1\right)\left(\xi^2-\eta^2\right)}$$

$$\Gamma^2_{11} = \dfrac{\eta\left(\eta^2-1\right)}{\left(\xi^2-1\right)\left(\eta^2-\xi^2\right)}$$

$$\Gamma^2_{12} = \dfrac{\xi}{\left(\xi^2-\eta^2\right)}$$

$$\Gamma^2_{22} = \dfrac{\eta\left(\xi^2-1\right)}{\left(\eta^2-1\right)\left(\eta^2-\xi^2\right)}$$

Unit basis

$$\mathbf{a}_1 = \dfrac{\eta \mathbf{u}_1}{\sqrt{\dfrac{-\eta^2+\xi^2}{-1+\xi^2}}} + \xi\sqrt{\dfrac{-1+\eta^2}{\eta^2-\xi^2}}\,\mathbf{u}_2$$

$$\mathbf{a}_2 = \dfrac{\xi \mathbf{u}_1}{\sqrt{\dfrac{\eta^2-\xi^2}{-1+\eta^2}}} - \eta\sqrt{\dfrac{-1+\xi^2}{-\eta^2+\xi^2}}\,\mathbf{u}_2$$

$$\mathbf{a}_3 = \mathbf{u}_3$$

2. Evaluate the gradient and the Laplacian of the function

$$F = x^2 + xy + y^2,$$

in the unit basis associated with the polar coordinates. To use the program `Operator` we input the following:

`tensor` = $\{x^2 + xy + y^2\}$;

`var` = $\{r, \varphi\}$;

`transform` = $\{r\text{Cos}[\varphi], r\text{Sin}[\varphi]\}$;

`characteristic` = `numeric`;

`option` = `operator`;

`simplifyoption` = `true`;

`Operator[tensor, var, transform, characteristic, option, simplifyoption]`

In output we obtain

The function $F = x^2 + xy + y^2$ in the coordinates $\{r, \varphi\}$ is written as $F = r^2(1 + \text{Cos}[\varphi]\text{Sin}[\varphi])$

Differential operators

Gradient

$$\nabla F = \begin{pmatrix} r\left(2 + \text{Sin}[2\varphi]\right) \\ r\text{Cos}[2\varphi] \end{pmatrix}$$

Laplacian

$$\triangle F = 4$$

3. Let $\{\xi, \eta, k\}$ be parabolic coordinates and let F be any function depending on these coordinates. The program `Operator` allows us to calculate both the geometric characteristic of the coordinates $\{\xi, \eta, k\}$ and the gradient and Laplacian of F by introducing the following data:

`tensor` = $\{F\}$;

`var` = $\{\xi, \eta, k\}$;

`transform` = $\{\frac{1}{2}(\xi^2 - \eta^2), \xi\eta, -k\}$;

`characteristic` = `symbolic`;

`option` = `all`;

`simplifyoption` = `true`;

`Operator[tensor, var, transform, characteristic, option, simplifyoption]`

We obtain this output:

2.9. The Program Operator

Jacobian matrix

$$\begin{pmatrix} \xi & -\eta & 0 \\ \eta & \xi & 0 \\ 0 & 0 & -1 \end{pmatrix}$$

Natural basis

$$\mathbf{e}_1 = \xi \mathbf{u}_1 + \eta \mathbf{u}_2$$
$$\mathbf{e}_2 = -\eta \mathbf{u}_1 + \xi \mathbf{u}_2$$
$$\mathbf{e}_3 = -\mathbf{u}_3$$

Metric matrix

$$\begin{pmatrix} \xi^2 + \eta^2 & 0 & 0 \\ 0 & \xi^2 + \eta^2 & 0 \\ 0 & 0 & 1 \end{pmatrix}$$

Inverse metric matrix

$$\begin{pmatrix} \dfrac{1}{\xi^2 + \eta^2} & 0 & 0 \\ 0 & \dfrac{1}{\xi^2 + \eta^2} & 0 \\ 0 & 0 & 1 \end{pmatrix}$$

Christoffel symbols

$$\Gamma^1_{11} = \frac{\xi}{\xi^2 + \eta^2}$$

$$\Gamma^1_{12} = \frac{\eta}{\xi^2 + \eta^2}$$

$$\Gamma^1_{22} = -\frac{\xi}{\xi^2 + \eta^2}$$

$$\Gamma^2_{11} = -\frac{\eta}{\xi^2 + \eta^2}$$

$$\Gamma^2_{12} = \frac{\xi}{\xi^2 + \eta^2}$$

$$\Gamma^2_{22} = \frac{\eta}{\xi^2 + \eta^2}$$

Unit basis

$$\mathbf{a}_1 = \frac{\xi}{\sqrt{\xi^2 + \eta^2}} \mathbf{u}_1 + \frac{\eta}{\sqrt{\xi^2 + \eta^2}} \mathbf{u}_2$$

$$\mathbf{a}_2 = -\frac{\eta}{\sqrt{\xi^2 + \eta^2}} \mathbf{u}_1 + \frac{\xi}{\sqrt{\xi^2 + \eta^2}} \mathbf{u}_2$$

$a_3 = -u_3$

Differential operators for the function F $[\xi, \eta, k]$

Gradient

$$\nabla F = \begin{pmatrix} \dfrac{1}{\sqrt{\xi^2 + \eta^2}} \partial_\xi [F] \\ \dfrac{1}{\sqrt{\xi^2 + \eta^2}} \partial_\eta [F] \\ \partial_k [F] \end{pmatrix}$$

Laplacian

$$\Delta F = \dfrac{(\xi^2 + \eta^2)\partial_{k^2}[F] + \partial_{\eta^2}[F] + \partial_{\xi^2}[F]}{\xi^2 + \eta^2}$$

Exercises

1. Apply the program Operator to the coordinates introduced in Section 2.7 to determine their geometric characteristics.

2. Using the program Operator, evaluate the differential operators of any scalar, vector, or 2-tensor field both in Cartesian coordinates and those of Exercise 1.

3. Evaluate the gradient and Laplacian of the function $F = x^2 + xy + y^2 + z$ in Cartesian and spherical coordinates.

4. Determine the differential operators of the vector field $\mathbf{v} = r\mathbf{e}_r + \cos\varphi\, \mathbf{e}_\varphi + \sin\theta\, \mathbf{e}_\theta$ in spherical coordinates.

5. Evaluate the differential operators of the tensor $\{\{x, y\}, \{x, y\}\}$ in polar coordinates.

Chapter 3

Finite and Infinitesimal Deformations

3.1 Deformation Gradient

Let S be a three-dimensional continuous body and let $R = (O, \mathbf{e}_i)$ be a rectangular coordinate system in the Euclidean space E_3 (see Chapter 1). Moreover, let C_* and C denote two **configurations** of S at different times. Any point of S in the **reference configuration** C_*, identified from now on by the label $\mathbf{X}$, is called a **material point**; the same point in the **actual** or **current configuration** C is identified by the position $\mathbf{x}$ and is called a **spatial point**. Furthermore, (x_i) and (X_L), with $i, L = 1, 2, 3$, are respectively the **spatial** and **material** coordinates of the particle $\mathbf{X}$ in $(O, \mathbf{e}_i)$.[1]

When a rectangular coordinate system is adopted ($g_{ij} \equiv \delta_{ij}$), the position of indices is immaterial since the covariant and contravariant components coincide.

A **finite deformation** from C_* to C is defined as the vector function

$$\mathbf{x} = \mathbf{x}(\mathbf{X}) \tag{3.1}$$

which maps any $\mathbf{X} \in C_*$ onto the corresponding $\mathbf{x} \in C$.

In order to preserve the basic property of matter that two particles cannot simultaneously occupy the same place, we require that $\mathbf{x}(\mathbf{X}) \neq \mathbf{x}(\mathbf{Y})$ if $\mathbf{X} \neq \mathbf{Y}$, for all $\mathbf{X}, \mathbf{Y} \in C_*$, so that (3.1) is a diffeomorphism preserving the topological properties of the reference configuration.

In terms of coordinates, (3.1) is equivalent to the three scalar functions

$$x_i = x_i(X_L), \tag{3.2}$$

---

[1] From now on, capital indices are used to identify quantities defined in C_*, while lower-case indices identify quantities in C.

which are assumed to be of class C^1, together with their inverse functions. It follows that the Jacobian must be different from zero:

$$J = \det\left(\frac{\partial x_i}{\partial X_L}\right) \neq 0. \tag{3.3}$$

More particularly, it is supposed that[2]

$$\det \mathbf{F} > 0. \tag{3.4}$$

Differentiation of (3.2) gives

$$dx_i = \frac{\partial x_i}{\partial X_L} dX_L, \tag{3.5}$$

so that, at any $\mathbf{X} \in C_*$, a linear mapping is defined, called the ***deformation gradient*** in $\mathbf{X}$. It is identified by the tensor

$$\mathbf{F} = (F_{iL}) = \left(\frac{\partial x_i}{\partial X_L}\right), \tag{3.6}$$

which maps the infinitesimal material vector $d\mathbf{X}$, emanating from $\mathbf{X}$, onto the corresponding vector $d\mathbf{x}$, emanating from $\mathbf{x}(\mathbf{X})$. In particular, if $\mathbf{F}$ does not depend on $\mathbf{X} \in C_*$, then the deformation is called *homogeneous*. Note that the linear mapping (3.6) locally describes the body deformation when passing from C_* to C. To make this more explicit, we start by noting that, due to condition (3.3), Cauchy's polar decomposition theorem 1.10 can be applied: it is always possible to find a proper *orthogonal matrix* $\mathbf{R}$ (i.e. $\mathbf{RR}^T = \mathbf{R}^T\mathbf{R} = \mathbf{I}$) and two *symmetric* and *positive definite* matrices $\mathbf{U}$ and $\mathbf{V}$ such that

$$\mathbf{F} = \mathbf{RU} = \mathbf{VR}. \tag{3.7}$$

In order to explore the meaning of this decomposition, we start with $(3.7)_1$. The matrix $\mathbf{U}$ is a linear mapping in $\mathbf{X}$; i.e., it maps the vector space at $\mathbf{X}$ into itself, whereas $\mathbf{R}$ defines an orthogonal mapping of vectors in $\mathbf{X}$ into vectors in $\mathbf{x}(\mathbf{X})$. For this reason, the second one is sometimes referred to as a two-point tensor (see Section 3.6).

Because $\mathbf{U}$ is symmetric and positive definite, its eigenvalues λ_L are real and positive, with corresponding mutually orthogonal eigenvectors $(\mathbf{u}_L)$. In addition, the mapping $\mathbf{R}$ transforms orthonormal bases in $\mathbf{X}$ into orthonormal bases in $\mathbf{x}(\mathbf{X})$, so that, if $d\mathbf{X} = dX_L \mathbf{u}_L$ and $\mathbf{e}_L = \mathbf{R}(\mathbf{u}_L)$, then

$$d\mathbf{x} = \mathbf{F} d\mathbf{X} = \mathbf{RU}(dX_L \mathbf{u}_L) = \lambda_L dX_L \mathbf{Ru}_L$$
$$= \lambda_1 dX_1 \mathbf{e}_1 + \lambda_2 dX_2 \mathbf{e}_2 + \lambda_3 dX_3 \mathbf{e}_3.$$

---

[2] This hypothesis will be motivated in the next chapter.

3.1. Deformation Gradient

Therefore, the vector $d\mathbf{x}$ has components $(\lambda_1 dX_1, \lambda_2 dX_2, \lambda_3 dX_3)$ relative to the basis $(\mathbf{e}_L)$ and the following is proved:

Theorem 3.1
In the deformation $C_* \to C$, the elementary parallelepiped dC_* which has a vertex at $\mathbf{X}$ and edges $(dX_1\mathbf{u}_1, dX_2\mathbf{u}_2, dX_3\mathbf{u}_3)$ parallel to the eigenvectors $(\mathbf{u}_L)$ of $\mathbf{U}$ is transformed into the parallelepiped at $\mathbf{x}(\mathbf{X})$ in three steps: stretching any dX_L of dC_* of the entity λ_L; transporting the element obtained with a rigid translation in $\mathbf{x}(\mathbf{X})$; and finally rotating $\lambda_L dX_L \mathbf{u}_L$ under $\mathbf{R}$, in order to superimpose this edge on the axis $(\mathbf{e}_L)$.

We observe that in any rigid translation the rectilinear components of a vector are not affected, whereas they change for a curvilinear coordinate change, due to the local character of the basis. This topic will be addressed in more detail in Section 3.9.

The decomposition expressed by $(3.7)_1$ allows the introduction of the following definitions:

$\mathbf{R} = $ ***rotation tensor***,
$\mathbf{U} = $ ***right stretching tensor***,
$\mathbf{V} = $ ***left stretching tensor***,
$\lambda_L = $ ***principal stretching in the direction of*** $\mathbf{u}_L$,
$\mathbf{u}_L = $ ***principal direction of stretching***.

We recall how the tensors $\mathbf{U}$ and $\mathbf{R}$ can be represented in terms of $\mathbf{F}$ (see Theorem 1.8):

$$\mathbf{U} = \sqrt{\mathbf{F}^T \mathbf{F}}, \qquad \mathbf{R} = \mathbf{F}\mathbf{U}^{-1}. \tag{3.8}$$

The meaning of $(3.8)_1$ is that the matrix $\mathbf{U}$ is represented, in the basis of the eigenvectors $(\mathbf{u}_L)$ of $\mathbf{F}^T\mathbf{F}$, by the diagonal matrix $\mathbf{U}'$, having as diagonal terms the square roots of the eigenvalues of $\mathbf{F}^T\mathbf{F}$.

The matrix $\mathbf{U}$ relative to any new basis $(\mathbf{e}_L)$ is given by the transformation rule

$$\mathbf{U} = \mathbf{A}\mathbf{U}'\mathbf{A}^T,$$

where $\mathbf{A}$ is the matrix representing the transformation $(\mathbf{u}_L) \longrightarrow (\mathbf{e}_L)$ (i.e., $\mathbf{e}_i = A_{ik}\mathbf{u}_k$; $U_{ij} = A_{ih}A_{jk}U'_{hk}$).

In any case, the evaluation of the matrix $\mathbf{U}$ is not easy, since its components are, in general, irrational (see Exercise 1). By inspecting $(3.8)_1$, we see that it may be more convenient to use the matrix $\mathbf{F}^T\mathbf{F}$ directly, rather than $\mathbf{U}$. Furthermore, we note that since $\mathbf{F}$ includes effects deriving from both the deformation and the rigid rotation $\mathbf{R}$, its use in constitutive laws is not appropriate, due to the fact that no changes in the state of stress are due to a rigid rotation.

These issues indicate the need to introduce other measures of deformation.

3.2 Stretch Ratio and Angular Distortion

Let $d\mathbf{x}$ be an infinitesimal vector in C corresponding to $d\mathbf{X}$ in C_*. The simplest (and most commonly used) measure of the extensional strain of an infinitesimal element is the **stretch ratio** (or simply **stretch**). This measure depends on the direction of the unit vector $\mathbf{u}_*$ of $d\mathbf{X}$ and is expressed by the scalar quantity

$$\delta_{\mathbf{u}_*} = \frac{|d\mathbf{x}|}{|d\mathbf{X}|}, \qquad (3.9)$$

while the quantity

$$\delta_{\mathbf{u}_*} - 1 = \frac{|d\mathbf{x}| - |d\mathbf{X}|}{|d\mathbf{X}|} \qquad (3.10)$$

defines the **longitudinal unit extension** (always along the direction of $\mathbf{u}_*$).

Moreover, the change in angle Θ_{12} between two arbitrary vectors $d\mathbf{X}_1$ and $d\mathbf{X}_2$, both emanating from $\mathbf{X} \in C_*$, is called the **shear** γ_{12} and it is defined as

$$\gamma_{12} = \Theta_{12} - \theta_{12}, \qquad (3.11)$$

where θ_{12} is the angle between $d\mathbf{x}_1$ and $d\mathbf{x}_2$ in C corresponding to $d\mathbf{X}_1$ and $d\mathbf{X}_2$.

The **right Cauchy–Green tensor** (Green, 1841) is defined as the symmetric tensor

$$\mathbf{C} = \mathbf{F}^T \mathbf{F}, \qquad C_{LM} = \frac{\partial x_k}{\partial X_L} \frac{\partial x_k}{\partial X_M}. \qquad (3.12)$$

By taking into account $(3.8)_1$, we see that it is also

$$\mathbf{C} = \mathbf{U}^2, \qquad (3.13)$$

so that $\mathbf{C}$ is a positive definite tensor with the following properties:

3.2. Stretch Ratio and Angular Distortion

Theorem 3.2

1. The stretch ratio in the direction of the unit vector $\mathbf{u}_*$ and the angle that two arbitrary unit vectors $\mathbf{u}_{*1}$ and $\mathbf{u}_{*2}$ assume during the deformation are given by

$$\delta_{\mathbf{u}_*} = \sqrt{\mathbf{u}_* \cdot \mathbf{C}\mathbf{u}_*}, \qquad (3.14)$$

$$\cos\theta_{12} = \frac{\mathbf{u}_{*1} \cdot \mathbf{C}\mathbf{u}_{*2}}{\sqrt{\mathbf{u}_{*1} \cdot \mathbf{C}\mathbf{u}_{*1}}\sqrt{\mathbf{u}_{*2} \cdot \mathbf{C}\mathbf{u}_{*2}}}; \qquad (3.15)$$

2. The tensors $\mathbf{C}$ and $\mathbf{U}$ have the same eigenvectors; the eigenvalues of $\mathbf{C}$ are equal to the squares of the eigenvalues of $\mathbf{U}$;

3. The components C_{LL} correspond to the squares of the stretch ratios in the direction of the unit vectors of the basis $(\mathbf{e}_L)$, whereas C_{LM}, $L \neq M$, are proportional to the sine of the shear strain between $\mathbf{e}_L$ and $\mathbf{e}_M$.

4. Finally, the eigenvalues of $\mathbf{C}$ are equal to the squares of the stretch ratios in the direction of the eigenvectors.

PROOF Let $d\mathbf{X} = |d\mathbf{X}|\mathbf{u}_*$; then from (3.9), (3.6), and (3.12), it follows

$$\delta_{\mathbf{u}_*}^2 = \frac{d\mathbf{x} \cdot d\mathbf{x}}{|d\mathbf{X}|^2} = \frac{|d\mathbf{X}|^2 \mathbf{u}_* \cdot \mathbf{C}\mathbf{u}_*}{|d\mathbf{X}|^2},$$

so that (3.14) is proved. In a similar manner, given $d\mathbf{X}_1 = |d\mathbf{X}_1|\mathbf{u}_{*1}$ and $d\mathbf{X}_2 = |d\mathbf{X}_2|\mathbf{u}_{*2}$, it follows that

$$\cos\theta_{12} = \frac{d\mathbf{x}_1 \cdot d\mathbf{x}_2}{|d\mathbf{x}_1||d\mathbf{x}_2|} = \frac{|d\mathbf{X}_1||d\mathbf{X}_2|\mathbf{u}_{*1} \cdot \mathbf{C}\mathbf{u}_{*2}}{|d\mathbf{X}_1||d\mathbf{X}_2|\sqrt{\mathbf{u}_{*1} \cdot \mathbf{C}\mathbf{u}_{*1}}\sqrt{\mathbf{u}_{*2} \cdot \mathbf{C}\mathbf{u}_{*2}}},$$

which proves (3.15).

The property (2) has been proved in Theorem 1.8.

The property (3) can easily be proved as a consequence of (3.14) and (3.15), if in these relations we assume that $\mathbf{u}_{*1} = \mathbf{e}_L$ and $\mathbf{u}_{*2} = \mathbf{e}_M$, and recall that $C_{LM} = \mathbf{e}_L \cdot \mathbf{C}\mathbf{e}_M$. ∎

We will also make use of the **left Cauchy–Green tensor** (Finger, 1894):

$$\mathbf{B} = \mathbf{F}\mathbf{F}^T = \mathbf{R}\mathbf{U}^2\mathbf{R}^T = \mathbf{B}^T = \mathbf{V}^2, \qquad B_{ij} = \frac{\partial x_i}{\partial X_K}\frac{\partial x_j}{\partial X_K}. \qquad (3.16)$$

It is of interest to prove the following theorem:

Theorem 3.3
The tensors $\mathbf{B}$ and $\mathbf{C}$ have the same eigenvalues; the eigenvectors of $\mathbf{C}$ are obtained by applying the rotation $\mathbf{R}^T$ to the eigenvectors of $\mathbf{B}$.

PROOF If $\mathbf{u}$ is an eigenvector of $\mathbf{C}$ corresponding to the eigenvalue Λ_C, then
$$\mathbf{Cu} = \mathbf{U}^2\mathbf{u} = \Lambda_C \mathbf{u},$$
so that
$$\mathbf{RU}^2\mathbf{R}^T(\mathbf{Ru}) = \Lambda_C(\mathbf{Ru})$$
and finally
$$\mathbf{B}(\mathbf{Ru}) = \Lambda_C(\mathbf{Ru}).$$
On the other hand, if $\mathbf{v}$ is an eigenvector of $\mathbf{B}$ corresponding to the eigenvalue Λ_B, then
$$\mathbf{Bv} = \mathbf{RU}^2\mathbf{R}^T\mathbf{v} = \Lambda_B \mathbf{v},$$
which proves the theorem. ∎

In order to highlight the physical meaning of the tensors $\mathbf{C}$ and $\mathbf{B}$, we observe that the Cauchy–Green tensor gives the square length of the element $d\mathbf{x}$ in which the original element $d\mathbf{X}$ is transformed:
$$dl^2 = d\mathbf{X} \cdot \mathbf{C} d\mathbf{X}. \tag{3.17}$$
Conversely, the tensor $\mathbf{B}^{-1} = (\mathbf{F}^{-1})^T \mathbf{F}^{-1}$, which has Cartesian components
$$B_{ij}^{-1} = \frac{\partial X_K}{\partial x_i} \frac{\partial X_K}{\partial x_j},$$
(Cauchy, 1827), gives the initial length of the element $d\mathbf{x}$
$$dl_*^2 = d\mathbf{x} \cdot \mathbf{B}^{-1} d\mathbf{x}. \tag{3.18}$$

In addition to the formulae (3.14) and (3.15) giving the stretch ratios and the angle changes, it is of practical interest to derive the following expressions, which relate the change of an infinitesimal oriented area $d\boldsymbol{\sigma}_*$, defined in the initial configuration, to the corresponding infinitesimal oriented $d\boldsymbol{\sigma}$ in the deformed configuration, as well as the change in volume from dC_* into dC:
$$d\boldsymbol{\sigma} \equiv \mathbf{N} d\sigma = J\left(\mathbf{F}^{-1}\right)^T \mathbf{N}_* d\sigma \equiv J\left(\mathbf{F}^{-1}\right)^T d\boldsymbol{\sigma}_*, \tag{3.19}$$
$$dc = J dc_*, \tag{3.20}$$
where $\mathbf{N}_*$ and $\mathbf{N}$ are the unit vectors normal to $d\sigma_*$ and $d\sigma$, respectively.

To prove (3.19), consider two infinitesimal vectors $\mathbf{dx}$ and $\mathbf{dy}$ both emanating from $\mathbf{x} \in C$. The area of the parallelogram with edges $\mathbf{dx}$ and $\mathbf{dy}$ is given by the length of the vector

$$\mathbf{d\sigma} \equiv \mathbf{N}\, d\sigma = \mathbf{dx} \times \mathbf{dy}.$$

In terms of components, this is

$$d\sigma_i = N_i d\sigma = \epsilon_{ijk} dx_j dy_k = \epsilon_{ijk} F_{jL} F_{kM} dX_L dY_M$$

and, multiplying by $\mathbf{F}$, we obtain

$$F_{iK} N_i d\sigma = \epsilon_{ijk} F_{iK} F_{jL} F_{kM} dX_L dY_M.$$

The definition of determinant allows us to write

$$J \epsilon_{KLM} = \epsilon_{ijk} F_{iK} F_{jL} F_{kM},$$

so that the previous relation becomes

$$F_{iK} N_i d\sigma = J \epsilon_{KLM} dX_L dY_M.$$

Finally, it follows that

$$N_i d\sigma = J\left(F^{-1}\right)_{Ki} \epsilon_{KLM} dX_L dY_M = J\left(F^{-1}\right)_{Ki} N_{*K} d\sigma_*,$$

where ϵ_{KLM} is the permutation symbol.

Relation (3.20) follows directly from the rule of change of variables in multiple integrals.

3.3 Invariants of C and B

As we showed in Chapter 1, we recall that

1. the eigenvalues of a symmetric tensor are real and they are strictly positive if the tensor is positive definite;

2. the characteristic spaces of a symmetric tensor are mutually orthogonal.

It follows that the eigenvalues of $\mathbf{C}$ and $\mathbf{B}$ are real and positive. Furthermore, the characteristic equation of $\mathbf{C}$ is written as

$$\Lambda^3 - I_C \Lambda^2 + II_C \Lambda - III_C = 0, \qquad (3.21)$$

where the coefficient I_C, II_C, III_C, are the **first**, **second**, and **third principal invariant** of the tensor $\mathbf{C}$, i.e., they are quantities independent of the basis in which $\mathbf{C}$ is expressed:

$$I_C = \text{tr}\,\mathbf{C} = C_{11} + C_{22} + C_{33},$$
$$II_C = \det\begin{pmatrix} C_{11} & C_{12} \\ C_{21} & C_{22} \end{pmatrix} + \det\begin{pmatrix} C_{11} & C_{13} \\ C_{31} & C_{33} \end{pmatrix} + \det\begin{pmatrix} C_{22} & C_{23} \\ C_{32} & C_{33} \end{pmatrix},$$
$$III_C = \det\mathbf{C}. \tag{3.22}$$

In particular, if the eigenvectors of $\mathbf{C}$ are selected as basis vectors, then the matrix representing $\mathbf{C}$ is diagonal, and (3.22) can also be written as

$$\begin{aligned} I_C &= \Lambda_1 + \Lambda_2 + \Lambda_3, \\ II_C &= \Lambda_1\Lambda_2 + \Lambda_1\Lambda_3 + \Lambda_2\Lambda_3, \\ III_C &= \Lambda_1\Lambda_2\Lambda_3. \end{aligned} \tag{3.23}$$

Moreover, $\mathbf{C}$ and $\mathbf{B}$ have the same eigenvalues so that:

$$\begin{aligned} I_C &= I_B, \\ II_C &= II_B, \\ III_C &= III_B. \end{aligned} \tag{3.24}$$

3.4 Displacement and Displacement Gradient

The deformation process of a continuous body S as it passes from the initial configuration C_* to the current configuration C can also be described by considering the **displacement field** $\mathbf{u}(\mathbf{X})$, defined as

$$\mathbf{u}(\mathbf{X}) = \mathbf{x}(\mathbf{X}) - \mathbf{X}. \tag{3.25}$$

If the **displacement gradient tensor**

$$\mathbf{H} = \nabla\mathbf{u}(\mathbf{X}) \equiv \left(\frac{\partial u_i}{\partial X_L}\right) \tag{3.26}$$

is introduced, then from (3.25) it follows that

$$\mathbf{F} = \mathbf{I} + \mathbf{H}. \tag{3.27}$$

A common measure of the deformation is given in this case by the **Green–St. Venant tensor** (Green, 1841; St.Venant, 1844)

$$\mathbf{G} = \frac{1}{2}(\mathbf{C} - \mathbf{I}) = \mathbf{G}^T. \tag{3.28}$$

3.4. Displacement and Displacement Gradient

This tensor gives the change in the squared length of the material vector $d\mathbf{X}$, i.e.,
$$(dl)^2 - (dl_*)^2 = 2dX_I G_{IJ} dX_J.$$
By taking into account the definition (3.12) of $\mathbf{C}$ and the equation (3.27), we find that the Green–St. Venant tensor also assumes the form
$$\mathbf{G} = \mathbf{E} + \frac{1}{2}\mathbf{H}^T\mathbf{H}, \tag{3.29}$$
where
$$\mathbf{E} = \frac{1}{2}\left(\mathbf{H} + \mathbf{H}^T\right) \tag{3.30}$$
is the ***infinitesimal strain tensor***.

The main properties of the tensor $\mathbf{G}$ are expressed by the following theorem:

Theorem 3.4

1. Both tensors $\mathbf{G}$ and $\mathbf{C}$ have the same eigenvectors; if Λ are the eigenvalues of $\mathbf{C}$, then $\mathbf{G}$ has eigenvalues equal to $(\Lambda - 1)/2$.

2. The squares of the stretch ratios of elements initially parallel to the coordinate axes are equal to $2G_{LL} + 1$; the components G_{LM} $(L \neq M)$ are proportional to the angle between the deformed elements that were initially parallel to the coordinate axes.

PROOF Property (1) follows from the definitions. To prove (2) it is enough to recall (3.14), (3.15), and (3.28), so that
$$\delta_{\mathbf{u}_1}^2 = C_{11} = 1 + 2G_{11},$$
$$\cos\theta_{12} = \frac{C_{12}}{\sqrt{C_{11}C_{22}}} = \frac{2G_{12}}{\sqrt{(1 + 2G_{11})(1 + 2G_{22})}}.$$

∎

Finally, the following relationships hold between the invariants of $\mathbf{G}$ and $\mathbf{B}$:
$$2I_G = I_B - 3,$$
$$4II_G = II_B - 2I_B + 3,$$
$$8III_G = III_B - II_B + I_B - 1,$$
$$I_B = 2I_G + 3,$$
$$II_B = 4II_G + 4I_G + 3,$$
$$III_B = 8III_G + 4II_G + 2I_G + 1.$$

3.5 Infinitesimal Deformation Theory

The deformation $C_* \to C$ is defined to be **infinitesimal** if both the components of $\mathbf{u}$ and $\nabla \mathbf{u}$ are first-order quantities, i.e., the squares and products of these quantities can be neglected.

We observe that $\mathbf{F}, \mathbf{C}$, and $\mathbf{B} \to \mathbf{I}$ when $\mathbf{u} \to \mathbf{0}$, whereas $\mathbf{H}, \mathbf{G}$, and $\mathbf{E} \to \mathbf{0}$ if $\mathbf{u} \to \mathbf{0}$.

From (3.29), in the case of infinitesimal deformations we have

$$\mathbf{G} \simeq \mathbf{E}, \qquad (3.31)$$

where the sign $\simeq$ indicates that the quantities on the left-hand side differ from the quantities on the right-hand side by an order greater than $|\mathbf{u}|$ and $|\nabla \mathbf{u}|$.

As far as the meaning of the linearized deformation tensor $\mathbf{E}$ are concerned, the following theorem holds:

Theorem 3.5
The tensors $\mathbf{E}$ and $\mathbf{U}$ have the same eigenvectors (and the same applies to $\mathbf{C}$ and $\mathbf{B}$)[3]; these eigenvectors identify the principal axes of deformation. The longitudinal unit extension in the direction parallel to the unit vector $\mathbf{e}_L$ of the axis X_L is given by

$$\delta_{u_L} - 1 \simeq E_{LL}, \qquad (3.32)$$

and the shear strain relative to the axes X_L and X_M is twice the off-diagonal component

$$\gamma_{LM} \simeq 2E_{LM}. \qquad (3.33)$$

PROOF To prove the first part of the theorem, we consider two unit vectors $\mathbf{u}_{*1}$ and $\mathbf{u}_{*2}$ emanating from $\mathbf{X} \in C_*$. By using (3.29), (3.30), and (3.31), we obtain

$$\mathbf{u}_{*1} \cdot \mathbf{C}\mathbf{u}_{*2} = \mathbf{u}_{*1} \cdot (2\mathbf{G} + \mathbf{I})\mathbf{u}_{*2} \simeq \mathbf{u}_{*1} \cdot 2\mathbf{E}\mathbf{u}_{*2} + \mathbf{u}_{*1} \cdot \mathbf{u}_{*2},$$

so that, if $\mathbf{u}_{*1} = \mathbf{u}_{*2} = \mathbf{u}_*$ and we take into account (3.14), we find that

$$\delta_{u_*} = \sqrt{1 + \mathbf{u}_* \cdot 2\mathbf{E}\mathbf{u}_*} \simeq 1 + \mathbf{u}_* \cdot \mathbf{E}\mathbf{u}_*. \qquad (3.34)$$

Relation (3.32) is then proved if $\mathbf{u}_* = \mathbf{e}_L$.

---
[3] Here and in the following discussion the equality is intended to be satisfied by neglecting second-order terms in $\mathbf{H}$.

3.5. Infinitesimal Deformation Theory

Relationship (3.34) replaces (3.14) in the linearized theory, showing that under such an assumption, the stretch ratios can be evaluated in any direction from the components of $\mathbf{E}$. For this reason, $\mathbf{E}$ is called the infinitesimal deformation tensor. Furthermore, from (3.15) and (3.34) and with $\mathbf{u}_{*1} = \mathbf{e}_L$, $\mathbf{u}_{*2} = \mathbf{e}_M$, it follows that

$$\cos\theta_{12} \simeq \frac{2E_{LM}}{(1+E_{LL})(1+E_{MM})} \simeq \frac{2E_{LM}}{1+(E_{LL}+E_{MM})}.$$

From (3.15) we have $\cos\theta_{12} = \cos(\frac{\pi}{2} - \gamma_{12}) = \sin\gamma_{12} \simeq \gamma_{12}$. But γ_{12} is a first-order quantity so that, by multiplying the numerator and denominator by $1-(E_{LL}+E_{MM})$ and neglecting terms of higher order, (3.33) is obtained. ∎

Theorem 3.6
In an infinitesimal deformation[4] we have

$$\mathbf{U} \simeq \mathbf{I} + \mathbf{E}, \qquad \mathbf{R} \simeq \mathbf{I} + \mathbf{W}, \tag{3.35}$$

where $\mathbf{W}$ is the skew-symmetric tensor of rotation for infinitesimal deformation

$$\mathbf{W} = \frac{1}{2}(\mathbf{H} - \mathbf{H}^T) = -\mathbf{W}^T. \tag{3.36}$$

Furthermore,
$$\mathbf{F} \cdot d\mathbf{X} = d\mathbf{X} + \mathbf{E} \cdot d\mathbf{X} + \boldsymbol{\varphi} \times d\mathbf{X}, \tag{3.37}$$

where
$$\varphi_i = -\frac{1}{2}\epsilon_{ijk} W_{jk} \tag{3.38}$$

is the vector of infinitesimal rotation.

PROOF By using (3.12) and (3.13) we obtain

$$\mathbf{U}^2 = \mathbf{F}^T\mathbf{F} \Rightarrow \mathbf{U} = \sqrt{\mathbf{C}}, \qquad \mathbf{R} = \mathbf{F}\mathbf{U}^{-1}. \tag{3.39}$$

It follows that $\mathbf{U} \simeq \sqrt{2\mathbf{E}+\mathbf{I}} \simeq \sqrt{2\mathbf{E}+\mathbf{I}} \simeq \mathbf{I}+\mathbf{E}$; moreover,

$$\mathbf{R} = \mathbf{F}\mathbf{U}^{-1} \simeq (\mathbf{I}+\mathbf{H})(\mathbf{I}+\mathbf{E})^{-1}.$$

Since $(\mathbf{I}+\mathbf{E})^{-1} \simeq \mathbf{I} - \mathbf{E} \simeq \mathbf{I} - \frac{1}{2}(\mathbf{H}+\mathbf{H})^T$, we have $\mathbf{R} \simeq \mathbf{I}+\mathbf{W}$ and (3.35) is proved. In addition,

$$\mathbf{F} \cdot d\mathbf{X} = \mathbf{R}\mathbf{U} \cdot d\mathbf{X} = (\mathbf{I}+\mathbf{E})(\mathbf{I}+\mathbf{W}) \cdot d\mathbf{X} \simeq (\mathbf{I}+\mathbf{E}+\mathbf{W}) \cdot d\mathbf{X}.$$

---
[4]Note that $\mathbf{I}+\mathbf{W}$ is orthogonal if we neglect 2nd-order terms in $\mathbf{H}$. In fact, $(\mathbf{I}+\mathbf{W})(\mathbf{I}+\mathbf{W})^T \simeq \mathbf{I}+\mathbf{W}+\mathbf{W}^T = \mathbf{I}$, since $\mathbf{W}$ is skew-symmetric.

In any case, it is always possible to express $\mathbf{W} \cdot d\mathbf{X}$ in the form $\varphi \times d\mathbf{X}$, where φ is the **adjoint** of $\mathbf{W}$, as defined by (3.38). ∎

To give a physical meaning to (3.37), at $\mathbf{X} \in C_*$ we consider a parallelepiped with edges that are parallel to the eigenvectors $\mathbf{u}_1, \mathbf{u}_2, \mathbf{u}_3$ of $\mathbf{E}$. Any edge $dX_L \mathbf{u}_L$ (no summation on L) is transformed into $dX_L \mathbf{u}_L + \lambda_L dX_L \mathbf{u}_L + \varphi \times dX_L \mathbf{u}_L$, i.e., the vector $\varphi \times dX_L \mathbf{u}_L$ is added to the vector $dX_L \mathbf{u}_L$, which rotates $dX_L \mathbf{u}_L$ by an angle $|\varphi|$ around the direction of φ, together with the deformed vector $\lambda_L dX_L \mathbf{u}_L$.

We can easily derive a further useful relation:

$$J \simeq 1 + \operatorname{tr} \mathbf{H} = 1 + \operatorname{tr} \mathbf{E}, \qquad (3.40)$$

so that the (3.20) can be written as

$$dc = (1 + \operatorname{tr} \mathbf{E}) dc_*. \qquad (3.41)$$

3.6 Transformation Rules for Deformation Tensors

In the previous analysis, the same Cartesian coordinate system has been used for both the initial configuration C_* and the current configuration C. For the sake of clarity, different symbols have been introduced for the coordinates (X^L) of material points $\mathbf{X} \in C_*$ and for the coordinates (x^i) of spatial points $\mathbf{x} \in C$.

At this stage it is useful to show how the deformation tensors transform when changing the coordinate system, both in the reference and the current configuration. The situation arises when dealing with properties related to the isotropy of the material (which in this case has an inherent reference system), a problem dealt with in Chapter 7, as well as when dealing with transformation rules of mechanical quantities for a change of the rigid reference frame (see Chapters 4 and 6).

As a starting point, first we note that the quantities F^i_L cannot be regarded as the components of a tensor in the sense specified in Chapter 1, due to the fact that $\mathbf{F}$ is a linear mapping between the space of vectors emanating from $\mathbf{X} \in C_*$ and the vector space defined at the corresponding point $\mathbf{x} \in C$ (rather than a linear mapping between vectors emanating from the same point).

If we consider rotations of the axes in C_* and C generated by the orthogonal matrices $\mathbf{Q}^*$ and $\mathbf{Q}$

$$\mathbf{X}^* = \mathbf{Q}^* \mathbf{X}, \qquad \mathbf{x}' = \mathbf{Q} \mathbf{x}, \qquad (3.42)$$

then from the transformation $(3.42)_1$ we derive

$$dx = \mathbf{F}^* d\mathbf{X}^* = \mathbf{F}^* \mathbf{Q}^* d\mathbf{X} = \mathbf{F} d\mathbf{X} \Rightarrow \mathbf{F} = \mathbf{F}^* \mathbf{Q}^*, \qquad (3.43)$$

and $(3.42)_2$ leads to

$$dx' = \mathbf{F}' d\mathbf{X} = \mathbf{Q}\mathbf{F} d\mathbf{X} \Rightarrow \mathbf{F}' = \mathbf{Q}\mathbf{F}. \qquad (3.44)$$

These results highlight the fact that the deformation tensor behaves as a vector for coordinate changes in only one configuration, and in this respect **F** is an example of a ***two-point tensor***. This definition is in fact used for a tensor whose components transform like those of a vector under rotation of only one of the two reference axes and like a two-point tensor when the two sets of axes are rotated independently. This aspect can be understood intuitively if we refer to the figurative picture suggested by Marsden and Hughes (1983), according to which the tensor **F** has two "legs": one in C_* and one in C.

It is also easy to verify that the tensors **H** and **G** satisfy the same transformation rule of **F**. Moreover,

$$\mathbf{C}^* = (\mathbf{F}^T)^* \mathbf{F}^* = \mathbf{Q}^* \mathbf{F}^T \mathbf{F} (\mathbf{Q}^*)^T = \mathbf{Q}^* \mathbf{C} (\mathbf{Q}^*)^T, \qquad (3.45)$$

$$\mathbf{C}' = \mathbf{F}'^T \mathbf{F}' = \mathbf{F}^T \mathbf{Q}^T \mathbf{Q} \mathbf{F} = \mathbf{C}, \qquad (3.46)$$

so that **C** behaves as a tensor for changes of coordinates in C_*, while it exhibits invariance for transformations in C.

Similarly, the following transformation rules can be proven to hold:

$$\mathbf{G}^* = \mathbf{Q}^* \mathbf{G} (\mathbf{Q}^*)^T, \qquad \mathbf{G}' = \mathbf{G}, \qquad (3.47)$$

$$\mathbf{B}^* = \mathbf{B}, \qquad \mathbf{B}' = \mathbf{Q} \mathbf{B} \mathbf{Q}^T. \qquad (3.48)$$

3.7 Some Relevant Formulae

In this section we will present and prove formulae that will be useful for topics developed later in this book.

Let us start by observing that, if **F** is the deformation gradient tensor, **a** is a nonsingular matrix, and $a = \det \mathbf{a}$, then

$$\frac{\partial}{\partial x_i} \left(\frac{1}{J} F_{iL} \right) = 0, \qquad (3.49)$$

$$\frac{\partial a}{\partial a_{ij}} = a(a^{-1})_{ji}, \qquad \frac{\partial (a^{-1})_{hk}}{\partial a_{ij}} = -(a^{-1})_{hi}(a^{-1})_{jk}. \qquad (3.50)$$

Furthermore, if $\mathbf{a}(\mathbf{h}) = \mathbf{I} + \mathbf{h}$, it follows that

$$a = 1 + I_h + II_h + O(\mathbf{h}^2), \qquad \mathbf{a}^{-1} = \mathbf{I} - \mathbf{h} + \mathbf{h}^2 + O(\mathbf{h}^2). \tag{3.51}$$

Finally, for the right Cauchy–Green tensor, the following relations hold:

$$\frac{\partial I_C}{\partial \mathbf{C}} = \mathbf{I}, \qquad \frac{\partial II_C}{\partial \mathbf{C}} = I_C \mathbf{I} - \mathbf{C}^T, \tag{3.52}$$

$$\frac{\partial III_C}{\partial \mathbf{C}} = III_C \left(\mathbf{C}^{-1}\right)^T = \left[\mathbf{C}^2 - I_C \mathbf{C} + II_C \mathbf{I}\right]^T, \tag{3.53}$$

where $\mathbf{I}$ is the unit matrix. Note that (3.49) and (3.50) apply to any non-singular tensor $\mathbf{C}$; if $\mathbf{C}$ is symmetric, then the sign of transpose can be omitted.

To prove $(3.50)_1$, it is enough to express the determinant a in terms of its first row elements, i.e., $a = a_{11}A_{11} + a_{12}A_{12} + a_{13}A_{13}$, where A_{ij} is the cofactor of a_{ij}. Since, $\partial a / \partial a_{12} = A_{12}$ and recalling that

$$\left(a^{-1}\right)_{21} = \frac{A_{12}}{a}, \tag{3.54}$$

we see that the relation $(3.50)_1$ is proved for $i = 1, j = 2$, and so for any other choice of indices.

To prove $(3.50)_2$, we differentiate the relation

$$a_{rl} \left(a^{-1}\right)_{lk} = \delta_{rk} \tag{3.55}$$

with respect to a_{ij} to obtain

$$\frac{\partial a_{rl}}{\partial a_{ij}} \left(a^{-1}\right)_{lk} + a_{rl} \frac{\partial \left(a^{-1}\right)_{lk}}{\partial a_{ij}} = 0.$$

By multiplying by $\left(a^{-1}\right)_{hr}$ and recalling (3.55), we get

$$\frac{\partial \left(a^{-1}\right)_{hk}}{\partial a_{ij}} = -\left(a^{-1}\right)_{hr} \frac{\partial a_{rl}}{\partial a_{ij}} \left(a^{-1}\right)_{lk} = -\left(a^{-1}\right)_{hr} \delta_{ri} \delta_{jl} \left(a^{-1}\right)_{lk}$$

and $(3.50)_2$ follows.

The relation (3.49) is verified through the sequence of identities

$$\frac{\partial}{\partial x_i}\left(\frac{1}{J}F_{iL}\right) = \frac{\partial F_{hM}}{\partial x_i} \frac{\partial}{\partial F_{hM}}\left(\frac{1}{J}F_{iL}\right)$$

$$= \frac{\partial F_{hM}}{\partial x_i}\left[-\frac{1}{J^2}J\left(F^{-1}\right)_{Mh}F_{iL} + \frac{1}{J}\delta_{ih}\delta_{ML}\right]$$

$$= \frac{1}{J}\left(F^{-1}\right)_{Ni}\frac{\partial F_{hM}}{\partial X_N}\left[\delta_{ih}\delta_{ML} - \delta_{ML}F_{iL}\right]$$

$$= \frac{1}{J}\left[\left(F^{-1}\right)_{Nh}\frac{\partial^2 x_h}{\partial X_L \partial X_N} - \left(F^{-1}\right)_{Nh}\frac{\partial^2 x_h}{\partial X_L \partial X_N}\right] = 0.$$

3.7. Some Relevant Formulae

Considering now (3.51), we first observe that, for a matrix $\mathbf{a}(\mathbf{h}) = \mathbf{I} + \mathbf{h}$, we have

$$a(\mathbf{h}) = a(\mathbf{0}) + \left(\frac{\partial a}{\partial h_{ij}}\right)_{\mathbf{h}=\mathbf{0}} h_{ij} + \frac{1}{2}\left(\frac{\partial^2 a}{\partial h_{ij}\partial h_{lm}}\right)_{\mathbf{h}=\mathbf{0}} h_{ij} h_{lm} + O(\mathbf{h}^2),$$

and since $a(\mathbf{0}) = 1$, from (3.50)$_1$ it follows that

$$\left(\frac{\partial a}{\partial h_{ij}}\right)_{\mathbf{h}=\mathbf{0}} = \left(\frac{\partial a}{\partial a_{ij}}\right)_{\mathbf{h}=\mathbf{0}} = a(0)(a^{-1})_{ji}(0) = \delta_{ij}.$$

By using (3.50), we can write

$$\frac{\partial^2 a}{\partial h_{ij}\partial h_{lm}} = \frac{\partial}{\partial h_{ij}} a(a^{-1})_{ml}$$

$$= a(a^{-1})_{ji}(a^{-1})_{ml} - a(a^{-1})_{mi}(a^{-1})_{jl},$$

$$\left(\frac{\partial^2 a}{\partial h_{ij}\partial h_{lm}}\right)_{\mathbf{h}=\mathbf{0}} h_{ij} h_{lm} = (\delta_{ji}\delta_{ml} - \delta_{mi}\delta_{jl}) h_{ij} h_{lm}$$

$$= h_{ii} h_{ll} - h_{mj} h_{jm} = 2II_{\mathbf{h}},$$

so that (3.51)$_1$ is proved.
Furthermore,

$$(a^{-1})_{ij} = \delta_{ij} + \left(\frac{\partial(\delta_{ij} + h_{ij})^{-1}}{\partial h_{lm}}\right)_{\mathbf{h}=\mathbf{0}} h_{lm}$$

$$+ \frac{1}{2}\left(\frac{\partial^2(\delta_{ij} + h_{ij})^{-1}}{\partial h_{lm}\partial h_{pq}}\right)_{\mathbf{h}=\mathbf{0}} h_{lm} h_{pq} + \cdots,$$

and, since (3.50)$_2$ gives

$$\left(\frac{\partial(\delta_{ij} + h_{ij})^{-1}}{\partial h_{lm}}\right)_{\mathbf{h}=\mathbf{0}} = -\left[(\delta_{il} + h_{il})^{-1}(\delta_{mj} + h_{mj})^{-1}\right]_{\mathbf{h}=\mathbf{0}} = -\delta_{il}\delta_{mj},$$

$$\left(\frac{\partial^2(\delta_{ij} + h_{ij})^{-1}}{\partial h_{lm}\partial h_{pq}}\right)_{\mathbf{h}=\mathbf{0}} = \delta_{il}\delta_{mp}\delta_{qj} + \delta_{ip}\delta_{ql}\delta_{mj},$$

(3.51)$_2$ is obtained.
To prove (3.52) and (3.53), consider the identity

$$\det(\Lambda\mathbf{I} + \mathbf{C}) = \Lambda^3 + I_C\Lambda^2 + II_C\Lambda + III_C. \tag{3.56}$$

Differentiating both sides of (3.56) with respect to $\mathbf{C}$ and recalling (3.50)$_1$, we have

$$\frac{\partial}{\partial \mathbf{C}}\det(\Lambda\mathbf{I} + \mathbf{C}) = \det(\Lambda\mathbf{I} + \mathbf{C})\left[(\Lambda\mathbf{I} + \mathbf{C})^{-1}\right]^{\mathbf{T}}, \tag{3.57}$$

$$\frac{\partial}{\partial \mathbf{C}}(\Lambda^3 + I_C\Lambda^2 + II_C\Lambda + III_C) = \frac{\partial I_C}{\partial \mathbf{C}}\Lambda^2 + \frac{\partial II_C}{\partial \mathbf{C}}\Lambda + \frac{\partial III_C}{\partial \mathbf{C}}. \quad (3.58)$$

The right-hand sides of (3.57) and (3.58) must be equal so that, multiplying the expression above by $(\Lambda \mathbf{I} + \mathbf{C})^T$ and considering (3.56), we get

$$(\Lambda^3 + I_C\Lambda^2 + II_C\Lambda + III_C)\mathbf{I}$$
$$= \frac{\partial I_C}{\partial \mathbf{C}}\Lambda^3 + \left(\frac{\partial II_C}{\partial \mathbf{C}} + \frac{\partial I_C}{\partial \mathbf{C}}\mathbf{C}^T\right)\Lambda^2$$
$$+ \left(\frac{\partial III_C}{\partial \mathbf{C}} + \frac{\partial II_C}{\partial \mathbf{C}}\mathbf{C}^T\right)\Lambda + \frac{\partial III_C}{\partial \mathbf{C}}\mathbf{C}^T.$$

By comparing the coefficients of Λ, (3.52) and (3.53) are easily derived. As a further result the **Cayley–Hamilton theorem** follows:

$$\mathbf{C}^3 - I_C\mathbf{C}^2 + II_C\mathbf{C} - III_C\mathbf{I} = 0, \quad (3.59)$$

which is valid for any 2-tensor $\mathbf{C}$.

3.8 Compatibility Conditions

The tensor field $C_{LM}(\mathbf{X})$ cannot be arbitrarily assigned, since the equations

$$\sum_{i=1}^{3} \frac{\partial x_i}{\partial X_L}\frac{\partial x_i}{\partial X_M} = C_{LM}, \quad (3.60)$$

due to the symmetry of $\mathbf{C}$, form a system of 6 nonlinear partial differential equations with the three unknowns $x_i(\mathbf{X})$, which describe the deformation of a continuous body S when passing from C_* to C.

The mathematical difficulty arises from the fact that equations (3.60) overdetermine the three unknowns, so that we have to search for necessary and sufficient conditions ensuring that a proposed set of functions C_{LM} can be regarded as the tensor components of a deformation field.

In order to find these integrability conditions on (3.60), we first observe that, since

$$d\mathbf{x} \cdot d\mathbf{x} = d\mathbf{X} \cdot \mathbf{C}d\mathbf{X}, \quad (3.61)$$

the coordinates X_L in C_* can also be regarded as *curvilinear coordinates* in C, in which the metric tensor components are C_{LM}. In addition, C is a subset of the Euclidean space and therefore the components C_{LM} must be such that they correspond to a Euclidean metric tensor. In this respect, **Riemann's theorem** states that a symmetric tensor is a metric tensor of a Euclidean space if and only if it is a nonsingular positive definite

3.8. Compatibility Conditions

tensor and the corresponding **Riemann–Christoffel tensor** **R** formed from C_{LM} identically vanishes:[5]

$$R_{NMLP} = \frac{1}{2}(C_{LM,NP} + C_{PN,LM} - C_{NL,MP} - C_{PM,LN})$$
$$+ C^{RS}(\Gamma_{PNR}\Gamma_{LMS} - \Gamma_{PMR}\Gamma_{LNS}) = 0, \qquad (3.62)$$

where
$$2\Gamma_{LMP} = C_{LP,M} + C_{MP,L} - C_{LM,P}$$

are Christhoffel symbols of the second kind.

These $3^4 = 81$ equations (3.62) are not independent. In fact, due to the symmetries of R_{NMLP} that can be deduced by inspection of (3.62),

$$R_{LMNP} = -R_{MLNP} = -R_{LMPN}; \qquad R_{LMNP} = R_{NPLM},$$

there are only 6 distinct and nonvanishing components R_{NMLP}. To prove this, let a pair of indices LP be fixed. Since R_{NMLP} is skew-symmetric with respect to the pair NM, we conclude that indices N and M can only assume the values 12, 13, and 23. The same argument holds for LP if the pair NM is given. Finally, the 9 nonvanishing components of R_{NMLP}

$$\begin{array}{ccc} R_{1212}, & R_{1213}, & R_{1223}, \\ R_{1312}, & R_{1313}, & R_{1323}, \\ R_{2312}, & R_{2313}, & R_{2323}, \end{array}$$

merge into the following 6 independent components, due to symmetry with respect to the first two indices:

$$\begin{array}{ccc} R_{1212}, & R_{1213}, & R_{1223}, \\ & R_{1313}, & R_{1323}, \\ & & R_{2323}. \end{array} \qquad (3.63)$$

As a special case, consider the linearized theory of elasticity. Since $\mathbf{C} \simeq \mathbf{I} + 2\mathbf{E}$, (3.62) assumes the form

$$R_{NMLP} \simeq E_{LM,NP} + E_{PN,LM} - E_{NL,MP} - E_{PM,LN} = 0, \qquad (3.64)$$

which is certainly true if $\mathbf{E}(\mathbf{X})$ is a linear function. The equations of interest deduced from (3.64) are the following, known as the **St. Venant compatibility conditions** (St.Venant, 1864):

$$R_{1212} \simeq 2E_{12,12} - E_{11,22} - E_{22,11} = 0,$$

---
[5]The Riemann–Christoffel tensor **R** is also referred as the curvature tensor and the condition (3.62) as Riemann's theorem.

$$\begin{aligned}
R_{1213} &\simeq E_{12,13} + E_{31,12} - E_{11,23} - E_{23,11} = 0, \\
R_{1223} &\simeq E_{22,13} + E_{31,22} - E_{12,23} - E_{23,21} = 0, \\
R_{1313} &\simeq 2E_{13,13} - E_{11,33} - E_{33,11} = 0, \\
R_{1323} &\simeq E_{13,23} + E_{23,13} - E_{12,33} - E_{33,21} = 0, \\
R_{2323} &\simeq 2E_{23,23} - E_{22,33} - E_{33,22} = 0.
\end{aligned} \qquad (3.65)$$

and (3.65) can be summarized by the following relation:

$$\epsilon_{QSR}\,\epsilon_{NML} E_{QN,SM} = 0 \Leftrightarrow \nabla \times \nabla \times \mathbf{E} = \mathbf{0}. \qquad (3.66)$$

If the divergence operator is applied to (3.66), it follows that

$$\epsilon_{QSR}\,\epsilon_{NML} E_{QN,SMR} = 0, \qquad (3.67)$$

which is identically satisfied by any second-order symmetric tensor $\mathbf{E}$, since the permutation symbol ϵ_{QSR} is skew-symmetric with respect to indices R and S and $E_{QN,SMR}$ is symmetric with respect to the same indices.

When $L = 1, 2, 3$, (3.67) gives the following three equations:

$$\begin{aligned}
-R_{1223,3} + R_{1323,2} - R_{2323,1} &= 0, \\
R_{1213,3} - R_{1313,2} + R_{1323,1} &= 0, \\
-R_{1212,3} + R_{1213,2} - R_{1223,1} &= 0.
\end{aligned} \qquad (3.68)$$

It is then possible to conclude that, if

$$R_{1213} = R_{1223} = R_{1323} = 0, \qquad (3.69)$$

it also holds that

$$R_{2323,1} = R_{1313,2} = R_{1212,3} = 0, \qquad (3.70)$$

i.e.,

$$\begin{aligned}
& R_{2323} \text{ is independent on } X^1, \\
& R_{1313} \text{ is independent on } X^2, \\
& R_{1212} \text{ is independent on } X^3.
\end{aligned} \qquad (3.71)$$

It is convenient to summarize all the above considerations as follows: *if C_* is linearly simply connected, system (3.60) can be integrated if and only if the 6 distinct components of the curvature tensor $\mathbf{R}(\mathbf{C})$ vanish. In the linearized theory, the tensor $\mathbf{E}$ must satisfy (3.65) or, equivalently, must satisfy (3.69) in C_* and (3.71) on the boundary ∂C_*.*

We note that, when dealing with infinitesimal deformations, (3.64) can be obtained by observing that, given a symmetric tensor field $\mathbf{E}(\mathbf{X})$, a displacement field $\mathbf{u}(\mathbf{X})$ has to be determined such that

$$u_{L,M} + u_{M,L} = 2E_{LM}.$$

By adding and subtracting $u_{L,M}$, we can write the above system of 6 equations with 3 unknowns $u_L(\mathbf{X})$ as, see (3.36)

$$u_{L,M} = E_{LM} + W_{LM}.$$

The 3 differential forms

$$du_L = u_{L,M} dX_M = (E_{LM} + W_{LM}) dX_M$$

can be integrated if and only if

$$E_{LM,N} + W_{LM,N} = E_{LN,M} + W_{LN,M}. \tag{3.72}$$

Cyclic permutation of indices in (3.72) gives two similar equations, and adding (3.72) to the first of these and subtracting the second one, leads to

$$E_{LM,N} - E_{MN,L} = W_{LN,M},$$

which is equivalent to the system of differential forms

$$dW_{LN} = (E_{LM,N} - E_{MN,L}) dX_M.$$

These forms can be integrated if and only if (3.64) holds.

3.9 Curvilinear Coordinates

This section extends the description of the deformation to the case in which both the initial and current configurations are given in curvilinear coordinates (y^i), which for sake of simplicity are still taken to be orthogonal (Chapter 2).

In this case, we have to distinguish contravariant components, relative to the natural basis $(\mathbf{e}_i)$, from covariant components, relative to the reciprocal or dual basis $(\mathbf{e}^i)$, furthermore, for physical reasons it is convenient to express vectors and tensors with respect to an orthonormal basis $(\mathbf{a}_i)$ (see Chapter 2, Section 2.6).

If (y^i) are the coordinates of a point in C whose coordinates in C_* are (Y^L), then the finite deformation from $C_* \longrightarrow C$ is given by

$$y^i = y^i(Y^L). \tag{3.73}$$

Differentiation gives

$$dy^i \mathbf{e}_i = \frac{\partial y^i}{\partial Y^L}(\mathbf{Y}) dY^L \mathbf{e}_i, \tag{3.74}$$

where $(\mathbf{e}_i)$ is the natural basis associated with (y^i) at the point $\mathbf{y}(\mathbf{Y}) \in C$. By recalling (2.59) and (2.60) and referring to the components $d\bar{y}^i$ in the basis of unit vectors $(\mathbf{a}_i)$, we find that

$$d\bar{y}^i \mathbf{a}_i = \overline{F}_L^i d\overline{Y}^L \mathbf{a}_i, \tag{3.75}$$

where

$$\overline{F}_L^i = \frac{\sqrt{g_{ii}}}{\sqrt{g_{LL}}} F_L^i. \tag{3.76}$$

The bar indicates that the quantities are evaluated in the basis $\mathbf{a}_i$. It is useful to observe that the metric coefficient g_{ii} is computed at $\mathbf{y}(\mathbf{Y}) \in C$, while g_{LL} is evaluated at $\mathbf{Y} \in C_*$. We can apply Cauchy's theorem of polar decomposition in the form (3.7) to the matrix $(\overline{F}_L^i)$ and, since $(\mathbf{a}_i)$ is an orthonormal basis, all considerations developed for Cartesian coordinates still apply to $(\overline{F}_L^i)$.

As an example, the components of tensors $\mathbf{C}$ and $\mathbf{B}$ relative to $(\mathbf{a}_i)$ are given by the elements of the matrices

$$\overline{\mathbf{C}} = \overline{\mathbf{F}}^T \overline{\mathbf{F}}, \quad \overline{\mathbf{B}} = \overline{\mathbf{F}} \overline{\mathbf{F}}^T,$$

and the eigenvalue equation is written as

$$\overline{C}_{LM} \bar{u}^M = \lambda \delta_{LM} \bar{u}^M.$$

The displacement gradient $\nabla \mathbf{u}$ and tensors $\mathbf{E}$ and $\mathbf{W}$ assume the expression given by the rules explained in Chapter 2. The program Deformation, discussed at the end of this chapter, can be used to advantage in this respect.

3.10 Exercises

1. Describe the deformation process corresponding to a one-dimensional extension or compression.

   The required deformation is characterized by two principal stretch ratios equal to unity, while the third one is different than unity. The vector function (3.1) is in this case equivalent to the following scalar functions:

   $$x_1 = \alpha X_1, \quad x_2 = X_2, \quad x_3 = X_3,$$

   where $\alpha > 0$ if it is an extension and $\alpha < 0$ if it is a compression.

3.10. Exercises

The deformation gradient, the right Cauchy–Green tensor, and the Green–St. Venant tensor are expressed by

$$\mathbf{F} = \begin{pmatrix} \alpha & 0 & 0 \\ 0 & 1 & 0 \\ 0 & 0 & 1 \end{pmatrix},$$

$$\mathbf{C} = \mathbf{F}^T\mathbf{F} = \begin{pmatrix} \alpha^2 & 0 & 0 \\ 0 & 1 & 0 \\ 0 & 0 & 1 \end{pmatrix},$$

$$\mathbf{G} = \frac{1}{2}(\mathbf{F}^T\mathbf{F} - \mathbf{I}) = \begin{pmatrix} \frac{\alpha^2 - 1}{2} & 0 & 0 \\ 0 & 0 & 0 \\ 0 & 0 & 0 \end{pmatrix}.$$

2. Given the plane pure shear deformation

$$x_1 = X_1, \quad x_2 = X_2 + kX_3, \quad x_3 = X_3 + kX_2,$$

determine $\mathbf{F}, \mathbf{C}$, and $\mathbf{G}$, the stretch ratios along the diagonal directions, and the angle θ_{23} in the current configuration, as shown in Figure 3.1.

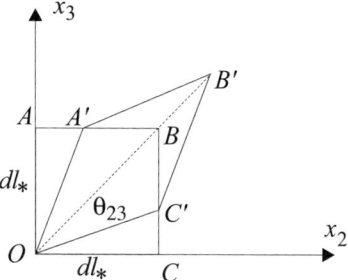

Figure 3.1

The required tensors are represented by the matrices

$$(F_{iL}) = \begin{pmatrix} 1 & 0 & 0 \\ 0 & 1 & k \\ 0 & k & 1 \end{pmatrix},$$

$$(C_{LM}) = \begin{pmatrix} 1 & 0 & 0 \\ 0 & 1 & k \\ 0 & k & 1 \end{pmatrix} \begin{pmatrix} 1 & 0 & 0 \\ 0 & 1 & k \\ 0 & k & 1 \end{pmatrix} = \begin{pmatrix} 1 & 0 & 0 \\ 0 & 1+k^2 & 2k \\ 0 & 2k & 1+k^2 \end{pmatrix},$$

$$(G_{LM}) = \begin{pmatrix} 0 & 0 & 0 \\ 0 & \dfrac{k^2}{2} & k \\ 0 & k & \dfrac{k^2}{2} \end{pmatrix}.$$

Given a line element parallel to the unit vector **N** identified in the reference configuration, the stretch ratio $\Lambda_\mathbf{N}$ is computed according to the relationship

$$\Lambda_\mathbf{N}^2 = \mathbf{N} \cdot \mathbf{CN}.$$

In the direction parallel to the diagonal OB we obtain

$$\Lambda_{OB}^2 = \begin{pmatrix} 0 & \dfrac{1}{\sqrt{2}} & \dfrac{1}{\sqrt{2}} \end{pmatrix} \begin{pmatrix} 1 & 0 & 0 \\ 0 & 1+k^2 & 2k \\ 0 & 2k & 1+k^2 \end{pmatrix} \begin{pmatrix} 0 \\ \dfrac{1}{\sqrt{2}} \\ \dfrac{1}{\sqrt{2}} \end{pmatrix} = (1+k)^2,$$

and in the direction CA we obtain

$$\Lambda_{CA}^2 = \begin{pmatrix} 0 & -\dfrac{1}{\sqrt{2}} & \dfrac{1}{\sqrt{2}} \end{pmatrix} \begin{pmatrix} 1 & 0 & 0 \\ 0 & 1+k^2 & 2k \\ 0 & 2k & 1+k^2 \end{pmatrix} \begin{pmatrix} 0 \\ -\dfrac{1}{\sqrt{2}} \\ \dfrac{1}{\sqrt{2}} \end{pmatrix} = (1-k)^2.$$

The angle θ_{23} in the current configuration is given by

$$\cos \theta_{23} = \dfrac{C_{23}}{\sqrt{C_{22}}\sqrt{C_{33}}} = \dfrac{2k}{1+k^2}.$$

3. Referring to the pure shear deformation of Figure 3.1, apply the polar decomposition to the tensor **F**.

Principal axes X_2' and X_3' are rotated $45°$ about X_1, so that the transformation matrix is

$$(A_{ij}) = (\cos(X_i', X_j)) = \begin{pmatrix} 1 & 0 & 0 \\ 0 & \dfrac{1}{\sqrt{2}} & \dfrac{1}{\sqrt{2}} \\ 0 & -\dfrac{1}{\sqrt{2}} & \dfrac{1}{\sqrt{2}} \end{pmatrix}.$$

The Cauchy–Green tensor **C** in the basis of principal axes is:

$$\mathbf{C}' = \mathbf{ACA}^\mathbf{T},$$

3.10. Exercises

with components

$$(C'_{ij}) = \begin{pmatrix} 1 & 0 & 0 \\ 0 & (1+k)^2 & 0 \\ 0 & 0 & (1-k)^2 \end{pmatrix}.$$

It is then proved, as stated by Theorem 3.2, that the eigenvalues of the tensor $\mathbf{C}$ are the squares of the stretch ratios along the eigenvectors.

Moreover, the same Theorem 3.2 also states that both $\mathbf{U}$ and $\mathbf{C}$ have the same eigenvectors, and the eigenvalues of $\mathbf{C}$ are the squares of the eigenvalues of $\mathbf{U}$, so that

$$(U'_{ij}) = \begin{pmatrix} 1 & 0 & 0 \\ 0 & 1+k & 0 \\ 0 & 0 & 1-k \end{pmatrix}.$$

It follows that

$$((U')^{-1}_{ij}) = \begin{pmatrix} 1 & 0 & 0 \\ 0 & \dfrac{1}{1+k} & 0 \\ 0 & 0 & \dfrac{1}{1-k} \end{pmatrix}.$$

The matrix $\mathbf{U}^{-1}$ in the initial frame is

$$\mathbf{U}^{-1} = \mathbf{A}^{T}(\mathbf{U}')^{-1}\mathbf{A},$$

so that

$$(U^{-1}_{ij}) = \begin{pmatrix} 1 & 0 & 0 \\ 0 & \dfrac{1}{1-k^2} & -\dfrac{k}{1-k^2} \\ 0 & -\dfrac{k}{1-k^2} & \dfrac{1}{1-k^2} \end{pmatrix}.$$

Finally, since we have proved that

$$\mathbf{R} = \mathbf{F}\mathbf{U}^{-1},$$

we have

$$(R_{ij}) = \begin{pmatrix} 1 & 0 & 0 \\ 0 & 1 & 0 \\ 0 & 0 & 1 \end{pmatrix}.$$

Thus, the required decomposition is

$$\begin{pmatrix} 1 & 0 & 0 \\ 0 & 1 & k \\ 0 & k & 1 \end{pmatrix} = \begin{pmatrix} 1 & 0 & 0 \\ 0 & 1 & 0 \\ 0 & 0 & 1 \end{pmatrix} \begin{pmatrix} 1 & 0 & 0 \\ 0 & 1 & k \\ 0 & k & 1 \end{pmatrix}.$$

and we observe that the matrix $\mathbf{R}$ rotates the principal axes of $\mathbf{C}$ in $\mathbf{X} \in \mathbf{C}_*$ in order to superimpose them on the principal axes of $\mathbf{B}^{-1}$ in $\mathbf{x} \in \mathbf{C}$. As these axes coincide in this case, $\mathbf{R} = \mathbf{I}$.

4. Given the deformation
$$x_i = X_i + A_{ij}X_j,$$
with A_{ij} constant, prove that plane sections and straight lines in the reference configuration correspond to plane sections and straight lines in the current configuration.

5. Verify that, if the deformation in Exercise 4 is infinitesimal, i.e., the quantities A_{ij} are so small that their products are negligible, then the composition of two subsequent deformations can be regarded as their sum.

6. Given the deformation
$$\begin{cases} x_1 = X_1 + AX_3, \\ x_2 = X_2 - AX_3, \\ x_3 = X_3 - AX_1 + AX_2, \end{cases}$$
find $\mathbf{F}, \mathbf{U}, \mathbf{R}, \mathbf{C}, \mathbf{B}$, principal directions, and invariants.

7. Determine under which circumstances the infinitesimal displacement field
$$\mathbf{u}(\mathbf{X}) = u_1(X_1)\mathbf{i} + u_2(X_2)\mathbf{j}$$
satisfies the compatibility conditions (3.66).

3.11 The Program Deformation

Aim of the Program

Given a transformation from *orthogonal curvilinear coordinates* var to Cartesian coordinates, the program Deformation allows the user to compute the tensors **F**, **C**, and **B**, the eigenvalues and eigenvectors of **C** and **B**, the deformation invariants, and the inverse of **B**, as well as the tensors **U** and **R**.

Furthermore, given two directions parallel to the unit vectors vers1 and vers2, the program defines the stretch ratios in these directions and the shearing angle, if vers1 and vers2 are distinct. We note that all tensor and vector components are relative to the basis of unit vectors associated with the holonomic basis of curvilinear coordinates var.

Description of the Algorithm and Instructions for Use

The program is based on the theoretical points presented in Sections 3.1–3.3 and 3.9. Curvilinear coordinates in both the reference and current configurations can be chosen, with only the restriction that they be orthogonal. The program makes vers1 and vers2 unit vectors, if they are not.

The caption option refers to tensors **R** and **U** and allows the following possibilities: symbolic, numeric, and null. In particular, if option is selected as null, the program does not compute **U** and **R**; by selecting symbolic or numeric the program computes **U** and **R**, preserving the symbolic or numeric structure.

If the input datum simplifyoption is chosen equal to true, then the program redefines the Mathematica routines Sqrt, Times, Plus, etc., in such a way that the quantities in irrational symbolic expressions can be treated as real. This procedure permits the simplification of all the irrational symbolic calculations inside the program. However, due to the complex form of the matrices U and R, it is not possible to use this option together with option=symbolic.[6]

---

[6] If both the choices option=symbolic and simplifyoption=true are input, then the program cannot give the requested outputs. In this situation, the computation can be stopped from the pop-up menu of Mathematica. To interrupt the computation relative to the last input line, choose Kernel→Interrupt Evaluation or Kernel→Abort Evaluation; to interrupt all the running processes, choose Kernel→Quit→Local→Quit. In this last case, the package Mechanics.m must be launched again before any other application.

Command Line of the Program Deformation

Deformation[func, var, transform, point, vers1, vers2, option, simplifyoption]

Parameter List

Input Data

func = r.h.s. elements of the vector field describing the deformation;

var = curvilinear or Cartesian coordinates;

transform = r.h.s. elements of the transformation from curvilinear var to Cartesian coordinates, or r.h.s. elements of the identity transformation;

point = coordinates at which we want to compute the stretch ratios along the directions vers1 and vers2 and the shearing angle;

vers1 = first vector;

vers2 = second vector;

option = options related to **U** and **R**; the possibilities are symbolic, numeric, and null;

simplifyoption = true or false. In the first case, the program computes many simplifications of symbolic expressions.

Output Data

deformation gradient **F**;

right Cauchy–Green tensor **C**;

eigenvalues, eigenvectors, and invariants of **C**;

left Cauchy–Green tensor **B**;

eigenvectors and inverse of **B**;

right stretching tensor;

rotation tensor;

stretching ratios in the directions vers1 and vers2, evaluated at point;

shearing angle between vers1 and vers2, evaluated at point.

3.11. The Program Deformation

Worked Examples

1. We consider the plane deformation

$$\begin{cases} x_1 = X_1 + kX_2, \\ x_2 = X_2, \end{cases} \quad \forall k \in \Re$$

which maps any $\mathbf{X} = (X_1, X_2) \in \mathbf{C}_*$ onto $\mathbf{x} \in \mathbf{C}$. Both the initial and the current configuration are relative to a Cartesian coordinate system. The input data are the following:

func = $\{X_1 + kX_2, X_2\}$;
var = $\{X_1, X_2\}$;
transform = $\{X_1, X_2\}$;
point = $\{0, 0\}$;
vers1 = $\{1, 0\}$;
vers2 = $\{0, 1\}$;
option = null;
simplifyoption = true;
Deformation[func, var, transform, point, vers1, vers2, option, simplifyoption]

The program Deformation gives the following results:

Deformation gradient

$$\mathbf{F} = \begin{pmatrix} 1 & k \\ 0 & 1 \end{pmatrix}$$

Right Cauchy-Green tensor

$$\mathbf{C} = \begin{pmatrix} 1 & k \\ k & 1 + k^2 \end{pmatrix}$$

Eigenvalues of C

$\lambda_1 = \dfrac{1}{2}(2 + k^2 - k\sqrt{4 + k^2})$: AlgMult = GeoMult = 1

$\lambda_2 = \dfrac{1}{2}(2 + k^2 + k\sqrt{4 + k^2})$: AlgMult = GeoMult = 1

An orthonormal basis of eigenvectors of C

$$\mathbf{u}_1 = \left\{ -\frac{k + \sqrt{4 + k^2}}{\sqrt{2}\sqrt{4 + k^2 + k\sqrt{4 + k^2}}}, \frac{\sqrt{2}}{\sqrt{4 + k^2 + k\sqrt{4 + k^2}}} \right\}$$

$$u_2 = \left\{ \frac{-k + \sqrt{4+k^2}}{\sqrt{2}\sqrt{4+k^2 - k\sqrt{4+k^2}}}, \frac{1}{\sqrt{1 + \frac{1}{4}(-k+\sqrt{4+k^2})^2}} \right\}$$

Principal invariants of C

$I_C = 2 + k^2$

$II_C = 1$

Left Cauchy–Green tensor

$$B = \begin{pmatrix} 1+k^2 & k \\ k & 1 \end{pmatrix}$$

An orthonormal basis of eigenvectors of B

$$v_1 = \left\{ \frac{k - \sqrt{4+k^2}}{\sqrt{2}\sqrt{4+k^2 - k\sqrt{4+k^2}}}, \frac{1}{\sqrt{1 + \frac{1}{4}(k-\sqrt{4+k^2})^2}} \right\}$$

$$v_2 = \left\{ \frac{k + \sqrt{4+k^2}}{2\sqrt{1 + \frac{1}{4}(k+\sqrt{4+k^2})^2}}, \frac{1}{\sqrt{1 + \frac{1}{4}(k+\sqrt{4+k^2})^2}} \right\}$$

Inverse of the left Cauchy–Green tensor

$$B^{-1} = \begin{pmatrix} 1 & -k \\ -k & 1+k^2 \end{pmatrix}$$

Stretch ratio in the direction $\{1, 0\}$

$\delta_1 = 1$

Stretch ratio in the direction $\{0, 1\}$

$\delta_2 = \sqrt{1+k^2}$

Shear angle between the directions $\{1, 0\}$ and $\{0, 1\}$

$$\cos\theta_{12} = \frac{k}{\sqrt{1+k^2}}$$

Initial value of the angle (in degrees)

$\theta_{12}{}^i = 90°$

Final value of the angle (in degrees)

3.11. The Program Deformation

$$\theta_{12}{}^f = \frac{180 \operatorname{ArcCos}\left[\frac{k}{\sqrt{1+k^2}}\right]}{\pi}{}^{\circ}$$

2. We consider the set of plane deformations

$$\begin{cases} r = R + k(R), \\ \phi = \Phi, \end{cases}$$

where $k = k(R)$ is an invertible function of class C^1 such that $(1 + k'(R))\left(\sqrt{(R+k(R))^2}/R\right) > 0$, for all R. Both the reference and the current configuration are relative to a polar curvilinear coordinate system. The program Deformation provides all the peculiar aspects of this deformation at a selected point, which we assume in this example is the point $\{1, 0\}$.

Input data:

```
func = {R + k[R], Φ};
var = {R, Φ};
transform = {RCos[Φ], RSin[Φ]};
point = {1, 0};
vers1 = {1, 0};
vers2 = {0, 1};
option = symbolic;
simplifyoption = true;
Deformation[func, var, transform, point, vers1, vers2,
option, simplifyoption]
```

Output data:

All tensor and vector components are relative to the basis of unit vectors associated with the natural basis of the curvilinear coordinates $\{R, \Phi\}$.

Deformation gradient

$$F = \begin{pmatrix} 1 + k'[R] & 0 \\ 0 & \dfrac{R + k[R]}{R} \end{pmatrix}$$

Right Cauchy–Green tensor

$$C = \begin{pmatrix} (1 + k'[R])^2 & 0 \\ 0 & \dfrac{(R + k[R])^2}{R^2} \end{pmatrix}$$

Eigenvalues of C

$$\lambda_1 = \frac{(R + k[R])^2}{R^2} \; : \; \text{AlgMult} = \text{GeoMult} = 1$$

$$\lambda_2 = (1 + k'[R])^2 \; : \; \text{AlgMult} = \text{GeoMult} = 1$$

An orthonormal basis of eigenvectors of C

$u_1 = \{0, 1\}$

$u_2 = \{1, 0\}$

Principal invariants of C

$$I_C = \frac{(R + k[R])^2}{R^2} + (1 + k'[R])^2$$

$$II_C = \frac{(R + k[R])^2 (1 + k'[R])^2}{R^2}$$

Left Cauchy–Green tensor

$$B = \begin{pmatrix} (1 + k'[R])^2 & 0 \\ 0 & \dfrac{(R + k[R])^2}{R^2} \end{pmatrix}$$

An orthonormal basis of eigenvectors of B

$v_1 = \{0, 1\}$

$v_2 = \{1, 0\}$

Inverse of the left Cauchy–Green tensor

$$B^{-1} = \begin{pmatrix} \dfrac{1}{(1 + k'[R])^2} & 0 \\ 0 & \dfrac{R^2}{(R + k[R])^2} \end{pmatrix}$$

Right stretching tensor

$$U = \begin{pmatrix} 1 + k'[R] & 0 \\ 0 & \dfrac{R + k[R]}{R} \end{pmatrix}$$

Rotation tensor

$$R = \begin{pmatrix} 1 & 0 \\ 0 & 1 \end{pmatrix}$$

Stretch ratio in the direction $\{1, 0\}$

$\delta_1 = 1 + k'[1]$

3.11. The Program Deformation

Stretch ratio in the direction $\{0, 1\}$
$\delta_2 = 1 + \text{k}[1]$

Shear angle between the directions $\{1, 0\}$ and $\{0, 1\}$
$\cos\theta_{12} = 0$

Initial value of the angle (in degrees)
$\theta_{12}{}^i = 90°$

Final value of the angle (in degrees)
$\theta_{12}{}^f = 90°$

Exercises

Apply the program **Deformation** to the following deformations:

1. $\begin{cases} x_1 = X_1 + kX_2, \\ x_2 = X_2, \\ x_3 = X_3. \end{cases} \quad \forall k \in \Re$

2. $\begin{cases} x_1 = \alpha X_1, \\ x_2 = \beta X_2, \\ x_3 = \gamma X_3. \end{cases} \quad \forall \alpha, \beta, \gamma \in \Re$

3. $\begin{cases} x_1 = X_1, \\ x_2 = X_2 - kX_3, \\ x_3 = X_3 + kX_2. \end{cases} \quad \forall k \in \Re$

4. $\begin{cases} x_1 = X_1, \\ x_2 = X_2 + \sqrt{2}X_3, \\ x_3 = X_3 + \sqrt{2}X_2. \end{cases}$

5. $\begin{cases} x_1 = X_1, \\ x_2 = X_2, \\ x_3 = X_3 + kX_2. \end{cases} \quad \forall k \in \Re$

6. $\begin{cases} x_1 = X_1, \\ x_2 = X_2 + kX_3, \\ x_3 = X_3 + kX_2. \end{cases} \quad \forall k \in \Re$

Chapter 4

Kinematics

4.1 Velocity and Acceleration

The aim of this section is to consider the motion of a continuous system S with respect to an orthonormal frame of reference $R = (O, \mathbf{e}_i)$ (see Chapter 1). As stated in Chapter 3, if C denotes the configuration of S in R at time t, then the coordinates (x_i) of the point $\mathbf{x} \in C$ are defined to be *Eulerian* or *spatial coordinates* of $\mathbf{x}$. In order to describe the motion of S, its points have to be labeled so that we can follow them during the motion. This requires the introduction of a reference configuration C_*, selected among all the configurations the system S can assume (the usual choice is the *initial* one), and the coordinates (X_L) of $\mathbf{X} \in C_*$ are defined as the *Lagrangian* or *material coordinates*.

The motion of S is accordingly defined by

$$x_i = x_i(\mathbf{X}, t), \qquad i = 1, 2, 3. \tag{4.1}$$

Equations (4.1) are required to satisfy, for any time instant, the assumptions introduced for (3.1), i.e., they are supposed to be one-to-one functions of class C^2. In particular, the requirement that one and only one $\mathbf{x} \in C$ corresponds to each $\mathbf{X} \in C_*$ guarantees that during the motion the system does not exhibit fractures or material discontinuities; the requirement that one and only one $\mathbf{X}$ corresponds to each $\mathbf{x}$ (existence of the inverse function) preserves the basic property of matter that two particles cannot simultaneously occupy the same position; finally, requiring that the functions are of class C^2 guarantees the regular and smooth behavior of the velocity and acceleration.

From the above properties it follows that at any $\mathbf{X} \in C_*$, the Jacobian J of (4.1) is different from zero.

Moreover, if C_* has been selected as the initial configuration, so that J is equal to 1 when $t = 0$, the continuity of J requires that, for all t,

$$J = \det\left(\frac{\partial x_i}{\partial X_j}\right) > 0. \tag{4.2}$$

Any quantity ψ of the continuous system S can be represented in **Lagrangian** or **Eulerian form**, depending on whether it is expressed as a function of $(\mathbf{X}, t)$ or $(\mathbf{x}, t)$, i.e., if it is a field assigned on the initial configuration C_* or on the current configuration C:

$$\psi = \psi(\mathbf{x}, t) = \psi(\mathbf{x}(\mathbf{X}, t), t) = \tilde{\psi}(\mathbf{X}, t). \tag{4.3}$$

When dealing with balance equations (see Chapter 5), it will be shown how it is important to correctly compute the rate of change of a physical quantity, occurring at a given material point. This rate of change is called the **material derivative**.

As an example, the **velocity** and the **acceleration** of the particle $\mathbf{X} \in C_*$ at time t are given by the partial derivative of the material representation

$$\mathbf{v} = \tilde{\mathbf{v}}(\mathbf{X}, t) = \frac{\partial \mathbf{x}}{\partial t} = \frac{\partial \mathbf{u}}{\partial t}, \quad \mathbf{a} = \tilde{\mathbf{a}}(\mathbf{X}, t) = \frac{\partial^2 \mathbf{x}}{\partial t^2} = \frac{\partial^2 \mathbf{u}}{\partial t^2}, \tag{4.4}$$

where $\mathbf{u} = \mathbf{x}(\mathbf{X}, t) - \mathbf{X}$ is the displacement field introduced in Chapter 3.

Because of the invertibility assumption of the motion, the formulation of the same fields in the spatial representation is allowed by the following identities:

$$\mathbf{v}(\mathbf{x}, t) = \mathbf{v}(\mathbf{x}(\mathbf{X}, t), t) = \tilde{\mathbf{v}}(\mathbf{X}, t),$$
$$\mathbf{a}(\mathbf{x}, t) = \mathbf{a}(\mathbf{x}(\mathbf{X}, t), t) = \tilde{\mathbf{a}}(\mathbf{X}, t), \tag{4.5}$$

and it follows that the material derivative of the spatial representation is given by the total derivative

$$\mathbf{a} = \frac{\partial \mathbf{v}}{\partial t} + \mathbf{v} \cdot \nabla \mathbf{v}, \tag{4.6}$$

where the gradient operator ∇ means differentiation with respect to spatial coordinates:

$$\mathbf{v} \cdot \nabla \mathbf{v} = v_j \frac{\partial \mathbf{v}}{\partial x_j}.$$

The general form for any field $\psi(\mathbf{x}, t) = \tilde{\psi}(\mathbf{X}, t)$ is

$$\dot{\psi} = \frac{\partial \tilde{\psi}}{\partial t} = \frac{\partial \psi}{\partial t} + \mathbf{v} \cdot \nabla \psi, \tag{4.7}$$

which shows that the material derivative in the spatial representation is composed of two contributions: the first one is a local change expressed by

4.1. Velocity and Acceleration

the partial time derivative and the second one is the **convective derivative** $\mathbf{v} \cdot \nabla \psi$.

Relevant features of the motion are often highlighted by referring to **particle paths** and **streamlines**. For this reason, the following definitions are introduced.

The vector field $\mathbf{v}(\mathbf{x}, t)$ at a fixed time t is called the **kinetic field**.

A **particle path** is the trajectory of an individual particle of S. In Lagrangian terms, particle paths can be obtained by integration of (4.4):

$$\mathbf{x} = \mathbf{x}_0 + \int_0^t \tilde{\mathbf{v}}(\mathbf{X}, t)\, dt. \tag{4.8}$$

If the velocity field is expressed in the Eulerian form $\mathbf{v} = \mathbf{v}(\mathbf{x}, t)$, then the determination of particle paths requires the integration of a **nonautonomous** system of first-order differential equations:

$$\frac{d\mathbf{x}}{dt} = \mathbf{v}(\mathbf{x}, t). \tag{4.9}$$

We briefly remind the reader that a system of first-order differential equations (4.9) is called **autonomous** if the functions on the right-hand side are real-valued functions that do not depend explicitly on t.

A **streamline** is defined as the continuous line, at a fixed time t, whose tangent at any point is in the direction of the velocity at that point.

Based on this definition, streamlines represent the integral curves of the kinetic field, i.e., the solution of the *autonomous* system

$$\frac{d\mathbf{x}}{ds} = \mathbf{v}(\mathbf{x}, t), \qquad t = const., \tag{4.10}$$

where s is a parameter along the curve. We note that two streamlines cannot intersect; otherwise we would have the paradoxical situation of a velocity going in two directions at the intersection point.

The motion is defined to be **stationary** if

$$\frac{\partial}{\partial t}\mathbf{v}(\mathbf{x}, t) = \mathbf{0}, \tag{4.11}$$

which is equivalent to saying that all the particles of S that cross the position $\mathbf{x} \in \mathbf{C}$ during the time evolution have the same velocity.

Since in the *stationary motion* the right-hand side of both (4.9) and (4.10) are independent of t, the two systems of differential equations are equivalent to each other, so that *particle paths and streamlines are coincident*.

4.2 Velocity Gradient

Two relevant kinematic tensors will be extensively used. The first one is the symmetric tensor defining the ***rate of deformation*** or ***stretching*** (Euler, 1770)

$$D_{ij} = \frac{1}{2}\left(\frac{\partial v_i}{\partial x_j} + \frac{\partial v_j}{\partial x_i}\right) = D_{ji}, \qquad (4.12)$$

and the second one is the skew-symmetric tensor defined as the ***spin*** or ***vorticity tensor***

$$W_{ij} = \frac{1}{2}\left(\frac{\partial v_i}{\partial x_j} - \frac{\partial v_j}{\partial x_i}\right) = -W_{ji}. \qquad (4.13)$$

By means of these two tensors, the ***gradient of velocity*** can be conveniently decomposed as

$$\nabla \mathbf{v} = \mathbf{D} + \mathbf{W}. \qquad (4.14)$$

By recalling the definition of differential

$$\mathbf{v}(\mathbf{x}+d\mathbf{x},t) = \mathbf{v}(\mathbf{x},t) + d\mathbf{x} \cdot \nabla \mathbf{v}, \qquad (4.15)$$

and using the decomposition (4.14), we obtain

$$\mathbf{v}(\mathbf{x}+d\mathbf{x},t) = \mathbf{v}(\mathbf{x},t) + \omega \times d\mathbf{x} + \mathbf{D}d\mathbf{x}, \qquad (4.16)$$

where ω is the vector such that $\omega \times d\mathbf{x} = \mathbf{W} d\mathbf{x}$. It can be verified that

$$\omega = \frac{1}{2}\nabla \times \mathbf{v}, \qquad (4.17)$$

which shows that, *in a neighborhood of* $\mathbf{x} \in C$, *the local kinetic field is composed of a rigid motion with angular velocity given by (4.17) that is a function of time and* $\mathbf{x}$, *and the term* $\mathbf{D}d\mathbf{x}$.

Later it will be useful to refer to the following expression for the Eulerian acceleration, derived from (4.6):

$$a_i = \frac{\partial v_i}{\partial t} + v_j \frac{\partial v_i}{\partial x_j} = \frac{\partial v_i}{\partial t} + v_j \frac{\partial v_i}{\partial x_j} + v_j \frac{\partial v_j}{\partial x_i} - v_j \frac{\partial v_j}{\partial x_i}$$

$$= \frac{\partial v_i}{\partial t} + 2W_{ij}v_j + \frac{1}{2}\frac{\partial v^2}{\partial x_j}.$$

Since $W_{ij}v_j = \epsilon_{ihj}\omega_h v_j$, we obtain

$$a_i = \frac{\partial v_i}{\partial t} + 2\epsilon_{ihj}\omega_h v_j + \frac{1}{2}\frac{\partial v^2}{\partial x_j},$$

and, referring to (4.17), we see that in vector terms it holds that

$$\mathbf{a} = \frac{\partial \mathbf{v}}{\partial t} + (\nabla \times \mathbf{v}) \times \mathbf{v} + \frac{1}{2}\nabla v^2. \qquad (4.18)$$

4.3 Rigid, Irrotational, and Isochoric Motions

For the analysis we are going to present, it is of interest to introduce Liouville's formula
$$\dot{J} = J\nabla \cdot \mathbf{v}. \tag{4.19}$$

To prove (4.19), first observe that by definition

$$\dot{J} = \frac{d}{dt}\det\left(\frac{\partial x_i}{\partial X_L}\right) = \begin{vmatrix} \frac{\partial v_1}{\partial X_1} & \frac{\partial v_1}{\partial X_2} & \frac{\partial v_1}{\partial X_3} \\ \frac{\partial x_2}{\partial X_1} & \frac{\partial x_2}{\partial X_2} & \frac{\partial x_2}{\partial X_3} \\ \frac{\partial x_3}{\partial X_1} & \frac{\partial x_3}{\partial X_2} & \frac{\partial x_3}{\partial X_3} \end{vmatrix} + \cdots + \begin{vmatrix} \frac{\partial x_1}{\partial X_1} & \frac{\partial x_1}{\partial X_2} & \frac{\partial x_1}{\partial X_3} \\ \frac{\partial x_2}{\partial X_1} & \frac{\partial x_2}{\partial X_2} & \frac{\partial x_2}{\partial X_3} \\ \frac{\partial v_3}{\partial X_1} & \frac{\partial v_3}{\partial X_2} & \frac{\partial v_3}{\partial X_3} \end{vmatrix}.$$

Moreover,
$$\frac{\partial v_h}{\partial X_L} = \frac{\partial v_h}{\partial x_j}\frac{\partial x_j}{\partial X_L},$$
and since each determinant can be written as $J\partial v_i/\partial x_i$ (no summation on i), (4.19) is proved.

Theorem 4.1
The motion of S is (globally) rigid if and only if
$$\mathbf{D} = \mathbf{0}. \tag{4.20}$$

PROOF To prove that (4.20) follows as a necessary condition, we observe that a rigid motion implies $v_i(\mathbf{x},t) = v_i(\mathbf{x_0},t) + \epsilon_{ijl}\omega_j(t)(x_l - x_{0l})$. The velocity gradient
$$\frac{\partial v_i}{\partial x_k} = \epsilon_{ijk}\omega_j$$
is then skew-symmetric, so that
$$2D_{ik} = \frac{\partial v_i}{\partial x_k} + \frac{\partial v_k}{\partial x_i} = (\epsilon_{ijk} + \epsilon_{kji})\omega_j = 0.$$

To prove that $\mathbf{D} = \mathbf{0}$ is a sufficient condition, by (4.15) and (4.16) we obtain
$$\frac{\partial v_i}{\partial x_j} = W_{ij}, \qquad W_{ij} = -W_{ji}. \tag{4.21}$$

The system (4.21) of 9 differential equations with the three unknown functions $v_i(\mathbf{x})$ can be written in the equivalent form
$$dv_i = W_{ij}dx_j, \tag{4.22}$$

so that (4.21) has a solution if and only if the differential forms (4.22) can be integrated. If the region C of the kinetic field is simply connected, then a necessary and sufficient condition for the integrability of (4.22) is

$$\frac{\partial W_{ij}}{\partial x_h} = \frac{\partial W_{ih}}{\partial x_j}.$$

By cyclic permutation of the indices, two additional conditions follow:

$$\frac{\partial W_{jh}}{\partial x_i} = \frac{\partial W_{ji}}{\partial x_h}, \qquad \frac{\partial W_{hi}}{\partial x_j} = \frac{\partial W_{hj}}{\partial x_i}.$$

Summing up the first two, subtracting the third one and taking into account $(4.21)_2$, we derive the condition

$$\frac{\partial W_{ij}}{\partial x_h} = 0,$$

which shows that the skew-symmetric tensor W_{ij} does not depend on the spatial variables and eventually depends on time. Then, integration of (4.22) gives

$$v_i(\mathbf{x}, t) = v_{0i}(t) + W_{ij}(t)(x_j - x_{0j}),$$

and the motion is rigid. ∎

The motion of S is defined to be *irrotational* if

$$\boldsymbol{\omega} = \frac{1}{2} \nabla \times \mathbf{v} = \mathbf{0}. \tag{4.23}$$

Again suppose that the region C is simply connected; then it follows that the motion is irrotational if and only if

$$\mathbf{v} = \nabla \varphi, \tag{4.24}$$

where φ is a **potential** for the velocity, also defined as the **kinetic potential**.

Let $c \subset C$ be a region which is the mapping of $c_* \subset C_*$ under the equations of the motion. This region is a **material volume** because it is always occupied by the same particles. If during the motion of S the volume of any arbitrary material region does not change, then the motion is called **isochoric** or **isovolumic**, i.e.,

$$\frac{d}{dt} \int_c dc = 0.$$

By changing the variables $(x_i) \longrightarrow (X_L)$, the previous requirement can be written as

$$\frac{d}{dt} \int_{c_*} J dc_* = 0,$$

and, since the volume c_* is fixed, differentiation and integration can be exchanged, and because of (4.19), we get

$$\int_{c_*} J\nabla \cdot \mathbf{v}\, dc_* = \int_c \nabla \cdot \mathbf{v}\, dc = 0 \qquad \forall c \subset C.$$

Next, we conclude that a motion is isochoric if and only if

$$\nabla \cdot \mathbf{v} = 0. \tag{4.25}$$

Finally, an irrotational motion is isochoric if and only if the velocity potential satisfies **Laplace**'s **equation**

$$\Delta \varphi = \nabla \cdot \nabla \varphi = 0, \tag{4.26}$$

whose solutions are known as **harmonic functions**.

4.4 Transformation Rules for a Change of Frame

As we will discuss in Chapter 7, constitutive equations are required to be invariant under change of frame of reference. Intuitively, this requirement means that two observers in relative motion with respect to each other must observe the same stress in a given continuous body. For this and other reasons it is of interest to investigate how the tensors $\dot{\mathbf{F}}$, $\nabla \mathbf{v}$, $\mathbf{D}$, and $\mathbf{W}$ transform *under a change of frame of reference*.

Let us briefly recall that a *frame of reference* can be considered as an *observer* measuring distances with a ruler and time intervals with a clock. In general, two observers moving relatively to each other will record different values of position and time of the *same event*. But within the framework of Newtonian mechanics, it is postulated that: *distance and time intervals between events have the same values in two frames of reference whose relative motion is rigid.*

Suppose that the first observer is characterized by the reference system Ox_i and time t and the second one by $O'x'_i$ and t'. The above requirement can be expressed in analytical terms as

$$\begin{aligned} x'_i &= Q_{ij}(t)x_j + c_i(t), \\ t' &= t + a, \end{aligned} \tag{4.27}$$

where $\mathbf{Q} = (Q_{ij})$, $\mathbf{c} = (c_i)$, and a are a proper orthogonal tensor, an arbitrary vector, and an arbitrary scalar quantity, respectively.

The change of frame of reference expressed by (4.27) is a time-dependent rigid transformation known as a *Euclidean transformation* (see Chapter 1).

A scalar field g, a vector field $\mathbf{q}$, and a tensor field $\mathbf{T}$ are defined to be **objective** if, under a change of frame of reference, they transform according to the rules

$$g'(\mathbf{x}', t') = g(\mathbf{x}, t),$$
$$\mathbf{q}'(\mathbf{x}', t') = \mathbf{Q}(t)\mathbf{q}(\mathbf{x}, t),$$
$$\mathbf{T}'(\mathbf{x}', t') = \mathbf{Q}(t)\mathbf{T}(\mathbf{x}, t)\mathbf{Q}^T(t). \tag{4.28}$$

In this case we say that two observers are considering the same event from two different points of view.

It was already observed in Section 3.6 that the deformation gradient $\mathbf{F}$ and the left and right Cauchy–Green tensors transform as follows:

$$\mathbf{F}' = \mathbf{QF}, \quad \mathbf{C}' = \mathbf{C}, \quad \mathbf{B}' = \mathbf{QBQ}^T. \tag{4.29}$$

Moreover, the following additional transformation rules hold under a change of frame (4.27):

$$\dot{\mathbf{F}}' = \mathbf{Q}\dot{\mathbf{F}} + \dot{\mathbf{Q}}\mathbf{F},$$
$$\nabla' \mathbf{v}' = \mathbf{Q}\nabla\mathbf{v}\mathbf{Q}^T + \dot{\mathbf{Q}}\mathbf{Q}^T,$$
$$\mathbf{D}' = \mathbf{Q}\mathbf{D}\mathbf{Q}^T,$$
$$\mathbf{W}' = \mathbf{Q}\mathbf{W}\mathbf{Q}^T + \dot{\mathbf{Q}}\mathbf{Q}^T. \tag{4.30}$$

The equation $(4.30)_1$ is obtained by differentiating $(4.29)_1$ with respect to time.

Furthermore, differentiating the relation $x'_i = Q_{ij}(t)x_j + c_i(t)$ with respect to time and x'_h, we get

$$\frac{\partial v'_i}{\partial x'_h} = Q_{ij}\frac{\partial v_j}{\partial x_k}\frac{\partial x_k}{\partial x'_h} + \dot{Q}_{ij}\frac{\partial x_j}{\partial x'_h}$$

and $(4.30)_2$ is proved. Moreover, the orthogonality condition $\mathbf{Q}\mathbf{Q}^T = \mathbf{I}$ allows us to write $\dot{\mathbf{Q}}\mathbf{Q}^T + \mathbf{Q}\dot{\mathbf{Q}}^T = \mathbf{0}$, i.e., $\dot{\mathbf{Q}}\mathbf{Q}^T = -\left(\dot{\mathbf{Q}}\mathbf{Q}^T\right)^T$, so that the skew-symmetry of $\dot{\mathbf{Q}}\mathbf{Q}^T$ is derived. By recalling the definitions (4.12) and (4.13) of $\mathbf{D}$ and $\mathbf{W}$, $(4.30)_{3,4}$ can easily be proved.

It follows that only the rate of deformation tensor $\mathbf{D}$ can be considered to be objective.

4.5 Singular Moving Surfaces

In continuum mechanics it is quite common to deal with a surface that is singular with respect to some scalar, vector, or tensor field and is moving

4.5. Singular Moving Surfaces

independently of the particles of the system. Typical phenomena include acceleration waves, shock waves, phase transitions, and many others. Due to the relevance of these topics, this section and the next one are devoted to the kinematics of singular surfaces.

Let
$$f(\mathbf{r}, t) = 0 \tag{4.31}$$

be a moving surface $\Sigma(t)$. Given a point $\mathbf{r} \in \Sigma(t)$, consider the straight line a of equation $\mathbf{r} + s\,\mathbf{n}$, where $\mathbf{n}$ is the unit vector normal to $\Sigma(t)$ at $\mathbf{r}$. If $f(\mathbf{r}, t + \tau) = 0$ is the equation of $\Sigma(t + \tau)$, let the intersection point of a with the surface $\Sigma(t + \tau)$ be denoted by $\mathbf{y}$. The distance between $\Sigma(t)$ and $\Sigma(t + \tau)$ measured along the normal at $\mathbf{r}$ is given by $s(\tau)$, and this parameter allows us to define the **normal speed** of $\Sigma(t)$ as the limit

$$c_n = \lim_{\tau \to 0} \frac{s(\tau)}{\tau} = s'(0). \tag{4.32}$$

The limit (4.32) can also be expressed in terms of (4.31) if it is observed that $s(\tau)$ is implicitly defined by equation

$$f(\mathbf{r} + s\,\mathbf{n}, t + \tau) \equiv \varphi(s, \tau) = 0,$$

so that, from Dini's theorem on implicit functions, it follows that

$$s'(0) = -\left(\frac{\partial \varphi / \partial \tau}{\partial \varphi / \partial s}\right)_{(0,0)} = -\frac{\partial f / \partial t}{\nabla f \cdot \mathbf{n}}.$$

Since ∇f is parallel to $\mathbf{n}$ and has the same orientation, it can be written as

$$c_n = -\frac{\partial f / \partial t}{|\nabla f|}. \tag{4.33}$$

The velocity of the surface with respect to the material particles instantaneously lying upon it is called the **local speed of propagation** and is given by $(c_n - \mathbf{v} \cdot \mathbf{n})$, if the current configuration is regarded as a reference configuration.

Note that we obtain (4.33) by using the implicit representation of the surface $\Sigma(t)$. If we adopt a parametric form of $\Sigma(t)$, i.e., $\mathbf{r} = \mathbf{g}(u_\alpha, t)$, where u_α, $\alpha = 1, 2$, are the parameters on the surface, then $f(\mathbf{g}(u_\alpha, t), t) = 0$ and (4.33) gives

$$c_n = \frac{1}{|\nabla f|} \nabla f \cdot \frac{\partial \mathbf{g}}{\partial t} = \mathbf{n} \cdot \frac{\partial \mathbf{r}}{\partial t}. \tag{4.34}$$

It can be proved that *the velocity of the surface is independent of the parametric representation of $\Sigma(t)$*.

In fact, if U_α with $u_\alpha = u_\alpha(U_\beta, t)$ are new parameters, then the parametric equations of the surface will be $\mathbf{R}(U_\alpha, t) = \mathbf{r}(u_\alpha(U_\beta, t))$ and (4.34) will be expressed in terms of $\partial \mathbf{R}/\partial t$ instead of $\partial \mathbf{r}/\partial t$. Since

$$\frac{\partial \mathbf{R}}{\partial t} = \frac{\partial \mathbf{r}}{\partial u_\alpha} \frac{\partial u_\alpha}{\partial t} + \frac{\partial \mathbf{r}}{\partial t},$$

and $\partial \mathbf{r}/\partial u^\alpha$ are tangent to the surface $\Sigma(t)$, it follows that

$$\mathbf{n} \cdot \frac{\partial \mathbf{r}}{\partial t} = \mathbf{n} \cdot \frac{\partial \mathbf{R}}{\partial t},$$

and both parametric representations lead to the same value of c_n.

Let $\mathbf{r} = \mathbf{g}(u_\alpha, t)$ be a surface $\Sigma(t)$ and let $\Gamma(t)$ be a moving curve on it, with parametric equations $\mathbf{r} = \mathbf{G}(s, t) = \mathbf{g}(u_\alpha(s, t), t)$, where s is the curvilinear abscissa on $\Gamma(t)$. The orientation of the tangent unit vector τ to the curve $\Gamma(t)$ is fixed according to the usual rule that τ is moving counterclockwise on $\Gamma(t)$ for an observer oriented along $\mathbf{n}$. Furthermore, the unit vector normal to $\Gamma(t)$ on the plane tangent to $\Sigma(t)$ is expressed by $\nu_\Sigma = \tau \times \mathbf{n}$.

According to these definitions, the velocity of the curve $\Gamma(t)$ is given by the scalar quantity

$$w_\nu = \frac{\partial \mathbf{G}}{\partial t} \cdot \nu_\Sigma \equiv \mathbf{w} \cdot \nu_\Sigma, \tag{4.35}$$

and it can be proved that $\mathbf{w} \cdot \nu_\Sigma$ is independent of the parametric representation of $\Sigma(t)$ and $\Gamma(t)$ (see Exercise 7).

A moving surface $\Sigma(t)$ is defined to be **singular of order k $\geq$ 0** with respect to the field $\psi(\mathbf{x}, t)$ if the same definition applies to the *fixed surface* S of $\Re^4$ of equation (4.31) (see Section 2.5).

It is convenient to write the relationships found in Section 2.5 in terms of variables (x_i, t). To this end, let $\Re^4$ be a four-dimensional space in which the coordinates (x_i, t) are introduced, and let the fixed surface S of $\Re^4$ be represented by (4.31) (see Figure 4.1).

The unit vector $\mathbf{N}$ normal to S has components

$$\mathbf{N} = \frac{\operatorname{grad} f}{|\operatorname{grad} f|} = \frac{1}{|\operatorname{grad} f|}\left(\nabla f, \frac{\partial f}{\partial t}\right), \tag{4.36}$$

where $\operatorname{grad} f = (\partial f/\partial x_i, \partial f/\partial t)$. Taking into account (4.33) and observing that the unit normal $\mathbf{n}$ in $\Re^3$ to the surface $f(\mathbf{r}, t) = 0$, with t fixed, has components $(\partial f/\partial x_i)/|\nabla f|$, we find that (4.36) can be written as

$$\mathbf{N} = \frac{|\nabla f|}{|\operatorname{grad} f|}(\mathbf{n}, -c_n) \equiv \beta(\mathbf{n}, -c_n). \tag{4.37}$$

4.6. Time Derivative of a Moving Volume

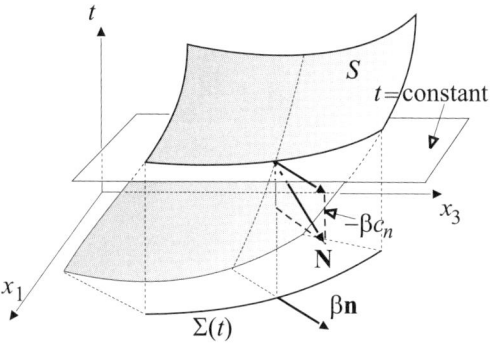

Figure 4.1

With this notation, if $\Sigma(t)$ is a surface of order 1 with respect to the tensor field $\mathbf{T}(\mathbf{x},t)$, then (2.48) and (4.37) give the following jump conditions:

$$[[\nabla \mathbf{T}]] = \mathbf{n} \otimes \mathbf{A}(\mathbf{x},t), \qquad (4.38)$$

$$\left[\left[\frac{\partial \mathbf{T}}{\partial t}\right]\right] = -c_n \mathbf{A}(\mathbf{x},t), \qquad (4.39)$$

where $\mathbf{A}(\mathbf{x},t) = \mathbf{a}\,|\nabla f|/|\mathrm{grad}\,f|$.

If the surface is of order 2, then (2.50) leads to the following conditions:

$$[[\nabla \nabla \mathbf{T}]] = \mathbf{n} \otimes \mathbf{n} \otimes \mathbf{A}(\mathbf{x},t), \qquad (4.40)$$

$$\left[\left[\nabla \frac{\partial \mathbf{T}}{\partial t}\right]\right] = -c_n \mathbf{n} \otimes \mathbf{A}(\mathbf{x},t), \qquad (4.41)$$

$$\left[\left[\frac{\partial^2 \mathbf{T}}{\partial t^2}\right]\right] = c_n^2 \mathbf{A}(\mathbf{x},t). \qquad (4.42)$$

4.6 Time Derivative of a Moving Volume

Let $V(t)$ be a moving volume with a smooth boundary surface $\partial V(t)$ having equation $g(\mathbf{x},t) = 0$ and unit outward normal $\mathbf{N}$. Suppose that its velocity component along the normal is given by

$$V_N = -\frac{1}{|\nabla g|}\frac{\partial g}{\partial t}.$$

When dealing with balance equations, it is useful to express the time derivative of a given quantity, defined over a volume which is not fixed but is changing with time. To do this, we need to use the following relation:

$$\frac{d}{dt}\int_{V(t)} \mathbf{f}(\mathbf{x},t)\,dc = \int_{V(t)} \frac{\partial \mathbf{f}}{\partial t}\,dc + \int_{\partial V(t)} \mathbf{f} V_N\,d\sigma. \qquad (4.43)$$

To prove (4.43), we observe that

$$\frac{d}{dt}\int_{V(t)} \mathbf{f}(\mathbf{x},t)\,dc$$

$$= \lim_{\Delta t \to 0} \frac{1}{\Delta t}\left(\int_{V(t+\Delta t)} \mathbf{f}(\mathbf{x}, t+\Delta t)\,dc - \int_{V(t)} \mathbf{f}(\mathbf{x}, t)\,dc\right).$$

The right-hand term can also be written in the form

$$\int_{V(t+\Delta t)} \mathbf{f}(\mathbf{x}, t+\Delta t)\,dc - \int_{V(t)} \mathbf{f}(\mathbf{x}, t+\Delta t)\,dc$$

$$+ \int_{V(t)} [\mathbf{f}(\mathbf{x}, t+\Delta t) - \mathbf{f}(\mathbf{x}, t)]\,dc,$$

and we see that the difference of the first two integrals gives the integral of $\mathbf{f}(\mathbf{x}, t+\Delta t)$ over the change of $V(t)$ due to the moving boundary $\partial V(t)$. By neglecting terms of higher order, we have

$$\int_{V(t+\Delta t)} \mathbf{f}(\mathbf{x}, t+\Delta t)\,dc - \int_{V(t)} \mathbf{f}(\mathbf{x}, t+\Delta t)\,dc$$

$$= \int_{\partial V(t)} \mathbf{f}(\mathbf{x}, t+\Delta t) V_N\,\Delta t\,d\sigma$$

so that in the limit $\Delta t \longrightarrow 0$, (4.43) is proved.

In the same context, let $\Sigma(t)$ be a singular surface of zero order with respect to the field $\mathbf{f}(\mathbf{x}, t)$ and $\sigma(t) = \Sigma(t) \cap V(t)$. If $f(\mathbf{x}, t) = 0$, $\mathbf{n}$, and c_n denote the equation of $\Sigma(t)$, its unit normal, and its advancing velocity, respectively, we have

$$c_n = -\frac{1}{|\nabla f|}\frac{\partial f}{\partial t}$$

and the following formula can be proved:

$$\frac{d}{dt}\int_{V(t)} \mathbf{f}(\mathbf{x},t)\,dc = \int_{V(t)} \frac{\partial \mathbf{f}}{\partial t}\,dc + \int_{\partial V(t)} \mathbf{f} V_N\,d\sigma - \int_{\sigma(t)} [[\mathbf{f}]] c_n\,d\sigma. \qquad (4.44)$$

To this end, suppose the surface $\Sigma(t)$ divides the volume $V(t)$ into two regions, V^- and V^+. In order to apply (4.43), we note that ∂V has velocity V_N, while $\sigma(t)$ is moving with velocity c_n if $\sigma(t)$ is considered to belong

4.6. Time Derivative of a Moving Volume

to $V^-(t)$, or with velocity $-c_n$ if it is part of V^+. Furthermore, the field $\mathbf{f}$ on $\sigma(t)$ will be $\mathbf{f}^-$ or $\mathbf{f}^+$, depending on whether $\sigma(t)$ belongs to V^- or V^+. We derive (4.44) by applying (4.43) to V^+ and V^- and subtracting the corresponding results.

As a particular case, if $V(t)$ is a fixed volume ($V_N = 0$), then from (4.44) it follows that

$$\frac{d}{dt} \int_V \mathbf{f}(\mathbf{x}, t) \, dc = \int_V \frac{\partial \mathbf{f}}{\partial t} \, dc - \int_{\sigma(t)} [[\mathbf{f}]] c_n \, d\sigma. \qquad (4.45)$$

As a further application of (4.44), consider a material volume $c(t) \subset C(t)$ that is the mapping of the initial volume $c_* \subset C_*$ by the equations of motion $\mathbf{x} = \mathbf{x}(\mathbf{X}, t)$. Since this moving volume is a collection of the same particles of $S(t)$, its boundary $\partial c(t)$ is moving at normal speed $\mathbf{v} \cdot \mathbf{N}$, and (4.44) can be written as

$$\frac{d}{dt} \int_{c(t)} \mathbf{f}(\mathbf{x}, t) \, dc = \int_{c(t)} \frac{\partial \mathbf{f}}{\partial t} \, dc + \int_{\partial c(t)} \mathbf{f} \otimes \mathbf{v} \cdot \mathbf{N} \, d\sigma - \int_{\sigma(t)} [[\mathbf{f}]] c_n \, d\sigma. \qquad (4.46)$$

By applying the generalized Gauss's theorem (2.42) to the second integral of the right-hand side of (4.46), we derive

$$\frac{d}{dt} \int_{c(t)} \mathbf{f}(\mathbf{x}, t) \, dc = \int_{c(t)} \left[\frac{\partial \mathbf{f}}{\partial t} + \nabla \cdot (\mathbf{v} \otimes \mathbf{f}) \right] dc - \int_{\sigma(t)} [[\mathbf{f}(c_n - v_n)]] \, d\sigma. \qquad (4.47)$$

In applications the need often arises to consider a **material open moving surface**, i.e., a surface which is the mapping, by the equations of motion $\mathbf{x} = \mathbf{x}(\mathbf{X}, t)$, of a surface S^* represented in the reference configuration by the equation $\mathbf{r} = \mathbf{r}(u^\alpha)$. In this case the following formula holds:

$$\frac{d}{dt} \int_{S(t)} \mathbf{u}(\mathbf{x}, t) \cdot \mathbf{N} \, d\sigma = \int_{S(t)} \left[\frac{\partial \mathbf{u}}{\partial t} + \nabla \times (\mathbf{u} \times \mathbf{v}) + \mathbf{v} \nabla \cdot \mathbf{u} \right] \cdot \mathbf{N} \, d\sigma, \qquad (4.48)$$

where $\mathbf{N}$ is the unit vector normal to $S(t)$.

To prove (4.48), we first observe that due to (3.19) we can write

$$\frac{d}{dt} \int_{S(t)} \mathbf{u}(\mathbf{x}, t) \cdot \mathbf{N} \, d\sigma = \frac{d}{dt} \int_{S(t)} u_i \, d\sigma_i$$
$$= \frac{d}{dt} \int_{S^*} u_i J \frac{\partial X_L}{\partial x_i} \, d\sigma_{*L} = \int_{S^*} \frac{d}{dt} \left(u_i J \frac{\partial X_L}{\partial x_i} \right) d\sigma_{*L}. \qquad (4.49)$$

In addition, from (4.19) it follows that

$$\frac{d}{dt} \left(J \frac{\partial X_L}{\partial x_i} \right) = J \nabla \cdot \mathbf{v} \frac{\partial X_L}{\partial x_i} + J \frac{d}{dt} \frac{\partial X_L}{\partial x_i}; \qquad (4.50)$$

and to evaluate the derivative on the right-hand side it is convenient to recall that
$$\frac{\partial X_L}{\partial x_i}\frac{\partial x_h}{\partial X_L}=\delta_{ih}.$$
Therefore,
$$\frac{d}{dt}\frac{\partial X_L}{\partial x_i}=-\frac{\partial \dot{x}_h}{\partial x_i}\frac{\partial X_L}{\partial x_h},$$
and (4.50) becomes
$$\frac{d}{dt}\left(J\frac{\partial X_L}{\partial x_i}\right)=J\nabla\cdot\mathbf{v}\frac{\partial X_L}{\partial x_i}-J\frac{\partial \dot{x}_h}{\partial x_i}\frac{\partial X_L}{\partial x_h}.$$
By using this expression in (4.50) we find that
$$\frac{d}{dt}\int_{S(t)}\mathbf{u}(\mathbf{x},t)\cdot\mathbf{N}\,d\sigma=\int_{S(t)}(\dot{\mathbf{u}}+\mathbf{u}\nabla\cdot\mathbf{v}-\mathbf{u}\cdot\nabla\mathbf{v})\cdot\mathbf{N}\,d\sigma$$
and (4.48) is proved by applying known vector identities.

Finally, it is of interest to investigate how (4.48) can be generalized to the case in which the material surface $S(t)$ intersects the moving surface $\Sigma(t)$, which is supposed to be singular with respect to the field $\mathbf{u}$. Let $\Gamma(t)=S(t)\cap\Sigma(t)$ be the intersection curve between $S(t)$ and $\Sigma(t)$, and let $S^-(t)$ and $S^+(t)$ be the regions into which $S(t)$ is subdivided by $\Gamma(t)$. If τ denotes the unit vector tangent to $\Gamma(t)$ and $\nu_{\mathbf{s}}$ the unit vector tangent to $S(t)$ such that $\nu_{\mathbf{s}},\tau,\mathbf{N}$ define a positive basis (see Figure 4.2), then we can easily prove the following generalization of (4.48):

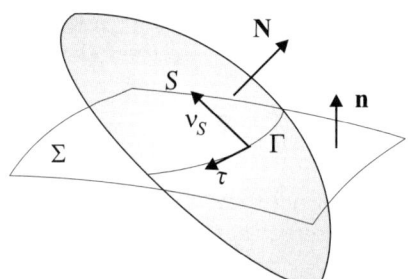

Figure 4.2

$$\frac{d}{dt}\int_{S(t)}\mathbf{u}(\mathbf{x},t)\cdot\mathbf{N}\,d\sigma=\int_{S(t)}\left[\frac{\partial \mathbf{u}}{\partial t}+\nabla\times(\mathbf{u}\times\mathbf{v})+\mathbf{v}\nabla\cdot\mathbf{u}\right]\cdot\mathbf{N}\,d\sigma$$
$$-\int_{\Gamma(t)}\mathbf{n}\times[[\mathbf{u}\times(\mathbf{w}-\mathbf{v})]]\cdot\nu_S\,ds,\qquad(4.51)$$

where $\mathbf{w}$ is the velocity of $\Gamma(t)$.

Remark When dealing with continuous systems characterized by interfaces (examples include shock waves, Weiss domains in crystals, phase transitions and many other phenomena), it is useful to evaluate the following derivative:

$$\frac{d}{dt} \int_{S(t)} \mathbf{\Phi}_S \, d\sigma,$$

where $\mathbf{\Phi}_S(\mathbf{x}, t)$ is a field transported by the *nonmaterial* surface $\Sigma(t)$ and $S(t) = c(t) \cap \Sigma(t)$, where $c(t)$ is a material or fixed volume. The formula (4.51) does not allow us to consider this case, and as a result many authors have paid attention to this problem (see for example Moeckel [13], Gurtin [14], Dell'Isola and Romano [15]). Recently, Marasco and Romano [16] have presented a simple and general formulation that will be further discussed in Volume II.

4.7 Worked Exercises

1. Given the motion

$$x_1 = \left(1 + \frac{t}{T}\right)^2 X_1, \quad x_2 = X_2, \quad x_3 = X_3,$$

find the Lagrangian and Eulerian representation of the velocity and acceleration.

The displacement components in the Lagrangian form are given by

$$u_1 = x_1 - X_1 = \left(2\frac{t}{T} + \frac{t^2}{T^2}\right) X_1, \quad u_2 = u_3 = 0,$$

and in Eulerian form they are

$$u_1 = \frac{2(t/T) + (t/T)^2}{\left(1 + \frac{t}{T}\right)^2} x_1, \quad u_2 = u_3 = 0.$$

The velocity components in the Lagrangian representation are

$$v_1 = \frac{\partial u_1(X, t)}{\partial t} = \frac{2}{T}\left(1 + \frac{t}{T}\right) X_1, \quad v_2 = v_3 = 0,$$

and in Eulerian form they are

$$v_1 = \frac{2}{T} \frac{1}{\left(1 + \frac{t}{T}\right)} x_1, \quad v_2 = v_3 = 0.$$

Finally, the acceleration components in Lagrangian form are
$$a_1 = \frac{2}{T^2}X_1, \quad a_2 = a_3 = 0,$$
while the Eulerian representation can be obtained by substituting in the previous relation the inverse motion
$$X_1 = \frac{x_1}{\left(1+\dfrac{t}{T}\right)^2}$$
or by applying the material derivative operator
$$\frac{d}{dt} = \frac{\partial}{\partial t} + \mathbf{v}\cdot\nabla,$$
so that
$$a_1 = \frac{2}{T^2}\frac{1}{\left(1+\dfrac{t}{T}\right)^2}x_1, \quad a_2 = a_3 = 0.$$

2. Prove that the motion is rigid if and only if the equations of motion have the structure
$$x_i = b_i(t) + Q_{ij}(t)X_j,$$
where Q_{ij} is a proper orthogonal tensor, i.e.,
$$Q_{ij}Q_{kj} = \delta_{ij}, \quad \det \mathbf{Q} = 1.$$

Let $\mathbf{X}^A$ and $\mathbf{X}^B$ label two particles in the reference configuration. If the motion is expressed by the previous relations, then
$$x_i^A - x_i^B = Q_{ij}(t)(X_j^A - X_j^B);$$
in addition, the properties of $\mathbf{Q}$ give
$$(x_i^A - x_i^B)(x_i^A - x_i^B) = Q_{ij}(t)Q_{ik}(X_j^A - X_j^B)(X_k^A - X_k^B)$$
$$= (X_j^A - X_j^B)(X_j^A - X_j^B),$$
so that it is proved that the distance between the two points does not change and the motion is rigid.

On the other hand, as shown in Section 3.2, in a rigid body
$$C_{LM} = F_{iL}F_{iM} = \delta_{LM}$$
and $\mathbf{F}$ is an orthogonal tensor. By differentiating with respect to X_P it can be shown that $\mathbf{F}$ is independent of $\mathbf{X}$, so that by integrating the equation
$$\frac{\partial x_i}{\partial X_L} = F_{iL}(t),$$
we obtain the equation of a rigid motion.

4.7. Worked Exercises

3. Prove that the kinetic field

$$\begin{cases} v_1 = -5x_2 + 2x_3, \\ v_2 = 5x_1 - 3x_3, \\ v_3 = -2x_1 + 3x_2, \end{cases}$$

corresponds to a rigid motion.

In fact, the spatial gradient of velocity is given by

$$\left(\frac{\partial v_i}{\partial x_j}\right) = \begin{pmatrix} 0 & -5 & 2 \\ 5 & 0 & -3 \\ -2 & 3 & 0 \end{pmatrix},$$

and it is skew-symmetric. Consequently, the rate of deformation

$$D_{ij} = \frac{1}{2}\left(\frac{\partial v_i}{\partial x_j} + \frac{\partial v_j}{\partial x_i}\right)$$

vanishes and the motion is rigid, according to (4.20).

4. Prove that a rigid motion is isochoric.

5. Find a class of isochoric but nonrigid motions (Hint: $\nabla \cdot \mathbf{v} = \mathbf{0}$ is equivalent to $\mathbf{v} = \nabla \times \mathbf{A}$).

6. Compute the angular velocity in an isochoric motion.

7. Prove that the velocity of propagation (4.35) of a curve on a surface is independent of their parametric representations.

4.8 The Program Velocity

Aim of the Program, Input and Output

Given a transformation from *orthogonal curvilinear coordinates* `var` to Cartesian coordinates and given the components of a velocity field in both the associated unit basis as well as in the holonomic basis of the coordinates `var`, the program `Velocity` computes the acceleration, the velocity gradient, and divergence, as well as the angular velocity.

Theoretical bases are provided in Sections 4.1 and 4.2. Velocity components must be given relative to the unit basis associated with the holonomic one of the curvilinear coordinates `var`.

The command line of the program `Velocity` is

`Velocity[vel, var, transform, characteristic, simplifyoption]`,

where the input data have the following meaning:

`vel` = velocity components;

`var` = orthogonal curvilinear or Cartesian coordinates;

`transform` = r.h.s. terms of the transformation from curvilinear `var` to Cartesian coordinates or r.h.s. terms of the transformation "identity";

`characteristic` = option, corresponding to possible choices `symbolic` or `numeric` (as previously defined for the program `Operator`).

`simplifyoption` = `true` or `false`. In the first case, the program computes many simplifications of symbolic expressions (as previously defined for the program `Deformation`).

From the previous data the corresponding output follows:

spatial gradient of velocity;

acceleration;

velocity divergence;

angular velocity.

4.8. The Program Velocity

Worked Examples

1. Let $\mathbf{v} = v_r \mathbf{e}_r + v_\varphi \mathbf{e}_\varphi + v_z \mathbf{e}_z$ be a velocity field in cylindrical coordinates. The required input data are the following:

   vel = $\{v_r, v_\varphi, v_z\}$;
   var = $\{r, \varphi, z\}$;
   transform = $\{R\text{Cos}[\varphi], R\text{Sin}[\varphi], z\}$;
   characteristic = symbolic;
   simplifyoption = true;
   Velocity[vel, var, transform, characteristic, simplifyoption]

   The program provides the following output data:[1]

   The components of any vector or tensor quantity are relative to the unit basis associated with the holonomic basis of the curvilinear coordinates $\{r, \varphi, z\}$.

   Spatial gradient of velocity

   $$\nabla \mathbf{v} = \begin{pmatrix} \partial_r[v_r] & \partial_r[v_\varphi] & \partial_r[v_z] \\ \dfrac{-v_\varphi + \partial_\varphi[v_r]}{r} & \dfrac{v_r + \partial_\varphi[v_\varphi]}{r} & \dfrac{\partial_\varphi[v_z]}{r} \\ \partial_z[v_r] & \partial_z[v_\varphi] & \partial_z[v_z] \end{pmatrix}$$

   Acceleration

   $$\mathbf{a} = \begin{pmatrix} v_r \partial_r[v_r] + v_z \partial_r[v_z] + v_\varphi \partial_r[v_\varphi] + \partial_t[v_r] \\ \partial_t[v_\varphi] + \dfrac{v_r \partial_\varphi[v_r] + v_z \partial_\varphi[v_z] + v_\varphi \partial_\varphi[v_\varphi]}{r} \\ \partial_t[v_z] + v_r \partial_z[v_r] + v_z \partial_z[v_z] + v_\varphi \partial_z[v_\varphi] \end{pmatrix}$$

   Divergence of velocity

   $$\nabla \cdot \mathbf{v} = \partial_r[v_r] + \partial_z[v_z] + \dfrac{v_r + \partial_\varphi[v_\varphi]}{r}$$

   Angular velocity

   $$\omega = \begin{pmatrix} \dfrac{1}{2}\left(-\partial_z[v_\varphi] + \dfrac{\partial_\varphi[v_z]}{r}\right) \\ \dfrac{1}{2}(-\partial_r[v_z] + \partial_z[v_r]) \\ \dfrac{1}{2}\left(\dfrac{v_\varphi}{r} + \partial_r[v_\varphi] - \dfrac{\partial_\varphi[v_r]}{r}\right) \end{pmatrix}$$

---

[1] It is suggested that the reader to compare these results to those provided by the program Operator.

2. Compute the dynamic characteristics of the velocity field

$$\mathbf{v} = \frac{x}{1+t}\mathbf{i} + \frac{2y}{1+t}\mathbf{j} + \frac{3z}{1+t}\mathbf{k}.$$

Input data:

```
vel = {x/(1+t), 2y/(1+t), 3z/(1+t)};
var = {x, y, z};
transform = {x, y, z};
characteristic = numeric;
simplifyoption = false;
Velocity[vel, var, transform, characteristic, simplify-
option]
```

Output:

Spatial gradient of velocity

$$\nabla \mathbf{v} = \begin{pmatrix} \frac{1}{1+t} & 0 & 0 \\ 0 & \frac{2}{1+t} & 0 \\ 0 & 0 & \frac{3}{1+t} \end{pmatrix}$$

Acceleration

$$\mathbf{a} = \begin{pmatrix} 0 \\ \frac{2y}{(1+t)^2} \\ \frac{6z}{(1+t)^2} \end{pmatrix}$$

Divergence of velocity

$$\nabla \cdot \mathbf{v} = \frac{6}{1+t}$$

Angular velocity

$$\omega = \begin{pmatrix} 0 \\ 0 \\ 0 \end{pmatrix}$$

Exercises

Apply the program Velocity to the following fields:

1. $\mathbf{v} = (-5y + 2z)\mathbf{i} + (5x - 3z)\mathbf{j} + (-2x + 3y)\mathbf{k}$

4.8. The Program Velocity

2. $\mathbf{v} = x^2 t \mathbf{i} + y t^2 \mathbf{j} + x z t \mathbf{k}$
3. $\mathbf{v} = \left(x^3 - xy^2\right) \mathbf{i} + \left(x^2 y + y\right) \mathbf{j}$
4. $\mathbf{v} = Bxzt \mathbf{i} + By^2 t^2 \mathbf{j} + Bzy \mathbf{k}$

5. Use the program Operator to obtain the differential operators for the velocity field in Exercise 3.

Chapter 5

Balance Equations

5.1 General Formulation of a Balance Equation

The fundamental laws of continuum mechanics are integral relations expressing conservation or balance of physical quantities: mass conservation, momentum balance, angular momentum balance, energy balance, and so on. These balance laws can refer to a *material volume*, intended as a collection of the same particles, or to a *fixed volume*.

In rather general terms, a balance law has the following structure:

$$\frac{d}{dt}\int_{c(t)} \mathbf{f}(\mathbf{x},t)\, dc = -\int_{\partial c(t)} \mathbf{\Phi} \cdot \mathbf{N}\, d\sigma + \int_{c(t)} \mathbf{r}\, dc, \qquad (5.1)$$

where $c(t)$ is an arbitrary *material volume* of the configuration $C(t)$ of the system.

The physical meaning of (5.1) is the following: the change that the quantity $\mathbf{f}$ exhibits is partially due to the flux $\mathbf{\Phi}$ across the boundary of $c(t)$ and partially due to the source term $\mathbf{r}$.

If we refer to a *fixed volume* V, then in addition to the flux $\mathbf{\Phi}$ we need to consider the transport of $\mathbf{f}$ throughout V with velocity $\mathbf{v}$, so that (5.1) assumes the form

$$\frac{d}{dt}\int_V \mathbf{f}(\mathbf{x},t)\, dc = -\int_{\partial V} (\mathbf{f}\otimes\mathbf{v} + \mathbf{\Phi}) \cdot \mathbf{N}\, d\sigma + \int_V \mathbf{r}\, dc. \qquad (5.2)$$

When the configuration $C(t)$ is subdivided in two regions by a moving singular surface $\Sigma(t)$, then the previous balance law in the integral form (5.1) or (5.2), the derivation rules (4.45) and (4.46), as well as the generalized Gauss theorem (2.46) allow us to derive the following **local form of the balance equation** and **jump condition**

$$\frac{\partial \mathbf{f}}{\partial t} + \nabla \cdot (\mathbf{f}\otimes\mathbf{v} + \mathbf{\Phi}) - \mathbf{r} = \mathbf{0} \qquad \text{in } C(t) - \Sigma(t),$$
$$[[\mathbf{f}(v_n - c_n) + \mathbf{\Phi}\cdot\mathbf{n}]] = \mathbf{0} \qquad \text{on } \Sigma(t). \qquad (5.3)$$

131

In many circumstances, on the basis of reasonable physical assumptions, we are led to the following integral balance equation:

$$\frac{d}{dt}\int_{c(t)} \mathbf{f}(\mathbf{x},t)\,dc = -\int_{\partial c(t)} \mathbf{s}\,d\sigma + \int_{c(t)} \mathbf{r}\,dc, \tag{5.4}$$

where there is no reason to conclude that $\mathbf{s} = \mathbf{\Phi} \cdot \mathbf{N}$.

In this case, Gauss's theorem cannot be applied to the surface integral on the right-hand side; consequently, (5.3) cannot be derived. Anyway, if $\mathbf{s}$ is supposed to depend on $(\mathbf{x},t)$ as well as on the unit vector $\mathbf{N}$ normal to $\partial c(t)$ (*Cauchy's hypothesis*), then **Cauchy's theorem** can be proved:[1]

Theorem 5.1
If in the integral momentum balance law the tensor of order r has the structure $\mathbf{s} = \mathbf{s}(\mathbf{x},t,\mathbf{N})$, then there is a tensor $\mathbf{\Phi}(\mathbf{x},t)$ of order $(r+1)$ such that

$$\mathbf{s} = \mathbf{\Phi} \cdot \mathbf{N}. \tag{5.5}$$

PROOF For sake of simplicity but without loss of generality, $\mathbf{s}$ is supposed to be a vector. If a material volume $c(t) \subset C(t)$ is considered such that $c(t) \cap \Sigma(t) = \emptyset$, then from (5.3) and (4.46) it follows that

$$\int_{c(t)}\left(\frac{\partial \mathbf{f}}{\partial t} + \nabla \cdot (\mathbf{f}\otimes\mathbf{v}) - \mathbf{r}\right)dc \equiv \int_{c(t)} \varphi\,dc = -\int_{\partial c(t)} \mathbf{s}\,d\sigma. \tag{5.6}$$

If A is the area of the surface ∂c and the functions under the integral are regular, then (5.6) gives

$$\frac{1}{A}\int_c \varphi_i\,dc = \frac{1}{A}\int_{\partial c} s_i(\mathbf{N})\,d\sigma.$$

By applying the mean-value theorem to the volume integral,[2] we have

$$\frac{1}{A}\mathrm{vol}(c)\varphi_i(\xi_i,t) = \frac{1}{A}\int_{\partial c} s_i(\mathbf{N})\,d\sigma.$$

where ξ_i, $i = 1, 2, 3$, are the coordinates of a suitable point internal to c. In the limit $A \to 0$ (or, which is the same for c merging into its internal point $\mathbf{x}$), the following result is obtained:

$$\lim_{A\to 0}\frac{1}{A}\int_{\partial c} s_i(\mathbf{N})\,d\sigma = 0. \tag{5.7}$$

---

[1] In [40], W. Noll proves that Cauchy's hypothesis follows from the balance of linear momentum under very general assumptions concerning the form of the function describing the surface source $\mathbf{s}$.

[2] The mean value theorem applies to each component of the vector function, and not to the vector function itself.

5.1. General Formulation of a Balance Equation

Now, if φ and **s** are regular functions, it must hold that

$$\mathbf{s}(\mathbf{N}) = -\mathbf{s}(-\mathbf{N}), \tag{5.8}$$

as we can prove by applying (5.7) to a small cylinder c_ϵ having height ϵ^2, bases with radius ϵ and by considering the surface σ through the internal point **x** (see Figure 5.1).

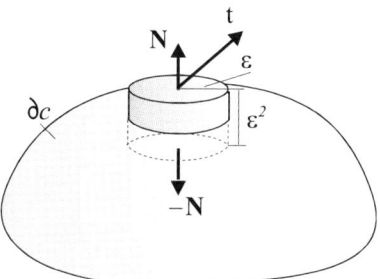

Figure 5.1

Let ξ_1, ξ_2, and ξ_3 be three points located on the surface of the cylinder, on the base whose normal is **N**, and on the other base of normal $-\mathbf{N}$, respectively; moreover, let $\mathbf{N}_1$ the unit normal to the lateral surface of c_ϵ.

By applying the mean-value theorem, it follows that

$$\int_{\partial c_\epsilon} s_i(\mathbf{x}, \mathbf{N})\, d\sigma = 2\pi\epsilon^3 s_i(\xi_1, \mathbf{N}_1) + \pi\epsilon^2 s_i(\xi_2, \mathbf{N}) + \pi\epsilon^2 s_i(\xi_3, -\mathbf{N}),$$

where the argument t has been omitted for sake of simplicity.

Since the term $2\pi\epsilon^3 + 2\pi\epsilon^2$ represents the area A of the cylinder c_ϵ, the previous relation gives (5.7) when $A \to 0$.

Now to prove (5.4), we use Cauchy's tetrahedron argument. At a point $\mathbf{x} \in C$ draw a set of rectangular coordinate axes, and for each direction **N** choose a tetrahedron such that it is bounded by the three coordinate planes through **x** and by a fourth plane, at a distance ϵ from **x**, whose unit outward normal vector is **N** (see Figure 5.2).

Let σ_0 be the area of the surface whose normal is **N** and let σ_i ($i = 1, 2, 3$) be the area of each of the three right triangles whose inward normal is given by the basis unit vector $\mathbf{e}_i$. Since $\sigma_i = \sigma_0|N_i|$ and the volume c of Δ is equal to $\sigma_0\epsilon/3$, if in (5.7) c denotes Δ, it follows that

$$\lim_{\epsilon \to 0} \frac{1}{\sigma_0(1 + \sum_{i=1}^{3}|N_i|)} \left[s_i(\xi_1, \mathbf{N})\sigma_0 + \sum_{i=1}^{3} s_i(\eta_j, -\mathbf{e}_j)|N_j|\sigma_0 \right] = 0,$$

and in addition, if $N_j > 0$ ($j = 1, 2, 3$), then from (5.8) it must be that

$$\mathbf{s}(\mathbf{x}, \mathbf{N}) = \sum_{i=1}^{3} \mathbf{s}(\mathbf{x}, \mathbf{e}_j) N_j, \qquad (5.9)$$

which is valid even if some $N_j = 0$, because of the continuity assumption on $\mathbf{s}(\mathbf{x}, \mathbf{N})$.

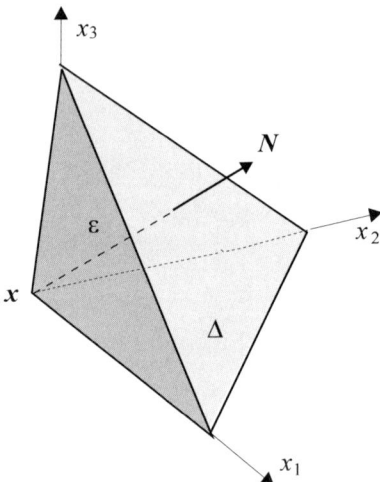

Figure 5.2

Conversely, if N_1 is negative, then $\sigma_1 = -\sigma_0 N_1$ and in (5.9) the term $-\mathbf{s}(\mathbf{x}, \mathbf{e}_1) N_1$ will replace $\mathbf{s}(\mathbf{x}, \mathbf{e}_1) N_1$; assuming the reference $(-\mathbf{e}_1, \mathbf{e}_2, \mathbf{e}_3)$ in place of $(\mathbf{e}_1, \mathbf{e}_2, \mathbf{e}_3)$ still gives (5.9).

If

$$\Phi_{ij} = \mathbf{e}_i \cdot \mathbf{s}(\mathbf{x}, \mathbf{e}_j), \qquad (5.10)$$

then (5.9) can be written as

$$s_i = \mathbf{s} \cdot \mathbf{e}_i = \Phi_{ij} N_j. \qquad (5.11)$$

Since $\mathbf{N}$ and $\mathbf{s}$ are vectors and $\mathbf{N}$ is arbitrary, (5.11) requires Φ_{ij} to be the components of a second-order tensor so that (5.5) is proved. ∎

In the presence of electromagnetic fields additional balance laws are needed. They have the following structure:

$$\frac{d}{dt}\int_{S(t)} \mathbf{u} \cdot \mathbf{N}\, d\sigma = \int_{\partial S(t)} \mathbf{a} \cdot \boldsymbol{\tau}\, ds + \int_{S(t)} \mathbf{g} \cdot \mathbf{N}\, d\sigma + \int_{\Gamma(t)} \mathbf{k} \cdot \boldsymbol{\nu}\, ds, \qquad (5.12)$$

5.1. General Formulation of a Balance Equation

where $\mathbf{u}, \mathbf{a}, \mathbf{g}$, and $\mathbf{k}$ are vector fields, $S(t)$ is any arbitrary material surface, and $\Gamma(t)$ is the intersection curve of $S(t)$ with the singular moving surface $\Sigma(t)$.

To obtain the local form in the region C as well as the jump conditions on $\Sigma(t)$, the rule (4.48) and the Stokes theorem (2.48) have to be applied to the first integral on the right-hand side, so that

$$\frac{\partial \mathbf{u}}{\partial t} + \mathbf{v}\nabla \cdot \mathbf{u} = \nabla \times (\mathbf{a} + \mathbf{v} \times \mathbf{u}) + \mathbf{g} \quad \text{in } C(t),$$
$$(\mathbf{n} \times [[\mathbf{u} \times (\mathbf{w} - \mathbf{v}) + \mathbf{a}]] + \mathbf{k}) \cdot \nu_\Sigma = 0 \quad \text{on } \Sigma(t). \quad (5.13)$$

Moreover, $\mathbf{w} \cdot \mathbf{N} = v_N$, $\mathbf{w} \cdot \mathbf{n} = c_n$, since $\Gamma(t)$ belongs to both the material surface $S(t)$ and the singular surface $\Sigma(t)$; therefore, $(5.13)_2$ can be written

$$([[(c_n - v_n)\mathbf{u} - u_n(\mathbf{w} - \mathbf{v}) + \mathbf{n} \times \mathbf{a}]] + \mathbf{k}) \cdot \nu_\Sigma = 0. \quad (5.14)$$

In the basis $(\tau, \mathbf{n}, \mathbf{N})$ it holds that

$$\mathbf{w} - \mathbf{v} = [w_\tau - \mathbf{v} \cdot \tau]\tau + \frac{c_n - v_n}{\sin^2 \alpha}\mathbf{n} - \frac{(c_n - v_n)\cos\alpha}{\sin^2 \alpha}\mathbf{N},$$

where $\cos\alpha = \mathbf{n} \cdot \mathbf{N}$, so that $(5.13)_2$ assumes the form

$$([[(c_n - v_n)\mathbf{u} + \mathbf{n} \times \mathbf{a}]] + \mathbf{k}) \cdot \nu_\Sigma + [[\mathbf{n} \cdot \mathbf{u}(c_n - v_n)]]\cot\alpha = 0. \quad (5.15)$$

Since the material surface $S(t)$ is arbitrary, the relation (5.15) must hold for any value of α. If, as an example, we first suppose that $\alpha = \pi/2$ and that it has an arbitrary value, we derive the following:

$$[[(c_n - v_n)\mathbf{u} + \mathbf{n} \times \mathbf{a}]] + \mathbf{k} = \mathbf{0},$$
$$[[u_n(c_n - v_n)]] = [[u_n]]c_n - [[u_n v_n]] = 0. \quad (5.16)$$

In literature, the following relation is usually proposed:

$$\mathbf{n} \times [[\mathbf{u} \times (c_n - v_n)\mathbf{n} + \mathbf{a}]] + \mathbf{k} = \mathbf{0} \quad \text{on } \Sigma(t). \quad (5.17)$$

We note that it is equivalent to $(5.16)_1$ if and only if $(5.16)_2$ is satisfied. Some consequences of $(5.16)_2$ are the following:

1. if $[[u_n]] = 0$, then $[[v_n]] = 0$;

2. if the surface is material, then $(5.16)_2$ is identically satisfied and $(5.16)_1$ reduces to

$$[[\mathbf{n} \times \mathbf{a}]] + \mathbf{k} = \mathbf{0}. \quad (5.18)$$

Remark The balance equations presented here are sufficiently general to allow the formulation of all the physical principles considered in this volume. However, they do not work for describing phenomena such as phase transitions (see [16]) or shock waves, with transport of momentum and energy on the wavefront. For these phenomena the reader is referred to Volume II.

5.2 Mass Conservation

In order to derive the mass conservation law for a continuous system S, we introduce the basic assumption that its mass is continuously distributed over the region $C(t)$ occupied by S at the instant t. In mathematical terms, there exists a function $\rho(\mathbf{x}, t)$ called the **mass density**, which is supposed to be of class C^1 on $C(t)$, except for the points of a singular surface $\Sigma(t)$ of order 0 (see Section 2) with respect to $\rho(\mathbf{x}, t)$ and the other fields associated with S.

According to this assumption, if $c(t) \subset C(t)$ is a material volume, i.e., the image under $\mathbf{x}(\mathbf{X}, t)$ of a given region $c_* \subset C_*$, then the mass of $c(t)$ at the instant t is given by

$$m(c_*) = \int_{c(t)} \rho(\mathbf{x}, t) \, dc. \tag{5.19}$$

The **mass conservation principle** postulates that during the motion the mass of any material region does not change in time:

$$\frac{d}{dt} \int_{c(t)} \rho(\mathbf{x}, t) \, dc = 0. \tag{5.20}$$

With reference to the general balance law (5.1), the mass conservation principle for an arbitrary fixed volume v assumes the form

$$\frac{d}{dt} \int_v \rho(\mathbf{x}, t) \, dc = -\int_{\partial v} \rho(\mathbf{x}, t) \, \mathbf{v} \cdot \mathbf{N} \, dc. \tag{5.21}$$

If the material or fixed volume is subdivided in two regions by a singular surface $\Sigma(t)$ of 0 order with respect to the fields of interest, then the general formulas (5.3) allow us to obtain the **local formulation of the mass balance principle and the jump condition**

$$\begin{aligned}\frac{\partial \rho}{\partial t} + \nabla \cdot (\rho \mathbf{v}) = \dot{\rho} + \rho \nabla \cdot \mathbf{v} = 0 & \quad \text{on } C(t) - \Sigma(t), \\ [[\rho(v_n - c_n)]] = 0 & \quad \text{on } \Sigma(t). \end{aligned} \tag{5.22}$$

5.3 Momentum Balance Equation

Let S be a continuous system and let $\rho(\mathbf{x},t)$ be its mass density in the current configuration. In the following discussion any dependence on time t will be omitted for the sake of simplicity.

A fundamental axiom of continuum mechanics is the ***momentum balance principle***:

For any arbitrary material volume c of S it holds that

$$\dot{\mathbf{Q}}(c) = \mathbf{F}(c, c^e), \qquad (5.23)$$

where

$$\mathbf{Q}(c) = \int_c \rho \mathbf{v}\, dc \qquad (5.24)$$

is the momentum of c and $\mathbf{F}(c, c^e)$ is the resultant of all the forces acting on c from its exterior c^e.

In a first simplified approach to continuum mechanics, the external actions on c are divided into ***mass forces***, continuously distributed over c, and ***contact forces***, acting on the boundary ∂c of c; therefore,

$$\mathbf{F}(c, c^e) = \int_c \rho \mathbf{b}\, dc + \int_{\partial c} \mathbf{t}\, d\sigma, \qquad (5.25)$$

where $\mathbf{b}$ is the ***specific force***, defined on c, and $\mathbf{t}$ is the ***traction*** or the ***stress***, defined on ∂c.

With such an assumption, the first integral in (5.25) represents all (gravitational, electromagnetic, etc.) forces acting on the volume c from the *exterior of S* and the vector $\mathbf{b}$ is an a priori assigned field depending on $\mathbf{x}$ and t. On the other hand, $\mathbf{t}$ is a field of *contact forces* acting on ∂c which come from the molecular attraction between particles of c^e and c at the boundary ∂c.

The contact forces, which are strictly influenced by the deformation of S, are unknown, as are reactions in rigid body mechanics. The main difference between reactions and contact forces is the fact that for the latter, we can provide *constitutive laws* that specify their link with the motion of the system.

At this stage, some remarks are necessary to show how restrictive the previous hypotheses are. First, the mass forces acting over c could originate from other portions of S, external to c. This is just the case for mutual gravitational or electromagnetic attractions among parts of S. In fact, mass forces are unknown a priori, and the need arises to add to the motion equations the other equations governing the behavior of those fields which generate such forces.

As an example, if **b** is the gravitational field produced by the system itself, it is necessary to introduce Poisson's equation for this field (see Volume II for other examples).

Furthermore, the assumption made about contact forces means that molecular actions can be *uniquely represented by the vector* **t** $d\sigma$. According to this assumption the approximation of a ***simple continuum*** is usually introduced. But when the need arises to capture essential features linked to the microstructure of the body, contact actions are better represented by a vector **t** $d\sigma$ as well as a torque **m** $d\sigma$. In this case, the approximation of a ***polar continuum*** (E. and F. Cosserat, 1907) is used, and the related model is particularly useful in describing liquid crystals, due to the fact that molecules or molecule groups behave as points of a polar continuum.

Anyway, in the following discussion and within the scope of this first volume, the classical assumption of $\mathbf{m} = \mathbf{0}$ will be retained, so that S will be modeled as a simple continuum. Since molecular actions have a reduced interaction distance, the force acting on the surface $d\sigma$ depends on the particles of S adjacent to $d\sigma$ and not on particles far away. This remark justifies the ***Euler–Cauchy postulate*** that the vector **t**, acting on the unit area at **x**, depends on the choice of the surface only through its *orientation*, i.e.,

$$\mathbf{t} = \mathbf{t}(\mathbf{N}), \tag{5.26}$$

where **N** is the outward unit vector normal to $d\sigma$.[3]

This assumption is equivalent to saying that the traction **t** depends on the boundary ∂c only to first order; i.e., **t** depends on the orientation of the tangent plane at $\mathbf{x} \in \partial c$ and not on the curvature of ∂c.

It is customary to introduce the decomposition

$$\mathbf{t} = t_n \mathbf{N} + \mathbf{t}_\sigma,$$

where the component $t_n = \mathbf{t} \cdot \mathbf{N}$ is called the ***normal stress*** and the component $\mathbf{t}_\sigma$ is called the ***shear stress***.

By applying (5.23) to a *material volume* c and using definitions of momentum and external forces, we obtain

$$\frac{d}{dt} \int_c \rho \mathbf{v}\, dc = \int_{\partial c} \mathbf{t}(\mathbf{N})\, d\sigma + \int_c \rho \mathbf{b}\, dc. \tag{5.27}$$

Then, applying Theorem 5.1 to (5.27), we have:

---

[3] See the footnote of Section 5.1.

5.3. Momentum Balance Equation

Theorem 5.2

If $\mathbf{t}, \mathbf{b}$ and the acceleration $\mathbf{a}$ are regular functions, then the action-reaction principle also holds for stresses, i.e.,

$$\mathbf{t}(\mathbf{N}) = -\mathbf{t}(-\mathbf{N}). \tag{5.28}$$

There exists a second-order tensor $\mathbf{T}$, called **Cauchy's stress tensor**, such that $\mathbf{t}$ is a linear function of $\mathbf{N}$, so that

$$\mathbf{t}(\mathbf{N}) = \mathbf{T}\mathbf{N}. \tag{5.29}$$

The tensor $\mathbf{T}$ depends on $(\mathbf{x}, t)$, but is independent of $\mathbf{N}$.

In the next section, it will proved that $\mathbf{T}$ is a symmetric tensor; therefore, its eigenvalues are real and there is at least one orthonormal basis of eigenvectors. Eigenvalues are known as **principal stresses** and eigenvectors of $\mathbf{T}$ give **principal directions of stress** characterized by the following property: the traction $\mathbf{t}$, acting on the area $d\sigma$ normal to the principal direction $\mathbf{u}$, is normal to $d\sigma$, or, equivalently, on this area there are no shear stress components (see Exercise 2).

When (5.29) is substituted into (5.27), the balance equation with the general structure (5.1) is recovered, where

$$\mathbf{f} = \rho \mathbf{v}, \quad \boldsymbol{\Phi} = -\mathbf{T}, \quad \mathbf{r} = \rho \mathbf{b}.$$

Taking into account (5.3) as well as the local form of the mass conservation (5.22), the *local expression of the momentum balance and the jump condition* can be easily derived:

$$\rho \dot{\mathbf{v}} = \nabla \cdot \mathbf{T} + \rho \mathbf{b}, \tag{5.30}$$

$$[\![\rho \mathbf{v}(c_n - v_n) + \mathbf{T}\mathbf{n}]\!] = \mathbf{0}. \tag{5.31}$$

In the case of a *material surface*, the jump condition (5.31) reduces to **Poisson's condition**

$$[\![\mathbf{T}]\!]\mathbf{n} = \mathbf{0},$$

on which are based the **stress boundary conditions**.

We observe, in fact, that one of the most interesting material surfaces is the boundary of a body, so that the mechanical interaction of two sub-bodies is uniquely determined by tractions on this surface, of equal magnitude and opposite sign.

5.4 Balance of Angular Momentum

Within the framework of nonpolar continua, in addition to (5.27), the **balance of the angular momentum** is supposed to hold:

For any material volume of c we have:

$$\dot{\mathbf{K}}_{x_0} = \mathbf{M}_{x_0}(c, c^e), \tag{5.32}$$

where

$$\mathbf{K}_{x_0} = \int_c \rho(\mathbf{x} - \mathbf{x}_0) \times \mathbf{v}\, dc \tag{5.33}$$

is the angular momentum of c with respect to $\mathbf{x}_0$ and $\mathbf{M}_{x_0}(c, c^e)$ is the moment with respect to $\mathbf{x}_0$ of all forces acting on c from its exterior c^e.

The assumption that S is a *simple continuum* implies that

$$\mathbf{M}_{x_0}(c, c^e) = \int_{\partial c} (\mathbf{x} - \mathbf{x}_0) \times \mathbf{TN}\, d\sigma + \int_c \rho(\mathbf{x} - \mathbf{x}_0) \times \mathbf{b}\, dc. \tag{5.34}$$

Here again, by using the expressions (5.33) and (5.34), the relation (5.32) assumes the general structure (5.1), where

$$\mathbf{f} = \rho(\mathbf{x} - \mathbf{x}_0) \times \mathbf{v}, \quad \boldsymbol{\Phi} = -(\mathbf{x} - \mathbf{x}_0) \times \mathbf{T}, \quad \mathbf{r} = \rho(\mathbf{x} - \mathbf{x}_0) \times \mathbf{b}.$$

As a consequence, (5.32) is equivalent to the local conditions (5.3); in addition, by recalling (5.30) and (5.31) and by considering the continuity of $(\mathbf{x} - \mathbf{x}_0)$ across Σ, we find the following local condition:

$$\epsilon_{ijh} T_{hj} = 0, \tag{5.35}$$

which shows the *symmetry of the stress tensor*[4]

$$\mathbf{T} = \mathbf{T}^T. \tag{5.36}$$

To summarize, mass conservation, momentum balance, and balance of angular momentum give the following 4 differential equations

$$\begin{aligned} \dot{\rho} + \rho \nabla \cdot \mathbf{v} &= 0, \\ \rho \dot{\mathbf{v}} &= \nabla \cdot \mathbf{T} + \rho \mathbf{b}, \end{aligned} \tag{5.37}$$

for 10 unknown fields: $\rho(\mathbf{x}, t)$, $\mathbf{v}(\mathbf{x}, t)$, and $\mathbf{T}(\mathbf{x}, t)$, where $\mathbf{T} = \mathbf{T}^T$.

---

[4]Cauchy's stress tensor is not symmetric in the case of a polar continuum.

5.5. Energy Balance

Thus, we reach the conclusion that (5.37), with jump conditions $(5.22)_2$ and (5.31), do not form a closed set of field equations for the unknown fields; therefore we need to introduce *constitutive laws* connecting the tensor **T** with the fields $\rho(\mathbf{x},t)$ and $\mathbf{v}(\mathbf{x},t)$. This conclusion could also be reached by observing that relations (5.37) are of general validity, in that they do not contain any information about the constitution of the continuous body. But it is also evident that different materials will suffer different stress states if subjected to the same action.

This aspect will be discussed in Chapter 7 in a rather general framework.

5.5 Energy Balance

Up to now, we have only considered purely mechanical aspects of the motion of the continuous system S. When dealing with relations between heat and work, we must introduce an additional field equation: the ***energy balance*** or the ***first law of thermodynamics***:

1. *To any arbitrary material volume c of S we can associate a scalar function $E(c)$, called the **internal energy**, such that if $T(c)$ is the **kinetic energy** of c, then the total energy content of c is given by $E(c) + T(c)$;*

2. *The total power exchanged by c with its exterior c^e is represented by the **mechanical power** $P(c, c^e)$, due to external forces acting on c, and the **thermal power** $Q(c, c^e)$;*

3. *The time change of energy of c is given by*

$$\dot{E}(c) + \dot{T}(c) = P(c, c^e) + Q(c, c^e). \tag{5.38}$$

As a matter of definition, we remark that the term **kinetic energy** indicates the *macroscopic* kinetic energy associated with the macroscopically observable velocity of the continuous system. Random thermal motion of molecules, associated with temperature, is part of the internal energy.

As a particular case, when $\dot{E}(c) = Q(c, c^e) = 0$, (5.38) reduces to the **kinetic energy theorem**

$$\dot{T}(c) = P(c, c^e), \tag{5.39}$$

which states that the rate change of the kinetic energy is equal to the power of the forces acting on c.

In continuum mechanics, quantities appearing in (5.38) have to be expressed as integrals of scalar or vector fields defined on c. To this end, for nonpolar continua we suppose that

$$E(c) = \int_c \rho\varepsilon\, dc, \qquad T(c) = \int_c \frac{1}{2}\rho v^2 dc, \tag{5.40}$$

where ε is the **specific internal energy**.

Moreover, making reference to (5.27) and (5.29), the mechanical power can be written as

$$P(c, c^e) = \int_c \rho \mathbf{b}\cdot\mathbf{v}\, dc + \int_{\partial c} \mathbf{v}\cdot\mathbf{T}\mathbf{N}\, d\sigma, \tag{5.41}$$

so that it assumes the meaning of the rate at which the body forces per unit mass and the traction per unit area are doing work.

Similarly, an additive decomposition is also assumed to hold for the thermal power $Q(c, c^e)$:

$$Q(c, c^e) = \int_c \rho r\, dc + \int_{\partial c} s\, d\sigma, \tag{5.42}$$

where r is a distributed **specific heat source** (supposed to be a given function of $(\mathbf{x}, t)$) and s represents the thermal power flux through the boundary ∂c.

According to the Euler–Cauchy postulate, we also have

$$s = s(\mathbf{x}, t, \mathbf{N}), \tag{5.43}$$

so that, referring to theorem (5.5), we write

$$s(\mathbf{x}, t, \mathbf{n}) = -\mathbf{h}(\mathbf{x}, t)\cdot\mathbf{N}, \tag{5.44}$$

where the vector $\mathbf{h}$ is the **heat flux vector** and the negative sign accounts for the fact that $\int_{\partial c} \mathbf{h}(\mathbf{x}, t)\cdot\mathbf{N}\, d\sigma$ represents the outward heat flux.

By virtue of the above considerations, formally expressed by (5.39), (5.41), and (5.44), (5.38) can be written as

$$\frac{d}{dt}\int_c \rho\left(\varepsilon + \frac{1}{2}v^2\right) dc = \int_{\partial c} (\mathbf{v}\cdot\mathbf{T}\mathbf{N} - \mathbf{h}\cdot\mathbf{n})\, d\sigma + \int_c \rho(\mathbf{v}\cdot\mathbf{b} + r)\, dc, \tag{5.45}$$

and usual arguments lead to the following local equation and jump condition:

$$\rho\dot{\varepsilon} = \mathbf{T}:\nabla\mathbf{v} - \nabla\cdot\mathbf{h} + \rho r \quad \text{on } C(t),$$

$$\left[\!\left[\rho\left(\frac{1}{2}v^2 + \epsilon\right)(c_n - v_n) + \mathbf{v}\cdot\mathbf{T}\mathbf{n} - \mathbf{h}\cdot\mathbf{n}\right]\!\right] = 0 \quad \text{on } \Sigma(t), \tag{5.46}$$

where $C(t)$ is the region occupied by S at the instant t and $\Sigma(t)$ is a singular surface.

Furthermore, it can be proved that

$$\mathbf{T} : \nabla \mathbf{v} = \mathbf{T} : \mathbf{D} = T_{ij} D_{ij} \tag{5.47}$$

by simply considering the decomposition of the velocity gradient tensor (4.14) and recalling that $\mathbf{T}$ is a symmetric tensor and $\mathbf{W}$ is skew-symmetric.

5.6 Entropy Inequality

A relevant concept of continuum mechanics is the **temperature**. As distances are measured by means of rulers, so the temperature of a body is measured by means of cylinders containing substances (mercury, alcohol, or others) whose expansion laws are well established. Such a device is called a **thermometer** and the measuring scale is rather arbitrary, as is the scale of a ruler. But independent of the device used to measure the temperature, there exists a lower bound, below which the body cannot be cooled further.

If this lower bound is assumed to be 0, then the corresponding temperature scale θ is called the **absolute temperature**, where

$$0 \leq \theta \leq \infty.$$

When $Q(c, c^e) > 0$, the region c is receiving thermal power from c^e; the reverse is true when $Q(c, c^e) < 0$, e.g. the region c is providing thermal power to c^e. It should be noted that the energy balance principle does not contain any restriction on the value of $Q(c, c^e)$. On the contrary, experience shows that work can be transformed into heat by friction and there is no negative limit to $Q(c, c^e)$. On the other hand, experience also shows that a body can spontaneously receive heat from its surroundings at constant ambient temperature θ until it reaches the same ambient temperature.

By using these arguments we conclude that there exists an upper bound of $Q(c, c^e)$, denoting the maximum heat power which can be absorbed by the region c during an isothermic transformation. This statement is formally expressed by the **entropy principle** or the **second law of thermodynamics** for isothermal processes: *For any material volume c whose evolution takes place at constant temperature, there exists an upper bound for the thermal power*

$$Q(c, c^e) \leq B(c). \tag{5.48}$$

If the **entropy** S of the material volume c is introduced through the definition

$$\theta \dot{S}(c) = B(c), \tag{5.49}$$

then the previous inequality becomes

$$\theta \dot{S}(c) \geq Q(c, c^e), \qquad (5.50)$$

which is known as the ***Clausius–Planck inequality***.

The extension of this principle to a continuous system S is not straightforward, as the temperature depends not only on time (as in the case for isothermic processes), but also on the point $\mathbf{x} \in C$.

At first glance, we might suppose that we could apply (5.50) to any element of the continuum system S. But a deeper insight would indicate that this assumption has serious shortcomings, as it cannot explain many phenomena occurring inside the system. For this reason, we state the following more general formulation of the ***entropy principle*** or the ***second law of thermodynamics***:

*Let S be a continuum system and let c be any arbitrary material volume of S. There exist two scalar functions $S(c)$ and $M(c, c^e)$, called **entropy** and **entropy flux**, respectively, such that*

$$\dot{S}(c) \geq M(c, c^e), \qquad (5.51)$$

where

$$S(c) = \int_c \rho \eta \, dc, \qquad (5.52)$$

and

$$M(c, c^e) = \int_{\partial c} i \, d\sigma + \int_c \rho k \, dc. \qquad (5.53)$$

In (5.52) and (5.53) η is the ***specific entropy***, i is the ***density of the conductive part of the entropy flux***, and k is the ***density of the entropy flux emanating by radiation***.

Assuming $i = i(\mathbf{x}, t, \mathbf{N})$, we can write $i = \mathbf{j}(\mathbf{x}, t) \cdot \mathbf{N}$ and the following local equations hold:

$$\begin{aligned} \rho \dot{\eta} &\geq \nabla \cdot \mathbf{j} + \rho k, \\ [[\rho \eta (c_n - v_n) + \mathbf{j} \cdot \mathbf{n}]] &\geq 0, \end{aligned} \qquad (5.54)$$

under the usual assumption of regularity of the introduced functions.

It is important to investigate the structure of the fields $\mathbf{j}$ and k. The Clausius–Plank inequality (5.50) applied to a material element dc of S,

$$\theta \frac{d}{dt}(\rho \eta \, dc) \geq (-\nabla \cdot \mathbf{h} + \rho) \, dc, \qquad (5.55)$$

5.6. Entropy Inequality

leads to the inequality

$$\rho\dot{\eta} \geq \frac{-\nabla \cdot \mathbf{h} + \rho r}{\theta}. \tag{5.56}$$

If, in addition to (5.56), we introduce the **Fourier inequality**

$$\mathbf{h} \cdot \nabla\theta \leq 0, \tag{5.57}$$

which states that heat flows from regions at higher temperature to regions at lower temperatures, then (5.56) and (5.57) imply the **Clausius–Duhem inequality**

$$\rho\dot{\eta} \geq -\nabla \cdot \frac{\mathbf{h}}{\theta} + \rho\frac{r}{\theta}, \tag{5.58}$$

which can be regarded as a generalization of the previous relations.

By comparing (5.58) to (5.54), we find that $\mathbf{j}$ and k assume the following expressions:

$$\mathbf{j} = -\frac{\mathbf{h}}{\theta}, \quad k = \frac{r}{\theta}, \tag{5.59}$$

and (5.54) becomes

$$\rho\dot{\eta} \geq -\nabla \cdot \frac{\mathbf{h}}{\theta} + \rho\frac{r}{\theta},$$
$$\left[\left[\rho\eta(c_n - v_n) - \frac{\mathbf{h}}{\theta} \cdot \mathbf{n}\right]\right] \geq 0. \tag{5.60}$$

But, owing to $(5.46)_1$, we can write $-\nabla \cdot \mathbf{h} + \rho r = \rho\dot{\epsilon} - \mathbf{T} : \nabla\mathbf{v}$ and substituting into $(5.60)_1$, we get the **reduced dissipation inequality**

$$-\rho(\dot{\psi} + \eta\dot{\theta}) + \mathbf{T} : \nabla\mathbf{v} - \frac{\mathbf{h} \cdot \nabla\theta}{\theta} \geq 0, \tag{5.61}$$

where the quantity

$$\psi = \epsilon - \theta\eta \tag{5.62}$$

is the **Helmholtz free energy**.

Finally, if the *temperature θ is supposed to be constant across the surface* $\Sigma(t)$, by eliminating the term $[[\mathbf{h} \cdot \mathbf{n}]]$ appearing in $(5.46)_2$ and $(5.60)_2$, we obtain the additional jump condition

$$\left[\left[\rho(\frac{1}{2}v^2 + \psi)(c_n - v_n) + \mathbf{v} \cdot \mathbf{Tn}\right]\right] \leq 0. \tag{5.63}$$

At this stage, the reader should be aware of the introductory nature of the material discussed here. In the next chapter the essential role played by the inequality (5.61) in continuum mechanics will be highlighted (in particular in Sections 6.1 and 6.2). Moreover, a critical review of the present formulation of thermodynamics will be presented in Section 6.5, and alternative formulations will also be discussed in Chapter 11.

At the moment, it is worth noting that there are applications where the energy flux of nonmechanical nature (represented by the term $-\mathbf{h} \cdot \mathbf{n}$) and the entropy flux ($-\mathbf{h} \cdot \mathbf{n}/\theta$) can require more complex relations. For instance, in the theory of mixtures I. Mueller (see [41]) showed that this form of the entropy flux has to be modified. This is not surprising if we consider continua where electrical charges and currents are present. In this case, in the energy balance equation there are flux terms of both thermal and electromagnetic energy, the latter through the Poynting vector, while in the entropy inequality the ratio between the heat flux vector and the absolute temperature still appears. Nevertheless, even when we limit our attention to thermal and mechanical phenomena, the need can arise for postulating the presence of an extra term of energy flux, which does not appear in the entropy inequality (see for example the interstitial theory of J. Serrin & J. Dunn [17], suggested to model capillarity phenomena). Finally the same need arises when dealing with ferromagnetic and dielectric continua (see [18], [19]).

5.7 Lagrangian Formulation of Balance Equations

It will be shown in the next chapter that (5.37) and (5.46)$_1$, under certain circumstances, allow us to compute the fields $\rho(\mathbf{x},t)$, $\mathbf{v}(\mathbf{x},t)$, and $\theta(\mathbf{x},t)$ in the current configuration C. But in many cases it is of interest, especially from a physical point of view, to compute the same fields as a function of $(\mathbf{X},t)$, e.g., in the reference configuration C_*.

Let $\Sigma_*(t)$ be the Lagrangian image in C_* of the singular surface $\Sigma(t) \subset C(t)$, having the equation

$$F(\mathbf{X},t) = f(\mathbf{x}(\mathbf{X},t),t) = 0. \tag{5.64}$$

If, in particular, $\Sigma(t)$ is material, then F is independent of t. The function $F(\mathbf{X},t)$ is continuous, as it is composed of continuous functions; in the following discussion, it will be supposed that *the function $F(\mathbf{X},t)$ is differentiable*, so that it is possible to define its normal speed c_N by the relation (see Figure 5.3, where the dotted lines denote the trajectories of the points of the continuum)

$$c_N = -\frac{1}{|\nabla F|}\frac{\partial F}{\partial t}. \tag{5.65}$$

When $\Sigma(t)$ is material, $c_N^\pm = 0$ since $v_n = c_n$.

We now devote our attention to the Lagrangian formulation of balance equations. For the mass balance equation, by considering (3.8) and (5.19)

5.7. Lagrangian Formulation of Balance Equations

we obtain
$$m(c_*) = \int_c \rho(\mathbf{x},t)\,dc = \int_{c_*} \rho_*(\mathbf{X},t)\,dc_*,$$
or
$$\int_{c_*} [\rho(\mathbf{x}(\mathbf{X},t)t)J - \rho_*(\mathbf{X},t)]\,dc_* = 0.$$

Since c_* is an arbitrary volume, taking into account the regularity of the density field on $(C_* - \Sigma_*)$, we derive the **local Lagrangian formulation of mass conservation**:
$$\rho J = \rho_*, \tag{5.66}$$
which allows us to compute the mass density once the motion is known; consequently, J is also determined, since ρ_* is an initially assigned value. We also observe that (5.66) implies that on $\Sigma(t)$ we have
$$(\rho J)^\pm = (\rho_*)^\pm.$$

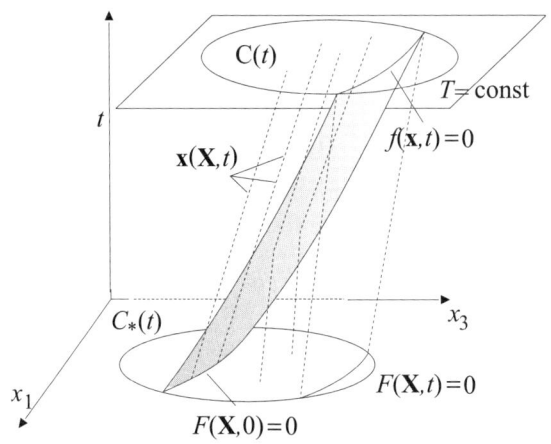

Figure 5.3

Similar arguments can be applied to the momentum equation
$$\frac{d}{dt}\int_c \rho \mathbf{v}\,dc = \int_{\partial c} \mathbf{T}\mathbf{N}\,d\sigma + \int_c \rho \mathbf{b}\,dc,$$
where $\mathbf{N}$ is the unit vector normal to ∂c, which, by using (3.7), (3.8), and (5.66), we can write as
$$\frac{d}{dt}\int_{c_*} \rho_* \mathbf{v}\,dc_* = \int_{\partial c_*} J\mathbf{T}(\mathbf{F}^{-1})^T \mathbf{N}_*\,d\sigma_* + \int_{c_*} \rho_* \mathbf{b}\,dc_*.$$

If we introduce the **first Piola–Kirchhoff tensor**,

$$\mathbf{T}_* = J\mathbf{T}(\mathbf{F}^{-1})^T \qquad (T_{*Li} = JT_{ij}(\mathbf{F}^{-1})_{Lj}), \tag{5.67}$$

then the previous global equation becomes

$$\frac{d}{dt}\int_{C_*} \rho_* \mathbf{v}\, dc_* = \int_{\partial C_*} \mathbf{T}_* \mathbf{N}_* \, d\sigma_* + \int_{C_*} \rho_* \mathbf{b}\, dc_*, \tag{5.68}$$

which, under the usual assumptions of regularity, allows us to derive the local conditions

$$\begin{aligned} \rho_* \mathbf{a} &= \nabla_{\mathbf{X}} \cdot \mathbf{T}_* + \rho_* \mathbf{b} &&\text{on } C(t) - \Sigma(t), \\ [[\rho_* \mathbf{v}\, c_N + \mathbf{T}_* \mathbf{n}_*]] &= \mathbf{0} &&\text{on } \Sigma(t), \end{aligned} \tag{5.69}$$

where $\nabla_{\mathbf{X}}\cdot$ is the divergence operator with respect to variables X_L.

It is also relevant to note that the definition (5.67) and the symmetry of the stress tensor (5.36) give

$$\mathbf{T}_* \mathbf{F}^T = \mathbf{F}\mathbf{T}^T. \tag{5.70}$$

If on the boundary ∂C or in some part of it there acts the traction $\mathbf{t}(\mathbf{x})$, $\mathbf{x} \in \partial C$, then Cauchy's stress tensor $\mathbf{T}$ satisfies the following boundary condition:

$$\mathbf{T}\mathbf{N} = \mathbf{t}. \tag{5.71}$$

By multiplying this relation for $d\sigma$ and taking into account (3.7), we obtain the traction $\mathbf{t}_* = \mathbf{T}_* \mathbf{N}_*$ acting on the boundary in the reference configuration:

$$\mathbf{t}_* = \frac{d\sigma}{d\sigma_*}\mathbf{t}. \tag{5.72}$$

Finally, it can be easily verified, starting from (3.7), that

$$d\sigma = J\sqrt{\mathbf{N}_* \cdot \mathbf{C}^{-1}\mathbf{N}_*}\, d\sigma_*. \tag{5.73}$$

We remark that the first Piola–Kirchhoff stress tensor has the shortcoming of being a nonsymmetric tensor, so that it is not so appropriate in constitutive laws involving symmetric deformations (as for example $\mathbf{G}$, see Section 3.4). This argument suggests the introduction of the **second Piola–Kirchhoff tensor** $\widetilde{\mathbf{T}}_*$, symmetric in nonpolar continua, defined in such a way that the force $\mathbf{t}\, d\sigma$ transforms into $\mathbf{t}_*\, d\sigma_*$ according to the transformation rule of material vectors (4.28):

$$\widetilde{\mathbf{T}}_* \mathbf{N}_* \, d\sigma_* = \mathbf{F}^{-1}(\mathbf{T}\mathbf{N})\, d\sigma. \tag{5.74}$$

By recalling the transformation formula of surfaces (3.7), we see that

$$\widetilde{\mathbf{T}}_* \mathbf{N}_* \, d\sigma_* = \mathbf{F}^{-1}(\mathbf{T}J(\mathbf{F}^{-1})^T \mathbf{N}_*)\, d\sigma_*,$$

5.7. Lagrangian Formulation of Balance Equations

and finally
$$\tilde{\mathbf{T}}_* = J\mathbf{F}^{-1}\mathbf{T}(\mathbf{F}^{-1})^T. \tag{5.75}$$

By comparing with (5.67), we also find that
$$\tilde{\mathbf{T}}_* = \mathbf{T}_*(\mathbf{F}^{-1})^T, \tag{5.76}$$

and the momentum equation $(5.69)_1$ becomes
$$\rho_* \mathbf{a} = \nabla_{\mathbf{X}}(\tilde{\mathbf{T}}_* \mathbf{F}^T) + \rho_* \mathbf{b}. \tag{5.77}$$

Similar considerations apply to the energy balance equation (5.45), which assumes the global Lagrangian formulation

$$\frac{d}{dt}\int_{c_*} \rho_* \left(\varepsilon + \frac{1}{2}v^2\right) dc_* = \int_{\partial c_*} (\mathbf{v} \cdot \mathbf{T}_* \mathbf{N}_* - \mathbf{h}_* \cdot \mathbf{N}_*) d\sigma_*$$
$$+ \int_{c_*} \rho_*(\mathbf{v} \cdot \mathbf{b} + r) dc_*, \tag{5.78}$$

where
$$\mathbf{h}_* = J\mathbf{h}(\mathbf{F}^{-1})^T. \tag{5.79}$$

From (5.78), the local equations follow:

$$\rho_* \dot{\varepsilon} = \mathbf{T} : \dot{\mathbf{F}}^T - \nabla_{\mathbf{X}} \cdot \mathbf{h}_* + \rho r \quad \text{in } C(t) - \Sigma(t),$$

$$\left[\!\left[\rho_* \left(\frac{1}{2}v^2 + \epsilon\right) c_N + \mathbf{v} \cdot \mathbf{T}_* \mathbf{n}_* - \mathbf{h}_* \cdot \mathbf{n}_*\right]\!\right] = 0 \quad \text{on } \Sigma(t). \tag{5.80}$$

Finally, the global form of entropy inequality

$$\frac{d}{dt}\int_{c(t)} \rho_* \eta \, dc_* \geq -\int_{\partial c_*(t)} \frac{\mathbf{h}_* \cdot \mathbf{N}_*}{\theta} d\sigma_* + \int_{c_*(t)} \rho_* \frac{r}{\theta} dc_*, \tag{5.81}$$

locally implies that

$$\rho_* \dot{\eta} \geq -\nabla_{\mathbf{X}} \left(\frac{\mathbf{h}_*}{\theta}\right) + \rho_* \frac{r}{\theta} \quad \text{in } C(t) - \Sigma(t),$$

$$\left[\!\left[\rho_* \eta c_N - \frac{\mathbf{h} \cdot \mathbf{n}_*}{\theta}\right]\!\right] \geq 0 \quad \text{on } \Sigma(t). \tag{5.82}$$

Similarly, to (5.61)–(5.63) correspond the Lagrangian expressions

$$-\rho_*(\dot{\psi} + \eta\dot{\theta}) + \mathbf{T}_* : \dot{\mathbf{F}}^T - \frac{\mathbf{h}_* \cdot \nabla_{\mathbf{X}} \theta}{\theta} \geq 0 \quad \text{in } C(t) - \Sigma(t),$$

$$\left[\!\left[\rho_* \left(\frac{1}{2}v^2 + \psi\right) c_N + \mathbf{v} \cdot \mathbf{T}_* \mathbf{n}_*\right]\!\right] \leq 0 \quad \text{on } \Sigma(t). \tag{5.83}$$

5.8 The Principle of Virtual Displacements

All the previously discussed balance laws refer to dynamical processes. As a particular case in static conditions they merge into the following equilibrium equations:

$$\nabla \cdot \mathbf{T} + \rho \mathbf{b} = \mathbf{0},$$
$$-\nabla \cdot \mathbf{h} + \rho r = 0. \qquad (5.84)$$

More specifically, if equilibrium is reached at uniform temperature, then the heat flux vector vanishes (also see next chapter) and from $(5.84)_2$ it follows that $r = 0$. In the following discussion we will focus on the case of constant uniform temperature, with the equilibrium condition expressed by $(5.84)_1$.

Let S be a system at equilibrium in the configuration C_0, subject to mass forces of specific density $\mathbf{b}$. Moreover, for boundary conditions we assume that a portion $\partial C_0'$ of the boundary ∂C_0 is subjected to prescribed tractions $\mathbf{t}$, while the portion $\partial C_0''$ is fixed. If $\delta \mathbf{u}(\mathbf{x})$ is a virtual infinitesimal displacement field, with the restriction that it be *kinematically admissible*, i.e., it is a vector field of class $C^1(C_0)$ and vanishing on $\partial C_0''$, then from $(5.84)_1$ we have the following global relation:

$$\int_{C_0} \nabla \mathbf{T} \cdot \delta \mathbf{u} \, dc + \int_{C_0} \mathbf{b} \cdot \delta \mathbf{u} \, dc = \mathbf{0},$$

which also can be written as

$$\int_{C_0} [\nabla (\mathbf{T} \cdot \delta \mathbf{u}) - \mathbf{T} : \nabla \delta \mathbf{u}] \, dc + \int_{C_0} \mathbf{b} \cdot \delta \mathbf{u} \, dc = 0.$$

Gauss's theorem, taking into account that $\delta \mathbf{u} = \mathbf{0}$ on $\partial C_0''$, leads to

$$\int_{C_0} \mathbf{T} : \nabla \delta \mathbf{u} \, dc = \int_{C_0} \mathbf{b} \cdot \delta \mathbf{u} \, dc + \int_{\partial C_0'} \mathbf{t} \cdot \delta \mathbf{u} \, dc. \qquad (5.85)$$

Note that, if (5.85) holds for any kinematically admissible field $\delta \mathbf{u}$, then (5.84) is satisfied. This remark proves that (5.85) is an expression totally equivalent to equilibrium conditions and allows us to state the following: *The continuous system S, whose boundary is partially fixed, is at equilibrium in the configuration C_0 under the external forces $\mathbf{b}$ and $\mathbf{t}$ if and only if the work done by these forces, given a kinematically admissible and infinitesimal displacement, is equal to the internal work done by stresses to produce the deformation field $\nabla \delta \mathbf{u}$ associated with the displacement field.*

5.9. Exercises

We remark that the previous global form of the equilibrium condition is just one of the possible expressions of the ***principle of virtual displacements*** (or ***principle of virtual work***, as it is often referred in the literature) for a continuous system. Early expressions are due to D'Alembert and Lagrange. In particular, Lagrange, in his *Mécanique Analytique* (1788) stated that *"any equilibrium law, which will be deduced in the future, will always be interpreted within the principle of virtual work: equivalently, it will be nothing else than a particular expression of this principle."*

In Chapter 1 of Volume II it will be shown that the principle of virtual work is a weak or variational formulation of the equilibrium problem. This different perspective, apparently not investigated after Cauchy introduced the variational formulation of differential equations (5.84) and related boundary conditions, is now considered to be the most convenient tool for deriving the equilibrium solutions (see also Chapter 10 and Appendix A).

5.9 Exercises

1. Given the stress tensor at the point P

$$T_{ij} = \begin{pmatrix} 300 & -50 & 0 \\ -50 & 200 & 0 \\ 0 & 0 & 100 \end{pmatrix},$$

derive the stress vector acting on a plane, through P, parallel to the plane shown in Figure 5.4.

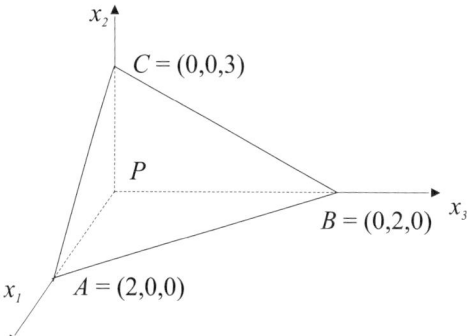

Figure 5.4

Since any vector normal to this plane is also normal to vectors $\overrightarrow{AB}$ and $\overrightarrow{BC}$, it can be expressed as

$$\mathbf{n} = \overrightarrow{AB} \times \overrightarrow{BC} = \begin{vmatrix} \mathbf{e}_1 & \mathbf{e}_2 & \mathbf{e}_3 \\ -2 & 2 & 0 \\ -2 & 0 & 3 \end{vmatrix} = 6\mathbf{e}_1 + 6\mathbf{e}_2 + 4\mathbf{e}_3,$$

and the unit normal vector $\hat{\mathbf{n}}$ has components

$$\left(\frac{3}{\sqrt{22}}, \frac{3}{\sqrt{22}}, \frac{2}{\sqrt{22}} \right).$$

Applying (5.29), we find the required stress vector to be

$$t_i^{(n)} = T_{ij} n_j = \begin{pmatrix} 300 & -50 & 0 \\ -50 & 200 & 0 \\ 0 & 0 & 100 \end{pmatrix} \begin{pmatrix} 3/\sqrt{22} \\ 3/\sqrt{22} \\ 2/\sqrt{22} \end{pmatrix} = \begin{pmatrix} 750/\sqrt{22} \\ 450/\sqrt{22} \\ 200/\sqrt{22} \end{pmatrix}.$$

2. Given
$$t_n = \mathbf{n} \cdot \mathbf{Tn} \quad \text{with } \mathbf{n} \cdot \mathbf{n} = 1,$$

verify that the maximum and minimum values of the normal component t_n act along the principal directions and coincide with the principal stresses.

In seeking the above constrained extremal values, we can use the Lagrangian multiplier method, which requires us to seek the extrema of the function

$$F(\mathbf{n}) = t_n - \lambda \left(\mathbf{n} \cdot \mathbf{n} - 1 \right) = \mathbf{n} \cdot \mathbf{Tn} - \lambda \left(\mathbf{n} \cdot \mathbf{n} - 1 \right),$$

where λ is a Lagrange multiplier. The method requires the derivative of $F(\mathbf{n})$ with respect to n_k to vanish, i.e.,

$$\frac{\partial}{\partial n_k} \left(T_{ij} n_j n_i - \lambda \left(n_i n_i - 1 \right) \right) = 0,$$

or equivalently,

$$\left(T_{kj} - \lambda \delta_{kj} \right) n_j = 0,$$

which shows that the problem is nothing other than the eigenvalue formulation for principal stresses, as already discussed in Section 5.2.

If the characteristic equation is written in the form (see Section 3.3)

$$\lambda^3 - I_T \lambda^2 + II_T \lambda - III_T = 0,$$

then it is plain that the algebraic nature of eigenvalues of $\mathbf{T}$, which are real owing to the symmetry of $\mathbf{T}$, depends on the principal invariants.

5.9. Exercises

If there are three distinct eigenvalues, then there are three distinct principal stresses and the stress state is called *triaxial*; if two eigenvalues are coincident, then the stress state is said to be *cylindrical* and there are no shear stresses on any plane parallel to the cylinder axis, which is the only determined direction; finally, if all three eigenvalues are coincident, then all directions are principal and the stress state is said to be *isotropic* or *spherical* or *hydrostatic*, because it is the stress state that exists in a fluid at rest.

3. Prove that, in a rigid body rotation where $\mathbf{v} = \omega \times \mathbf{r}$, equations (5.32) and (5.40)$_2$ reduce to the balance equation of angular momentum and to the kinetic energy for a rigid body.

By referring to the expression of double vector product, from (5.32) we find that

$$\mathbf{M}_O = \dot{\mathbf{K}}_O = \frac{d}{dt}\int_V \mathbf{r} \times \rho \mathbf{v}\, dc = \frac{d}{dt}\int_V \mathbf{r} \times \rho\,(\omega \times \mathbf{r})\, dc$$

$$= \frac{d}{dt}\int_V \rho\left(|\mathbf{r}|^2 \omega - (\mathbf{r} \cdot \omega)\mathbf{r}\right) dc = \frac{d}{dt}\int_V \rho\left(|\mathbf{r}|^2 \mathbf{1} - \mathbf{r} \otimes \mathbf{r}\right) dc\,(\omega),$$

where $\mathbf{1}$ is the identity mapping and $\mathbf{r} \otimes \mathbf{r}$ is a linear mapping such that $\mathbf{r} \otimes \mathbf{r}\,(\omega) = (\mathbf{r} \cdot \omega)\mathbf{r}$.

If

$$\mathbf{I}_O = \int_V \rho\left(|\mathbf{r}|^2 \mathbf{1} - \mathbf{r} \otimes \mathbf{r}\right) dc$$

denotes the inertia tensor with respect to a fixed pole O, then

$$\mathbf{M}_O = \frac{d}{dt}\,(\mathbf{I}_O \omega).$$

Finally, substituting $\mathbf{v} = \omega \times \mathbf{r}$ into (5.40)$_2$, we obtain

$$T = \frac{1}{2}\int_V \rho v^2\, dc = \frac{1}{2}\int_V \rho\,(\omega \times \mathbf{r}) \cdot \mathbf{v}\, dc$$

$$= \frac{1}{2}\int_V \rho \mathbf{r} \times \mathbf{v} \cdot \omega\, dc = \frac{1}{2}\mathbf{K}_O \cdot \omega = \frac{1}{2}\omega \cdot \mathbf{I}_O \omega,$$

where it has been taken into account that ω is independent of $\mathbf{r}$ and $\mathbf{K}_O = \mathbf{I}_O \omega$.

Chapter 6

Constitutive Equations

6.1 Constitutive Axioms

It has already been stated in Chapters 4 and 5 that balance equations are general relations whose validity does not depend on body properties. However, we also know from experience that two bodies with the same dimensions and shape may react differently when subjected to the same load and thermal conditions.

Thus we intuitively conclude that, in contrast to rigid body dynamics, the evolution of a continuous deformable body cannot be completely predicted by knowing the equations of motion, the massa distribution, and the external forces acting on the body. We will now use the results of Chapter 5 tho express this statement in a formal way.

In the Eulerian description, $\{\rho, \mathbf{v}, \mathbf{T}, \epsilon, \mathbf{h}, \theta\}$ represent the unknown fields, and the balance equations for mass conservation, momentum balance, and energy balance are the equations we use to determine these fields. Similarly, in the Lagrangian description, the unknowns become $\{\mathbf{v}, \mathbf{T}_*, \epsilon_*, \mathbf{h}_*, \theta\}$ and the balance equations are the momentum balance and energy balance.

In both cases the balance equations do not form a closed set of *field equations* for the above mentioned fields, so that we must add relations that connect the *stress tensor*, the *internal energy*, and the *heat flux* with the basic fields.

These relations are called **constitutive equations** because they describe the material constitution of the system from a macroscopic view point.

In order to obtain such relationships, we presume that the macroscopic response of a body, as well as any macroscopic property of it, depends on its molecular structure, so that response functions could in principle be obtained from statistical mechanics, in terms of the average of microscopic quantities.

As a matter of fact, such an approach, although promising from a theoretical point of view, is not straightforward if applied to the complex materials that interest us in continuum mechanics.

In addition, the basic assumption of continuum mechanics consists of erasing the discrete structure of the matter, so that constitutive equations are essentially based on *experimental evidence*. Again, this is not an easy task, but continuum mechanics greatly simplifies this task through the introduction of general rules, called **constitutive axioms**. In fact, they represent constraints for the structure of constitutive equations.

In order to discuss these axioms, we introduce some definitions.

The **history of a thermokinetic process up to time** t is defined by the two functions

$$\mathbf{x}_S^t = \mathbf{x}(\mathbf{Y}, \tau), \theta_S^t = \theta(\mathbf{Y}, \tau) \qquad (\mathbf{Y}, \tau) \in C_* \times (-\infty, t]. \qquad (6.1)$$

According to this definition, two processes are **locally equivalent** at $\mathbf{X}$ if there is at least a neighborhood of $\mathbf{X}$ in which the two histories coincide.

By **dynamic process** we mean the set of fields $\mathbf{x}(\mathbf{Y}, t)$, $\theta(\mathbf{Y}, t)$, $\mathbf{T}_*(\mathbf{Y}, t)$, $\epsilon(\mathbf{Y}, t)$, $\mathbf{h}_*(\mathbf{Y}, t)$, $\eta(\mathbf{Y}, t)$, where $\mathbf{T}_* \mathbf{F}^T = \mathbf{F} \mathbf{T}_*^T$, which are solutions of equations $(5.69)_1$–$(5.80)_1$, under a given body density $\mathbf{b}$ and a source r.

Taking into account all the previous remarks, it turns out that the **material response** is formally expressed by the set of fields $\mathbf{A} = \{\mathbf{T}_*, \epsilon, \mathbf{h}_*, \eta\}$, which depend on the history of the thermokinetic process. Constitutive axioms, listed below, deal with such functional dependence.

1. ***Principle of Determinism*** (Noll, 1958).

    At any instant t, the value of $\mathbf{A}$ at $\mathbf{X} \in C_*$ depends on the whole history of the thermokinetic process $\mathbf{x}_S^t$, θ_S^t up to the time t.

    This is equivalent to stating that the material response at $\mathbf{X} \in C_*$ and at time t is influenced by the history of S through a functional

    $$\mathbf{A} = \boldsymbol{\Upsilon}(\mathbf{x}_S^t, \theta_S^t, \mathbf{X}), \qquad (6.2)$$

    which depends on $\mathbf{X} \in C_*$ (nonhomogeneity) *as well as on the reference configuration*.

    If the equations (6.2) are substituted into $(5.69)_1$, $(5.80)_1$, with $\mathbf{T}_* \mathbf{F}^T = \mathbf{F} \mathbf{T}_*^T$, then a system of 4 scalar equations for the 4 unknown functions $\mathbf{x}(\mathbf{X}, t)$, $\theta(\mathbf{X}, t)$ is obtained. If $\mathbf{b}$ and the history of the process are given up to the instant t_0 under suitable boundary and initial conditions, then this system allows us to predict, at least in principle, the process from t_0 on.

    Note that (6.2) assumes that the material has a memory of the whole history of the motion. We make two remarks about this assumption: first, it is unrealistic to have experience of the whole history of the thermokinetic process; moreover, the results are quite reliable if we presume that the response of the system S is mainly influenced by its

6.1. Constitutive Axioms

recent history. Depending on the kind of memory selected, there are different classes of materials (e.g., materials with **fading memory**, **plastic memory**, and so on). Furthermore, if only the recent history (how recent has to be specified) is supposed to be relevant, then for suitably regular motions in $C_* \times (-\infty, t]$, the history of the motion can be expressed, at least in a neighborhood of $(\mathbf{X}, t)$, in terms of a Taylor expansion at the initial point $(\mathbf{X}, t)$ of order n. Such an assumption allows us to consider a class of materials for which $\mathbf{A}$ depends on $\mathbf{F}$ as well as on its spatial and time derivatives evaluated at $(\mathbf{X}, t)$:

$$\mathbf{A} = \mathbf{A}(\mathbf{F}, \dot{\mathbf{F}}, \nabla_X \mathbf{F}, ..., \mathbf{X}). \qquad (6.3)$$

The materials described by a constitutive relation (6.3) are said to be of **grade** n if n is the maximum order of the derivatives of motion and temperature.

It is worthwhile to note that this principle includes, as special cases, both *classical elasticity* and the *history-independent Newtonian fluids* (see Chapter 7).

2. **Principle of Local Action** (Noll, 1958).

   Fields $\mathbf{A}(\mathbf{X}, t)$ depend on the history of the thermokinetic process through a local class of equivalence at $\mathbf{X}$.

   According to the notion of contact forces, this principle states that the thermokinetic process of material points at a finite distance from $\mathbf{X}$ can be disregarded in computing the fields $\mathbf{A}(\mathbf{X}, t)$ at $\mathbf{X}$.

   The previous two principles (i.e., determinism and local action), when combined, imply that the response at a point depends on the history of the thermokinetic process relative to an arbitrary small neighborhood of the particle. Materials satisfying these two principles are called **simple materials**.

3. **Principle of Material Frame-Indifference.**

   Constitutive equations (6.2) must be invariant under changes of frame of reference.

   We first remark that the principle of **material frame-indifference** or **material objectivity** is not to be confused with the term *objectivity* in the sense of transformation behavior as discussed in Chapter 4.

   In fact, the term *objectivity* denotes transformation properties of given quantities, whereas the principle of *material objectivity* discussed here postulates the complete independence of the material response from the frame of reference.

In other words, since constitutive equations represent a mathematical model of material behavior, they are supposed to be independent of the observer. To make this principle clear, consider a point P with mass m moving in an inertial frame I under the action of an elastic force $-k(\mathbf{x} - \mathbf{x}_O)$, where O is the force center. The motion equation of P in I is written as

$$m\ddot{\mathbf{x}} = -k(\mathbf{x} - \mathbf{x}_O).$$

If a frame I' is considered to be in rigid motion with respect to I, so that the transformation $\mathbf{x}' = \mathbf{x}_\Omega(t) + \mathbf{Q}(\mathbf{t})\mathbf{x}$ holds, where $\mathbf{x}_\Omega(t)$ is the position vector fixing the origin of I with respect to I' and $\mathbf{Q}(t)$ is an orthogonal matrix, then the motion equation assumes the form

$$m\ddot{\mathbf{x}}' = -k(\mathbf{x}' - \mathbf{x}'_O) - m\mathbf{a}_\tau - 2m\omega_\tau \times \dot{\mathbf{x}}',$$

where $\mathbf{a}_\tau$ and ω_τ are the acceleration and the angular velocity of I' with respect to I. The motion of P in I' can be obtained by integrating this equation or, alternatively, by integrating the previous one and by applying the rigid transformation rule to the result. In this example it is relevant to observe that, when writing the motion equation in I', it has been assumed that the elastic force is an invariant vector and the *constitutive elastic law* is invariant when passing from I to I'.

In summary, the principle of material frame-indifference states that constitutive quantities transform according to their nature (e.g. ϵ is invariant, $\mathbf{T}_*$ is a tensor, and so on), and, in addition, their dependence on the thermokinetic process is invariant under changes of frame of reference.

4. **Principle of Dissipation** (Coleman e Noll, 1963).

   Constitutive equations satisfy the reduced dissipation inequality $(5.83)_1$ *in any thermokinetic process compatible with momentum and energy balance* $(5.69)_1, (5.80)_1$.

   To investigate the relevance of this principle, we first observe that, given (6.2) with $\mathbf{T}_*\mathbf{F}^T = \mathbf{F}\mathbf{T}^T$, the momentum and energy balance equations can always be satisfied by conveniently selecting $\mathbf{b}$ and r. This remark implies that momentum and energy balance equations $(5.69)_1, (5.80)_1$ do not, *as a matter of principle*, play any role of constraint for the thermokinetic process; in other words, any thermokinetic process is a solution of the balance equations provided that $\mathbf{b}$ and r are conveniently selected.

   Moreover, we note that the reduced dissipation inequality is equivalent to the entropy principle, presuming that the energy balance is

6.1. Constitutive Axioms

satisfied. Historically, the entropy inequality has been considered to be a *constraint for the processes*: i.e., only those processes that are solutions of the balance equations and satisfy the second law of thermodynamics for given body forces **b**, energy sources, and prescribed boundary conditions are admissible. It is apparent how complex such a requirement is, since explicit solutions can only be obtained in a few cases.

A different point of view was introduced by Coleman and Noll. They entrust the reduced dissipation inequality with the task that it be a further *constraint for the constitutive equations*, i.e., these equations must satisfy the reduced inequality in any process that is compatible with the momentum and energy balance. In a rather effective way, one could say that *the material itself is required to satisfy the second law of thermodynamics in responding to any thermokinetic process*.

A further remark is concerned with the fact that there is no general agreement on the possibility of arbitrarily selecting the thermodynamic process and then finding the functions $\mathbf{b}(\mathbf{X},t)$ and $r(\mathbf{X},t)$ from the momentum and balance equations. In fact, it is argued that such a choice could be physically very difficult or even impossible. As a consequence, it should be more appropriate to require (see [39] and [41]) that the dissipation principle be satisfied for those processes in which the momentum and energy balance are satisfied with arbitrary but prescribed forces and energy sources.[1]

5. ***Principle of Equipresence*** (Truesdell and Toupin, 1960).

    All constitutive quantities depend a priori on the same variables, i.e., on the history of the thermokinetic process.

    According to this principle, response functions that depend on only one of the two components of the thermokinetic process cannot be postulated, i.e., the stress tensor cannot be supposed to depend only on the motion and the heat flux on the temperature field. This remark is especially relevant for composite materials and in the presence of interaction phenomena, because in such cases it ensures that coupled phenomena are properly taken into account.

The mathematical theory of the constitutive equations is a rather broad branch of continuum mechanics, so that it is almost impossible to give a complete review of this subject in a textbook. For this reason, here only one class of materials will be considered with the twofold aim: to give the reader examples of applications of the general principles discussed above and to introduce the behavior of the most common materials that are considered in

---

[1] This problem will analyzed more extensively in Chapter 11.

this volume. Materials exhibiting a more complex behavior will be analyzed in Volume II.

6.2 Thermoviscoelastic Behavior

Material behavior described in this section exhibits the following features:

- the stress state depends on temperature and on local deformations, i.e. on $\mathbf{F}$;

- *internal friction* or *dissipation* phenomena are generated when a part of the system is in relative shearing motion with respect to other parts, so that the material response depends on temperature and on the velocity gradient $\nabla \mathbf{v} = \dot{\mathbf{F}} \mathbf{F}^{-1}$.

Let S be a continuous system, with C_* its reference configuration and C the current configuration, associated with the motion $\mathbf{x} = \mathbf{x}(\mathbf{X}, t)$ at the time instant t. We consider a material response such that

$$\mathbf{A} = \mathbf{A}_{C_*}(\mathbf{F}, \dot{\mathbf{F}}, \theta, \Theta, \mathbf{X}), \tag{6.4}$$

where $\Theta = \nabla_\mathbf{X} \theta$.

We note that, if $\mathbf{A}$ is independent of $\mathbf{X}$, then the system S is *homogeneous* in the reference configuration C_*.

Material response, as expressed by (6.4), depends on both the local thermal state, i.e., the temperature and its gradient in a neighborhood of $\mathbf{X}$, and on the local state of deformation and its time changes. This dependence justifies the definition of **thermoviscoelastic behavior**, given to responses exhibiting the structure (6.4).

Particular cases include **thermoelastic behavior** if (6.4) does not depend on $\dot{\mathbf{F}}$, so that there are no dissipation effects due to internal friction, and **elastic behavior** if thermal effects are also disregarded.

Making reference to the *constitutive axioms*, we first remark that (6.4) satisfies the determinism and local action principles owing to its dependence on the thermokinetic process, locally on space and on time.[2]

---
[2]It should be remarked that a material behavior of order 3 is represented by a constitutive response having the structure

$$\mathbf{A} = \mathbf{A}_{C_*}(\mathbf{v}, \mathbf{F}, \nabla_\mathbf{X} \mathbf{F}, \dot{\mathbf{F}}, \theta, \Theta, \dot{\theta}, \ddot{\theta}, d^3\theta/dt^3, \nabla_\mathbf{X} \Theta, \mathbf{X}),$$

where *all* the third-order derivatives are included. In any case, as shown in Exercise 1, the objectivity principle rules out the dependence on $\mathbf{v}$ and the principle of dissipation does not allow any dependence on $\nabla_\mathbf{X} \mathbf{F}$, $d^3\theta/dt^3$, $\nabla_\mathbf{X} \Theta$.

6.2. Thermoviscoelastic Behavior

As far as the *principle of dissipation* is concerned, the following theorem can be proved:

Theorem 6.1
A constitutive equation (6.4) satisfies the reduced dissipation inequality in any thermokynetic process if and only if the following relations hold:

$$\psi = \psi(\mathbf{F}, \theta),$$

$$\eta = -\frac{\partial \psi}{\partial \theta} = \eta(\mathbf{F}, \theta),$$

$$\mathbf{T}_*^{(e)} = \rho_* \frac{\partial \psi}{\partial \mathbf{F}} = \mathbf{T}_*^{(e)}(\mathbf{F}, \theta),$$

$$\mathbf{T}_*^{(d)} : \dot{\mathbf{F}}^T - \frac{\mathbf{h}_* \cdot \mathbf{\Theta}}{\theta} \geq 0, \tag{6.5}$$

where

$$\mathbf{T}_*^{(e)} = \mathbf{T}_*^{(e)}(\mathbf{F}, \mathbf{0}, \theta, \mathbf{0}) \tag{6.6}$$

is the Piola–Kirchhoff stress tensor at **equilibrium** *and*

$$\mathbf{T}_*^{(d)} = \mathbf{T}(\mathbf{F}, \dot{\mathbf{F}}, \theta, \mathbf{\Theta}) - \mathbf{T}(\mathbf{F}, \mathbf{0}, \theta, \mathbf{0}) \tag{6.7}$$

is its **dynamic component**.

PROOF If the time derivative of $\psi(\mathbf{F}, \dot{\mathbf{F}}, \theta, \mathbf{\Theta})$ is substituted into $(5.83)_1$, then we get the relation

$$-\rho_* \frac{\partial \psi}{\partial \dot{F}_{iL}} \ddot{F}_{iL}^T - \rho_* \frac{\partial \psi}{\partial \Theta_i} \dot{\Theta}_i - \rho_* \left(\eta + \frac{\partial \psi}{\partial \theta} \right) \dot{\theta}$$
$$+ \left(T_{*Li} - \rho_* \frac{\partial \psi}{\partial F_{iL}} \right) \dot{F}_{iL}^T - \frac{h_{*i} \cdot \Theta_i}{\theta} \geq 0, \tag{6.8}$$

which can be written in the compact form

$$\mathbf{a} \cdot \mathbf{u} + b \geq 0, \tag{6.9}$$

by assuming that

$$\mathbf{a} = \left(-\rho_* \frac{\partial \psi}{\partial \dot{\mathbf{F}}}, -\rho_* \frac{\partial \psi}{\partial \mathbf{\Theta}}, -\rho_* \left(\eta + \frac{\partial \psi}{\partial \theta} \right) \right), \quad \mathbf{u} = \left(\ddot{\mathbf{F}}, \dot{\mathbf{\Theta}}, \dot{\theta} \right),$$
$$b = \mathrm{tr} \left(\left(\mathbf{T}_* - \rho_* \frac{\partial \psi}{\partial \mathbf{F}} \right) \dot{\mathbf{F}}^T \right) - \frac{\mathbf{h}_* \cdot \mathbf{\Theta}}{\theta}. \tag{6.10}$$

Given that **a** and b are *independent* of **u**, the inequality (6.9) is satisfied, for any arbitrary **u**, if and only if $\mathbf{a} = \mathbf{0}$ and $b \geq 0$. This is certainly true because of the choice of the constitutive relations (6.4), so that $(6.5)_{1,2}$ are proved if it is verified that $\mathbf{u}$ can be arbitrary. This is made possible by virtue of the statement that *any* thermokinetic process is a solution of the balance equations, for a convenient choice of source terms $\mathbf{b}$ and r. Because the thermokinetic process is arbitrary, it follows that $\ddot{\mathbf{F}}, \dot{\boldsymbol{\Theta}}, \dot{\theta}$ are also arbitrary, for a given $\mathbf{X} \in C_*$ and time t.

From all the above remarks it follows that (6.8) reduces to

$$\left(\mathbf{T}_*^{(e)} - \rho_* \frac{\partial \psi}{\partial \mathbf{F}}\right) : \dot{\mathbf{F}}^T + \mathbf{T}_*^{(d)} : \dot{\mathbf{F}} - \frac{\mathbf{h}_* \cdot \boldsymbol{\Theta}}{\theta} \geq 0, \tag{6.11}$$

where $(6.5)_1$ and $(6.5)_2$ have been taken into account. The previous considerations cannot be applied to this inequality since (6.11) does not exhibit the form (6.9). The factor of $\dot{\mathbf{F}}^T$ does not depend on $\dot{\mathbf{F}}$, whereas the last two terms do. By substituting $\alpha \dot{\mathbf{F}}$ for $\dot{\mathbf{F}}$ and $\alpha \boldsymbol{\Theta}$ for $\boldsymbol{\Theta}$, with α a positive real number, (6.11) can be written as

$$\left(\mathbf{T}_*^{(e)} - \rho_* \frac{\partial \psi}{\partial \mathbf{F}}\right) : \alpha \dot{\mathbf{F}}^T + \mathbf{T}_*^{(d)} : \alpha \dot{\mathbf{F}} - \frac{\mathbf{h}_* \cdot \boldsymbol{\Theta}}{\theta} \geq 0, \tag{6.12}$$

and, by observing that

$$\lim_{\alpha \to 0} \mathbf{T}_*^{(d)}(\mathbf{F}, \alpha \dot{\mathbf{F}}, \theta, \alpha \boldsymbol{\Theta}) = \mathbf{0},$$

from (6.12), by dividing by α and considering the limit $\alpha \longmapsto 0$, we find that

$$\left(\mathbf{T}_*^{(e)} - \rho_* \frac{\partial \psi}{\partial \mathbf{F}}\right) : \dot{\mathbf{F}}^T - \frac{\mathbf{h}_*(\mathbf{F}, \mathbf{0}, \theta, \mathbf{0}) \cdot \boldsymbol{\Theta}}{\theta} \geq 0.$$

This inequality implies $(6.5)_3$, so that the theorem is proved. ∎

Thus we conclude that the principle of dissipation implies that the *free energy* of a thermoviscoelastic material

- only depends on the deformation gradient and temperature;

- acts as a thermodynamic potential for the entropy and the Piola–Kirchhoff stress tensor at equilibrium, and these quantities both depend on $\mathbf{F}$ and θ. In particular, when considering a thermoelastic material, the free energy acts as a thermodynamic potential for the whole Piola–Kirchhoff stress tensor.

It follows that, given the free energy, the constitutive equations for $\mathbf{T}_*^{(e)}$ and η are known. Furthermore, the dynamic component of the stress tensor and the heat flux vector must be defined in such a way that $(6.5)_4$ is satisfied in any process.

6.2. Thermoviscoelastic Behavior

In the Eulerian formulation, by taking into account (5.67) and (5.79) and observing that

$$\mathbf{T}_* : \dot{\mathbf{F}}^T = T_{*Li}\dot{F}_{iL} = JT_{ij}F_{Lj}^{-1}\dot{F}_{iL} = JT_{ij}\frac{\partial v_i}{\partial x_j} = JT_{ij}D_{ij},$$

$$\mathbf{h}_* \cdot \mathbf{\Theta} = h_{*L}\Theta_L = Jh_i F_{iL}^{-1}\Theta_L = Jh_i g_i = J\mathbf{h} \cdot \mathbf{g},$$

where $\mathbf{D}$ is the rate of deformation tensor and $\mathbf{g} = \nabla\theta$, we find that relations (6.5) reduce to

$$\psi = \psi(\mathbf{F}, \theta),$$

$$\eta = -\frac{\partial \psi}{\partial \theta} = \eta(\mathbf{F}, \theta),$$

$$\mathbf{T}^{(e)} = \rho \frac{\partial \psi}{\partial \mathbf{F}}\mathbf{F}^T = \mathbf{T}^{(e)}(\mathbf{F}, \theta),$$

$$\mathbf{T}^{(d)} : \mathbf{D} - \frac{\mathbf{h} \cdot \mathbf{g}}{\theta} \geq 0. \tag{6.13}$$

The constraints derived from the material objectivity principle are expressed by the following theorem:

Theorem 6.2
Constitutive equations (6.13) satisfy the principle of material frame-indifference if and only if

$$\psi = \psi(\mathbf{C}, \theta),$$

$$\eta = -\frac{\partial \psi}{\partial \theta} = \eta(\mathbf{C}, \theta),$$

$$\mathbf{T}^{(e)} = 2\rho \mathbf{F}\frac{\partial \psi}{\partial \mathbf{C}}\mathbf{F}^T,$$

$$\mathbf{T}^{(d)} = \mathbf{F}\hat{\mathbf{T}}^{(d)}(\mathbf{C}, \mathbf{F}^T\mathbf{D}\mathbf{F}, \theta, \mathbf{\Theta})\mathbf{F}^T,$$

$$\mathbf{h} = \mathbf{F}\hat{\mathbf{h}}(\mathbf{C}, \mathbf{F}^T\mathbf{D}\mathbf{F}, \theta, \mathbf{\Theta}), \tag{6.14}$$

where $\mathbf{C} = \mathbf{F}^T\mathbf{F}$ is the right Cauchy–Green tensor (see Chapter 3).

PROOF To prove the theorem, we recall that in a rigid change of frame of reference $I \longrightarrow I'$, tensors $\mathbf{F}, \mathbf{C},$ and $\dot{\mathbf{F}}$ transform according to rules (see (3.46) and (4.30)):

$$\mathbf{F}' = \mathbf{Q}\mathbf{F}, \quad \mathbf{C}' = \mathbf{C}, \quad \mathbf{D}' = \mathbf{Q}\mathbf{D}\mathbf{Q}^T,$$

$$\mathbf{W}' = \mathbf{Q}\mathbf{W}\mathbf{Q}^T + \dot{\mathbf{Q}}\mathbf{Q}^T, \quad \mathbf{G}' = \mathbf{G}, \tag{6.15}$$

where $\mathbf{Q}$ is the orthogonal matrix of the change of frame.

To prove the necessary condition, we observe that the objectivity of ψ implies that
$$\psi(\mathbf{F}, \theta) = \psi(\mathbf{QF}, \theta) \qquad \forall \mathbf{Q} \in O(3).$$
Resorting to the polar decomposition theorem (see Chapter 1) we can write that $\mathbf{F} = \mathbf{RU}$ and assuming that $\mathbf{Q}$ is equal to $\mathbf{R}^T$, we find that
$$\psi(\mathbf{F}, \theta) = \psi(\mathbf{U}, \theta),$$
and since $\mathbf{U}^2 = \mathbf{C}$, $(6.14)_1$ and $(6.14)_2$ are proved.

Moreover,
$$\frac{\partial \psi}{\partial F_{iQ}} F_{jQ} = \frac{\partial \psi}{\partial C_{LM}} \frac{\partial C_{LM}}{\partial F_{iQ}} F_{jQ} = \frac{\partial \psi}{\partial C_{LM}} (\delta_{hi}\delta_{LQ} F_{hM} + F_{hL}\delta_{hi}\delta_{MQ}) F_{jQ},$$
so that $(6.14)_3$ is also proved.

For $(6.14)_4$, since $\nabla_{\mathbf{X}} \mathbf{v} = \mathbf{F}^T \nabla \mathbf{v}$, it holds that
$$\mathbf{T}^{(d)}(\mathbf{F}, \dot{\mathbf{F}}, \theta, \Theta) = \tilde{\mathbf{T}}^{(d)}(\mathbf{F}, \nabla \mathbf{v}, \theta, \Theta) = \bar{\mathbf{T}}^{(d)}(\mathbf{F}, \mathbf{D}, \mathbf{W}, \theta, \Theta).$$

Furthermore, since Θ is invariant, the objectivity of $\bar{\mathbf{T}}^{(d)}(\mathbf{F}, \mathbf{D}, \mathbf{W}, \theta, \Theta)$ implies that
$$\bar{\mathbf{T}}^{(d)}(\mathbf{F}', \mathbf{D}', \mathbf{W}', \theta, \Theta) = \mathbf{Q} \bar{\mathbf{T}}^{(d)}(\mathbf{F}, \mathbf{D}, \mathbf{W}, \theta, \Theta) \mathbf{Q}^T,$$
i.e.,
$$\bar{\mathbf{T}}^{(d)}(\mathbf{F}, \mathbf{D}, \mathbf{W}, \theta, \Theta) = \mathbf{Q}^{\mathbf{T}} \bar{\mathbf{T}}^{(\mathbf{d})}(\mathbf{QF}, \mathbf{QDQ}^{\mathbf{T}}, \mathbf{QWQ}^{\mathbf{T}} + \dot{\mathbf{Q}}\mathbf{Q}^{\mathbf{T}}, \theta, \Theta)\mathbf{Q}. \tag{6.16}$$

If $\mathbf{Q} = \mathbf{I}$ and $\dot{\mathbf{Q}}\mathbf{Q}^T = \dot{\mathbf{Q}} = -\mathbf{W}$, then (6.16) gives
$$\bar{\mathbf{T}}^{(d)}(\mathbf{F}, \mathbf{D}, \mathbf{W}, \theta, \Theta) = \bar{\mathbf{T}}^{(d)}(\mathbf{F}, \mathbf{D}, 0, \theta, \Theta),$$
which proves that $\bar{\mathbf{T}}^{(d)}$ is independent of $\mathbf{W}$. It follows that (6.16) reduces to
$$\bar{\mathbf{T}}^{(d)}(\mathbf{F}, \mathbf{D}, \theta, \Theta) = \mathbf{Q}^T \bar{\mathbf{T}}^{(d)}(\mathbf{QF}, \mathbf{QDQ}^T, \theta, \Theta) \mathbf{Q}.$$
If again the relations $\mathbf{F} = \mathbf{RU}$ and $\mathbf{Q} = \mathbf{R}^T$ are used, we deduce that:
$$\bar{\mathbf{T}}^{(d)}(\mathbf{F}, \mathbf{D}, \theta, \Theta) = \mathbf{R}\bar{\mathbf{T}}^{(d)}(\mathbf{U}, \mathbf{QDQ}^T, \theta, \Theta)\mathbf{R}^T = \mathbf{F}\hat{\mathbf{T}}^{(d)}(\mathbf{C}, \mathbf{QDQ}^T, \theta, \Theta)\mathbf{F}^T$$
and $(6.14)_4$ is also proved.

Equation $(6.14)_5$ can be verified from similar arguments.

It is easy to check that $(6.14)_{1,2}$ are also sufficient conditions. As far as $(6.14)_3$ is concerned, it holds that
$$\mathbf{T}^{(e)}(\mathbf{F}', \theta, \Theta') = 2\rho \mathbf{F}' \frac{\partial \psi}{\partial \mathbf{C}'} \mathbf{F}'^T = 2\rho \mathbf{QF} \frac{\partial \psi}{\partial \mathbf{C}} \mathbf{F}^T \mathbf{Q}^T = \mathbf{Q} \mathbf{T}^{(e)}(\mathbf{F}, \theta, \Theta) \mathbf{Q}^T.$$

6.3. Linear Thermoelasticity

The same procedure proves the other relations. ∎

A thermoviscoelastic material is *incompressible* if

$$J = 1 \iff \nabla \cdot \mathbf{v} = 0 \iff V(c) = const, \tag{6.17}$$

where $V(c)$ is the volume of any material region c (see (3.20) and (4.25)).

It is relevant to observe that in this case the inequality of the reduced dissipation no longer implies (6.13), since (6.8) must be satisfied for any process for which $\nabla \cdot \mathbf{v} = 0$. To take this constraint into account, the inequality $R + p\nabla \cdot \mathbf{v} \geq 0$, instead of (6.8), must be considered, where R represents (6.8) and p is a Lagrangian multiplier. It can be verified that $(6.13)_3$ reduces to

$$\mathbf{T}^{(e)}(\mathbf{F}, \theta, \Theta) = -p\mathbf{I} + \rho \frac{\partial \psi}{\partial \mathbf{F}} \mathbf{F}^T, \tag{6.18}$$

where $p(\mathbf{x}) = \tilde{p}(\mathbf{X})$ is an undetermined function.

If a thermoelastic behavior, or a thermoviscoelastic behavior at rest, is supposed, i.e., $\mathbf{D} = \mathbf{0}$, then from $(6.13)_4$ we derive the condition

$$\mathbf{h} \cdot \mathbf{g} \leq 0, \tag{6.19}$$

which allows us to say that *when the system is at rest or there are no viscous effects, heat flows from higher temperature regions to lower temperature regions.*

In addition, observing that the function $\mathbf{f}(\mathbf{F}, \mathbf{0}, \theta, \mathbf{g}) = \mathbf{h} \cdot \mathbf{g}$ attains its maximum when $\mathbf{g} = \mathbf{0}$, we find that

$$\nabla_{\mathbf{g}} \mathbf{f}(\mathbf{F}, \mathbf{0}, \theta, \mathbf{g})_{\mathbf{g}=\mathbf{0}} = \mathbf{h}(\mathbf{F}, \mathbf{0}, \theta, \mathbf{0}) = \mathbf{0} \tag{6.20}$$

and it can be said that *in a thermoviscoelastic system at rest, as well as in an elastic system, we cannot have heat conduction without a temperature gradient, i.e., when the temperature is uniform.*

6.3 Linear Thermoelasticity

In this section, we consider a thermokinetic process of viscoelastic materials satisfying the following assumptions:

1. the reference configuration C_* is a state of equilibrium in which the stress tensor is $\mathbf{T}_0$ and the temperature field θ_0 is uniform;

2. the displacement field $\mathbf{u}(\mathbf{X}, t)$, its gradient $\mathbf{H} = \nabla_{\mathbf{X}} \mathbf{u}(\mathbf{X}, t)$, the temperature difference $\theta - \theta_0$, and its gradient $\boldsymbol{\Theta} = \nabla_{\mathbf{X}} \theta$ are first-order quantities.

The linearity assumption is equivalent to requiring that powers and products of these quantities appearing in the constitutive equation of the stress or powers greater than two in the thermodynamic potentials can be disregarded.

In order to illustrate the physical relevance of the above assumptions, we present the following two examples:

- Let S be a hollow cylinder with inner radius r_1, outer radius r_2, and height l, constituted by a thermoelastic homogeneous material in equilibrium under uniform temperature and in the absence of forces. Suppose now that a sector (as shown in Figure 6.1) is extracted and the extremities A and A' are welded in such a way that a new cylinder, whose cross section is no longer circular, is obtained. If we assume this configuration as a reference configuration C_*, then the stress state is not zero *even in absence of applied forces*, and the system will reach a new equilibrium temperature θ_0. Any perturbation of the system not far from C_* will satisfy only the assumption (2).

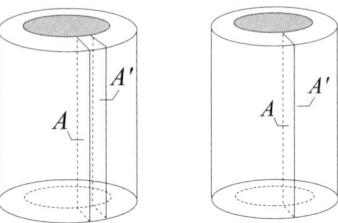

Figure 6.1

- As a second example, consider a cylinder of height l filled with fluid. Due to the weight of the fluid, there is a radial pressure acting on the internal cylinder wall, whose intensity increases linearly with depth from the surface (Stevino's law, Chapter 9). This pressure will produce a deformation of the cylinder, which defines the reference configuration C_*. If a piston acting on the surface will produce an additional small pressure changing in time, then there will be an evolution of the system satisfying only assumption (2).

By taking advantage of the fact that all the quantities describing the evolution from the reference configuration C_* under the assumptions (1) and (2) are first-order quantities, the stress constitutive relation

6.3. Linear Thermoelasticity

can be replaced with its Taylor expansion, truncated to the first-order terms, whereas the constitutive equations of the thermodynamic potentials can be approximated with their Taylor expansions, truncated to the second-order terms.

From $(6.14)_{3,5}$, since in the reference configuration $\mathbf{C} = \mathbf{I}$ and $\boldsymbol{\Theta} = \mathbf{0}$, it follows that

$$\mathbf{T}_0 = 2\rho_* \left(\frac{\partial \psi}{\partial \mathbf{C}}\right)_*, \qquad (6.21)$$

where the notation $(A)_*$ emphasizes that the quantity A is evaluated for $\mathbf{C} = \mathbf{I}$, $\theta = \theta_0$, and $\boldsymbol{\Theta} = \mathbf{0}$, i.e., in C_*.

In particular, if the reference configuration is stress free, then

$$\mathbf{T}_0 = \mathbf{0}. \qquad (6.22)$$

Under the assumption (2), the constitutive equations can be proved to have the structure

$$\mathbf{T} \cong \mathbf{T}_0 + \mathbf{H}\mathbf{T}_0 + \mathbf{T}_0\mathbf{H}^T - \mathrm{tr}\mathbf{H}\mathbf{T}_0 + \mathbb{C}\mathbf{E} + \mathbb{A}(\theta - \theta_0),$$
$$\mathbf{h} \cong -\mathbf{K}\mathbf{g}, \qquad (6.23)$$

where $\mathbf{E}$ is the tensor of infinitesimal strain (see (3.30)), $\mathbf{g} = \nabla_\mathbf{x} \theta$, and

$$\mathbb{C}_{ijnp} = 4\rho_* \left(\frac{\partial^2 \psi}{\partial C_{ij} \partial C_{np}}\right)_*, \quad \mathbb{A}_{ij} = 2\rho_* \left(\frac{\partial^2 \psi}{\partial \theta \partial C_{ij}}\right)_*,$$
$$K_{ij} = -\frac{\partial \hat{h}_i}{\partial \Theta_j}(\mathbf{I}, \theta_0, \mathbf{0}). \qquad (6.24)$$

Tensors $\mathbb{C}$ and $\mathbf{K}$ are called the **tensor of linear elasticity** and the **tensor of thermal conductivity**, relative to C_*.

Owing to (6.24), $\mathbb{C}$ and $\mathbb{A}$ have the following properties of symmetry:

$$\mathbb{C}_{ijnp} = \mathbb{C}_{jinp} = \mathbb{C}_{ijpn} = \mathbb{C}_{npij}, \quad \mathbb{A}_{ij} = \mathbb{A}_{ji}; \qquad (6.25)$$

furthermore, from (6.19) and $(6.23)_2$, it follows that the tensor $\mathbf{K}$ is positive definite:

$$\mathbf{g} \cdot \mathbf{K}\mathbf{g} \geq 0. \qquad (6.26)$$

Moreover, since

$$\mathbf{T}_* \cong \mathbf{T} + \mathrm{tr}\,\mathbf{H}\mathbf{T}_0 - \mathbf{T}_0\mathbf{H}^T,$$
$$\mathbf{h}_* \cong \mathbf{h}, \qquad (6.27)$$

if the reference configuration C_* is stress free, it follows that

$$\mathbf{T} \cong \mathbf{T}_* \cong \mathbb{C}\mathbf{E} + \mathbb{A}(\theta - \theta_0),$$
$$\mathbf{h} \cong \mathbf{h}_* \cong -\mathbf{K}\mathbf{g}, \qquad (6.28)$$

so that *there is no difference between the Lagrangian and Eulerian descriptions.*

Finally, we observe that $(6.23)_1$ and (6.27) give the following approximate expression of $\mathbf{T}_*$:

$$\mathbf{T}_* \cong \mathbf{T}_0 + \mathbf{H}\mathbf{T}_0 + \mathbb{C}\mathbf{E} + A(\theta - \theta_0).$$

To prove previous formulae, we will made use of lowercase Latin indices for both the Eulerian and Lagrangian formulations. In addition, since the aim is to find a set of linearized constitutive equations for the stress, the free energy ψ will be represented by a Taylor expansion truncated to second-order terms:

$$\psi \cong \psi_* + \left(\frac{\partial \psi}{\partial \theta}\right)_* (\theta - \theta_0) + \left(\frac{\partial \psi}{\partial C_{lm}}\right)_* (C_{lm} - \delta_{lm})$$
$$+ \frac{1}{2}\left(\frac{\partial^2 \psi}{\partial \theta^2}\right)_* (\theta - \theta_0)^2 + \left(\frac{\partial^2 \psi}{\partial \theta \partial C_{lm}}\right)_* (\theta - \theta_0)(C_{lm} - \delta_{lm})$$
$$+ \frac{1}{2}\left(\frac{\partial^2 \psi}{\partial C_{lm} \partial C_{np}}\right)_* (C_{lm} - \delta_{lm})(C_{np} - \delta_{np}). \qquad (6.29)$$

From (6.29) it also follows that

$$\frac{\partial \psi}{\partial C_{lm}} \cong \left(\frac{\partial \psi}{\partial C_{lm}}\right)_* + \left(\frac{\partial^2 \psi}{\partial \theta \partial C_{lm}}\right)_* (\theta - \theta_0) + \left(\frac{\partial^2 \psi}{\partial C_{lm} \partial C_{np}}\right)_* (C_{np} - \delta_{np}), \qquad (6.30)$$

and since $\mathbf{C} = \mathbf{F}^T\mathbf{F} = \mathbf{I} + 2\mathbf{E} + \mathbf{H}^T\mathbf{H}$, by neglecting second-order terms we get

$$\frac{\partial \psi}{\partial C_{lm}} \cong \left(\frac{\partial \psi}{\partial C_{lm}}\right)_* + \left(\frac{\partial^2 \psi}{\partial \theta \partial C_{lm}}\right)_* (\theta - \theta_0) + 2\left(\frac{\partial^2 \psi}{\partial C_{lm} \partial C_{np}}\right)_* E_{np}. \qquad (6.31)$$

In addition, from (5.66), (3.40), and (3.27) it follows that

$$\rho \cong \rho_*(1 - tr\mathbf{H}), \quad \mathbf{F} = \mathbf{I} + \mathbf{H},$$

and $(6.14)_3$ becomes

$$T_{ij} \cong 2\rho_*(1 - H_{hh})(\delta_{il} + H_{il})\frac{\partial \psi}{\partial C_{lm}}(\delta_{jm} + H_{jm}). \qquad (6.32)$$

By substituting (6.31) into (6.32), neglecting terms of higher order, and taking into account (6.21) and $(6.24)_{1,2}$, $(6.23)_1$ is proved.

In a similar way, first-order terms of the heat flux vector give

$$h_i \cong \frac{\partial h_i}{\partial \Theta_j}(\mathbf{C}, \theta, 0)\Theta_j \cong \frac{\partial h_i}{\partial \Theta_j}(\mathbf{I}, \theta, 0)\Theta_j,$$

6.4. Exercises

so that $(6.23)_2$ and $(6.24)_3$ are also proved, since $\Theta = \mathbf{F}^T \mathbf{g} = (\mathbf{I} + \mathbf{H}^T)\mathbf{g}$.

Finally, to prove (6.27) we recall (3.40) as well as the Lagrangian definitions (5.67) and (5.79), which, when expressed as a Taylor series, become

$$T_{*ij} \cong (1 + H_{hh})T_{ih}(\delta_{jh} - H_{jh}) \cong T_{ij} + H_{hh}T_{ij} - T_{ih}H_{jh},$$
$$h_{*i} = (1 + H_{hh})h_j(\delta_{ij} - H_{ij}) \cong h_j + H_{hh}h_j - h_j H_{ij}.$$

The introduction of equations (6.23) into the latter gives (6.27).

6.4 Exercises

1. Given the constitutive equations of order 3

$$\mathbf{A} = \mathbf{A}_{C_*}(\mathbf{v}, \mathbf{F}, \nabla_\mathbf{X}\mathbf{F}, \dot{\mathbf{F}}, \theta, \mathbf{G}, \dot{\theta}, \ddot{\theta}, d^3\theta/dt^3, \nabla_\mathbf{X}\mathbf{G}, \nabla_\mathbf{X}\dot{\mathbf{G}}, \mathbf{X}),$$

prove that the objectivity principle does not allow any dependence on the velocity $\mathbf{v}$ of $\mathbf{X}$.

Consider the change of reference frame from I to I', the latter having a constant velocity $\mathbf{u}$ with respect to I. The objectivity principle requires that

$$\mathbf{A}_{C_*}(\mathbf{v} + \mathbf{u}, \xi) = \mathbf{A}_{C_*}(\mathbf{v}, \xi),$$

where ξ is the set of all other variables. The choice $\mathbf{u} = -\mathbf{v}$ proves the independence of $\mathbf{v}$.

Using the principle of dissipation, other constraints for the constitutive quantities are obtained.

2. Let $\mathbf{d}(\mathbf{X},t)$ be a vector field and consider the constitutive equation for the heat flux vector in the form

$$\mathbf{h} = \mathbf{h}(\mathbf{F}, \dot{\mathbf{d}}, \theta, \Theta).$$

Prove that the objectivity principle requires that

$$\mathbf{h} = \tilde{\mathbf{h}}(\mathbf{C}, \hat{\mathbf{d}}, \theta, \Theta),$$

where

$$\hat{\mathbf{d}} = \dot{\mathbf{d}} - \omega \times \mathbf{d}$$

and ω is the angular velocity at $\mathbf{x} \in \mathbf{C}$.

Chapter 7

Symmetry Groups: Solids and Fluids

7.1 Symmetry

The experimental determination of the constitutive equation of a material is not an easy task. It requires special apparatuses that must satisfy very restrictive requirements, the most important being the capability of inducing a uniform state of stress and strain in the material element being tested. In addition, even in the simplest cases of elastic behavior, the experimental procedure is far from straightforward due to the number of stress (or strain) states required to determine the material response. For these reasons, it is relevant to ask if *symmetry properties* of materials can to some extent simplify this task, because of the additional restrictions they introduce into the constitutive equations.

Within this framework, we will first turn our attention to *isotropic solids* and to *viscous fluids* and then we will highlight the general properties of *elastic anisotropic solids* in isothermal conditions.

Before we proceed, let us introduce some intuitive considerations that will help us to understand a more formal definition of symmetry.

Let C_* be a sphere of radius r exhibiting the same elastic properties in all directions and subjected to a uniform pressure $-p\mathbf{N}$, where $\mathbf{N}$ is the unit forward vector normal to ∂C_*. We can predict, and it is experimentally proved, that C_* deforms into a sphere C of radius $r_1 < r$, whose state of stress is given by $-p\mathbf{I}$. This final configuration is also obtained if C_* is rotated around an axis passing through its center, before applying the uniform pressure $-p\mathbf{N}$ on its boundary.

If the material of C_* does not have the same elastic properties in all directions, then the deformed state of C will be an ellipsoid whose minor axis is in the direction of lower stiffness.

Suppose now that the sphere C_* is first subjected to a rigid rotation $\mathbf{Q}$ about its center, so that its configuration changes from C_* to C'_*, and it is subsequently deformed through the application of $-p\mathbf{IN}$. Its final configuration C' will be an ellipsoid rotated with respect to C, according to the matrix $\mathbf{Q}$.

More generally, consider an elastic body S in the reference configuration C_* subjected to mass forces $\rho\mathbf{b}$ and surface tractions $\mathbf{t}$, and suppose it reaches the deformed configuration C, in which the strain and the stress state in a neighborhood of $\mathbf{x}$ are given by the deformation gradient $\mathbf{F}$ and the stress tensor $\mathbf{T}$. If the material exhibits a symmetry expressed by the orthogonal matrix $\mathbf{Q}$,[1] then we could rotate it by $\mathbf{Q}$, so that C_* rotates into C'_*, and we could apply a force system, conveniently modified in such a way that the deformation gradient in $\mathbf{Qx}$ is still represented by $\mathbf{F}$. In this case, the stress state will be the same.

In the following discussion the notations $Unim$ and $O(3)$ identify respectively the group of 3×3 matrices whose determinant is ± 1 and the group of orthogonal matrices (see Chapter 1):

$$Unim = \{\mathbf{H} : E_3 \to E_3, \quad \mathbf{H} \in Lin(E_3), \quad \det \mathbf{H} = \pm 1\},$$
$$O(3) = \{\mathbf{Q} : E_3 \to E_3, \quad \mathbf{Q}^T\mathbf{Q} = \mathbf{I}, \quad \det \mathbf{Q} = \pm 1\}. \tag{7.1}$$

Accordingly, $O(3)$ is a subgroup of $Unim$.

Any matrix $\mathbf{H}$ belonging to one of these groups represents a transformation of the vector space $\mathcal{E}_3$ that leaves unchanged the mass density, as can be proved by noting that $\rho = \rho_* |\det \mathbf{H}|$.

If not otherwise stated, $\mathbf{H}$ and $\mathbf{Q}$ will denote unimodular and orthogonal matrices, respectively.

Let C_* be a reference configuration of a thermoelastic system S and let $\mathbf{X_0}$ be an arbitrary selected point of C_*, and consider the transformation $C_* \longrightarrow C'_*$

$$\mathbf{X}' = \mathbf{L}(\mathbf{X}) = \mathbf{X}_0 + \mathbf{H}(\mathbf{X} - \mathbf{X}_0), \quad \mathbf{H} \in Unim,$$

which, due to the properties of $\mathbf{H}$, preserves volume and density and leaves $\mathbf{X_0}$ a fixed point.

Now consider an arbitrary deformation $C_* \longrightarrow C$, represented by $\mathbf{x} = \mathbf{f}(\mathbf{X})$, and a second one $C_* \longrightarrow C'_* \longrightarrow C$, obtained by first applying $\mathbf{L}(\mathbf{X})$ and then $\mathbf{f}'(\mathbf{X}')$, so that $\mathbf{f} = \mathbf{f}'\mathbf{H}$. If S is a thermoviscoelastic material and its response is specified by $\mathbf{A}(\mathbf{F}, \dot{\mathbf{F}}, \theta, \Theta)$ (see (6.4)), then $\mathbf{H} \in Unim$ represents a ***symmetry*** at $\mathbf{X_0}$ and in the configuration C_* if and only if

$$\mathbf{A}_{C_*}(\mathbf{F}, \dot{\mathbf{F}}, \theta, \Theta) = \mathbf{A}_{C_*}(\mathbf{FH}, \dot{\mathbf{F}}\mathbf{H}, \theta, \Theta\mathbf{H}); \tag{7.2}$$

i.e., the material response in $\mathbf{X_0}$ is the same whether we consider the deformation $\mathbf{f}(\mathbf{X})$ or $\mathbf{f}'(\mathbf{L}(\mathbf{X}))$.

---
[1] A linear transformation is represented here by a matrix in a given orthonormal basis, as discussed in Chapter 1.

7.1. Symmetry

Theorem 7.1
Symmetry transformations at the point $\mathbf{X}_0$ of C_* are a subgroup G_{C_*} of Unim.

PROOF   To prove that if $\mathbf{H}_1, \mathbf{H}_2 \in G_{C_*}$ then also $\mathbf{H}_1\mathbf{H}_2 \in G_{C_*}$, we first observe that, if $\mathbf{F}$ is nonsingular and $\mathbf{H}_1 \in G_{C_*}$, it follows that $\mathbf{FH}_1$ is nonsingular. If we replace $\mathbf{F}$ with $\mathbf{FH}_1$ and $\boldsymbol{\Theta}$ with $\boldsymbol{\Theta}\mathbf{H}_1$ in (7.2), then we obtain

$$\mathbf{A}_{C_*}(\mathbf{FH}_1, \dot{\mathbf{F}}\mathbf{H}_1, \boldsymbol{\Theta}\mathbf{H}_1) = \mathbf{A}_{C_*}(\mathbf{FH}_1\mathbf{H}_2, \mathbf{FH}_1, \dot{\mathbf{F}}\mathbf{H}_1\mathbf{H}_2, \boldsymbol{\Theta}\mathbf{H}_1\mathbf{H}_2),$$

where, for the sake of simplicity, the dependence on θ has been omitted. In addition, if $\mathbf{H}_1 \in G_{C_*}$ and therefore $\mathbf{A}_{C_*}(\mathbf{F}, \dot{\mathbf{F}}, \boldsymbol{\Theta}) = \mathbf{A}_{C_*}(\mathbf{FH}_1, \dot{\mathbf{F}}\mathbf{H}_1, \boldsymbol{\Theta}\mathbf{H}_1)$, then it holds that

$$\mathbf{A}_{C_*}(\mathbf{F}, \dot{\mathbf{F}}, \boldsymbol{\Theta}) = \mathbf{A}_{C_*}(\mathbf{FH}_1\mathbf{H}_2, \mathbf{FH}_1, \dot{\mathbf{F}}\mathbf{H}_1\mathbf{H}_2, \boldsymbol{\Theta}\mathbf{H}_1\mathbf{H}_2), \qquad (7.3)$$

and since $\mathbf{H}_1\mathbf{H}_2 \in Unim$, we have $\mathbf{H}_1\mathbf{H}_2 \in G_{C_*}$. Furthermore, if $\mathbf{H} \in G_{C_*}$, then $\mathbf{H}^{-1}$ exists and by replacing $\mathbf{F}$ with $\mathbf{FH}^{-1}$ and $\boldsymbol{\Theta}$ with $\boldsymbol{\Theta}\mathbf{H}^{-1}$ in (7.2), it is proved that $\mathbf{H}^{-1} \in G_{C_*}$. ∎

The subgroup G_{C_*} is called the ***symmetry group*** of S at $\mathbf{X}_0$ and in the configuration C_*.

It can also be proved that symmetry groups in two configurations C_* and C'_* and at the point $\mathbf{X}_0$, where these are related to each other through a unimodular transformation $\mathbf{P}$, are *conjugate*; i.e., there exists the relation,

$$G_{C'_*} = \mathbf{P} G_{C_*} \mathbf{P}^{-1}. \qquad (7.4)$$

At this stage, we can give the following definitions:

- the material is ***homogeneous*** in the configuration C_* if its symmetry group is *independent* of $\mathbf{X}_0 \in C_*$;

- a thermoelastic solid S is called ***anisotropic*** in the configuration C_* and at $\mathbf{X}_0 \in C_*$ if

$$G_{C_*} \subset O(3), \qquad (7.5)$$

  i.e., if its symmetry group at $\mathbf{X}_0$ is a *proper* subgroup of all possible rotations about $\mathbf{X}_0$;

- a thermoelastic system S represents a ***thermoelastic isotropic solid*** at $\mathbf{X}_0$, *if there exists a configuration C_*, called the **natural** or **undistorted configuration**, relative to which the group of isotropy at $\mathbf{X}_0$ is the orthogonal group*:

$$G_{C_*} = O(3). \qquad (7.6)$$

Finally we leave as an exercise to verify the following formulae

$$\mathbf{F} = \mathbf{F}'\mathbf{H}, \quad \mathbf{\Theta} = \mathbf{H}^T\mathbf{\Theta}', \quad \mathbf{C} = \mathbf{H}^T\mathbf{C}'\mathbf{H}, \tag{7.7}$$

where quantities without primes refer to the deformation $C_* \longrightarrow C$ and those with primes to $C'_* \longrightarrow C$.

7.2 Isotropic Solids

We now proceed to explore the restrictions on the constitutive equations of a thermoelastic material deriving from the assumption of isotropy. To do this, the following definitions are necessary.

A scalar function $f(\mathbf{B})$ of the second-order tensor $\mathbf{B}$ is defined to be **isotropic** if

$$f(\mathbf{B}) = f(\mathbf{Q}\mathbf{B}\mathbf{Q}^T) \quad \forall \mathbf{Q} \in O(3). \tag{7.8}$$

Similarly, a vector function $\mathbf{f}(\mathbf{v}, \mathbf{B})$ and a tensor function $\mathbf{T} = \mathbf{T}(\mathbf{v}, \mathbf{B})$, whose arguments are the vector $\mathbf{v}$ and the tensor $\mathbf{B}$, are defined to be isotropic if

$$\begin{aligned} \mathbf{Q}\mathbf{f}(\mathbf{v}, \mathbf{B}) &= \mathbf{f}(\mathbf{Q}\mathbf{v}, \mathbf{Q}\mathbf{B}\mathbf{Q}^T) & \forall \mathbf{Q} \in O(3), \\ \mathbf{Q}\mathbf{T}(\mathbf{v}, \mathbf{B})\mathbf{Q}^T &= \mathbf{T}(\mathbf{Q}\mathbf{v}, \mathbf{Q}\mathbf{B}\mathbf{Q}^T) & \forall \mathbf{Q} \in O(3). \end{aligned} \tag{7.9}$$

We refer the reader to [20], [21], [22] for more insight into the structure of isotropic functions. For the applications we are going to discuss, it is relevant to observe that the above theorems for isotropic functions allow us to write (7.8) and (7.9)$_1$ in the following form:

$$\begin{aligned} f(\mathbf{B}) &= f(I_B, II_B, III_B), \\ \mathbf{f}(\mathbf{v}, \mathbf{B}) &= (K_0\mathbf{I} + K_1\mathbf{B} + K_2\mathbf{B}^2)\mathbf{v}, \end{aligned} \tag{7.10}$$

where I_B, II_B, III_B are the principal invariants of the second-order tensor $\mathbf{B}$ and K_0, K_1, K_2 depend on $I_B, II_B, III_B, g^2, \mathbf{g} \cdot \mathbf{Bg}$, and $\mathbf{g} \cdot \mathbf{B}^2\mathbf{g}$.

We can now prove the following fundamental theorem:

Theorem 7.2
A thermoelastic system S with constitutive equations that satisfy the principle of material frame-indifference and the principle of dissipation is an isotropic solid if and only if

$$\psi = \psi(I, II, III, \theta),$$

7.2. Isotropic Solids

$$\eta = -\frac{\partial \psi}{\partial \theta} = \eta(I, II, III, \theta),$$
$$\mathbf{T} = f_0 \mathbf{I} + f_1 \mathbf{B} + f_2 \mathbf{B}^2,$$
$$\mathbf{h} = (K_0 \mathbf{I} + K_1 \mathbf{B} + K_2 \mathbf{B}^2)\mathbf{g}, \tag{7.11}$$

where I, II, III are the principal invariants of $\mathbf{B}$ or $\mathbf{C}$ and the functions f are defined as

$$f_0 = 2\rho III \frac{\partial \psi}{\partial III}, \quad f_1 = 2\rho \left(\frac{\partial \psi}{\partial I} + I \frac{\partial \psi}{\partial II} \right), \quad f_2 = -2\rho \frac{\partial \psi}{\partial II}, \tag{7.12}$$

and the functions K depend on θ and on $I, II, III, g^2, \mathbf{g} \cdot \mathbf{Bg}$, and $\mathbf{g} \cdot \mathbf{B}^2 \mathbf{g}$.

PROOF If the constitutive equations of S satisfy both the principles of material frame-indifference and dissipation and S is thermoelastic (i.e., it does not exhibit frictional dissipation), then (6.14) becomes

$$\psi = \psi(\mathbf{C}, \theta),$$
$$\eta = -\frac{\partial \psi}{\partial \theta} = \eta(\mathbf{C}, \theta),$$
$$\mathbf{T}^{(e)} = 2\rho \mathbf{F} \frac{\partial \psi}{\partial \mathbf{C}} \mathbf{F}^T,$$
$$\mathbf{h} = \mathbf{F}\hat{\mathbf{h}}(\mathbf{C}, \theta, \boldsymbol{\Theta}). \tag{7.13}$$

Referring to (7.13)$_1$, we find that the isotropy assumption (7.2) applied to the free energy ψ,

$$\psi(\mathbf{F}, \theta) = \psi(\mathbf{FQ}, \theta) \quad \forall \mathbf{Q} \in O(3), \tag{7.14}$$

gives (see (7.7))

$$\psi(\mathbf{C}, \theta) = \psi(\mathbf{Q}^T \mathbf{CQ}, \theta) \quad \forall \mathbf{Q} \in O(3). \tag{7.15}$$

This relation, despite the immaterial exchange of $\mathbf{Q}^T$ with $\mathbf{Q}$, is coincident with (7.8), so that the function $\psi(\mathbf{C}, \theta)$ is isotropic.

From (7.10)$_1$, (7.11)$_{1,2}$ follow.

From (6.14)$_2$, (3.52), and (3.53) it follows that

$$\mathbf{T} = 2\rho \mathbf{F} \left(\frac{\partial \psi}{\partial I} \frac{\partial I}{\partial \mathbf{C}} + \frac{\partial \psi}{\partial II} \frac{\partial II}{\partial \mathbf{C}} + \frac{\partial \psi}{\partial III} \frac{\partial III}{\partial \mathbf{C}} \right) \mathbf{F}^T$$
$$= 2\rho \mathbf{F} \left(\frac{\partial \psi}{\partial I} \mathbf{I} + \frac{\partial \psi}{\partial II} (II - \mathbf{C}) + \frac{\partial \psi}{\partial III} III \left(\mathbf{C}^2 - I_C \mathbf{C} + II_C \mathbf{I} \right)^T \right) \mathbf{F}^T, \tag{7.16}$$

and since $\mathbf{C} = \mathbf{F}^T \mathbf{F}$, $\mathbf{B} = \mathbf{FF}^T$, and $\mathbf{B}^2 = \mathbf{FCF}^T$, (7.11)$_3$ is proved, also taking into account (7.12).

To prove (7.11)$_4$, we see that the isotropy requirement (7.2) applied to **h** gives (7.13)$_4$ the form (see (7.7))

$$\mathbf{F}\widehat{\mathbf{h}}(\mathbf{C}, \theta, \boldsymbol{\Theta}) = \mathbf{F}\mathbf{Q}\widehat{\mathbf{h}}(\mathbf{Q}^T\mathbf{C}\mathbf{Q}, \theta, \mathbf{Q}^T\boldsymbol{\Theta}),$$

from which it follows that

$$\mathbf{Q}^T\widehat{\mathbf{h}}(\mathbf{C}, \theta, \boldsymbol{\Theta}) = \widehat{\mathbf{h}}(\mathbf{Q}^T\mathbf{C}\mathbf{Q}, \theta, \mathbf{Q}^T\boldsymbol{\Theta}).$$

This expression, despite the exchange of $\mathbf{Q}^T$ with $\mathbf{Q}$, is coincident with (7.9)$_1$. From (7.10) we can derive

$$\mathbf{h} = \mathbf{F}(h_0\mathbf{I} + h_1\mathbf{C} + h_2\mathbf{C}^2)\boldsymbol{\Theta},$$

and by substituting $\boldsymbol{\Theta} = \mathbf{F}^T\mathbf{g}$ and $\mathbf{B}^2 = \mathbf{F}\mathbf{C}\mathbf{F}^T$, relations (7.11)$_4$ are obtained.

Finally, we observe that it is an easy exercise to verify that equations (7.11) satisfy the isotropy requirement (7.2). ∎

If

- the transformation $C_* \to C$ is infinitesimal, and
- in the reference configuration C_* the system is elastic, homogeneous at a constant and uniform temperature, and
- C_* is a stress-free state,

then (7.11)$_3$ allows us to derive the constitutive equation of a ***linear isotropic solid***.

To do this, we first note that the following relations hold: $\mathbf{B} = \mathbf{F}\mathbf{F}^T = (\mathbf{I} + \mathbf{H})(\mathbf{I} + \mathbf{H})^T \simeq \mathbf{I} + 2\mathbf{E}$, $\mathbf{B}^2 = \mathbf{I} + 4\mathbf{E}$.

If, in addition, we assume that f_0, f_1, f_2 can be expanded in power series of principal invariants in the neighborhood of the identity matrix $\mathbf{I}$, then by neglecting second-order terms of $\mathbf{H}$ we get the condition

$$f_i \simeq a_i + b_i(I - 3) \simeq a_i + 2b_i \mathrm{tr}\mathbf{E}, \quad (a_1 + a_2 + a_3 = 0), \tag{7.17}$$

from which it follows that

$$\mathbf{T} = \lambda(\mathrm{tr}\,\mathbf{E})\mathbf{I} + 2\mu\mathbf{E}, \tag{7.18}$$

where the coefficients λ and μ are called the ***Lamé coefficients***.

We also have

$$\mathrm{tr}\,\mathbf{T} = (3\lambda + 2\mu)\mathrm{tr}\,\mathbf{E},$$

so that, if $(3\lambda + 2\mu) \neq 0$, combining with (7.18) leads to the inverse relation

$$\mathbf{E} = \frac{1}{2\mu}\mathbf{T} - \frac{\lambda}{2\mu(3\lambda + 2\mu)}(\mathrm{tr}\,\mathbf{T})\mathbf{I}. \tag{7.19}$$

7.3. Perfect and Viscous Fluids

To give these coefficients a physical interpretation, we suppose the system S to be subjected to a uniform traction of intensity t along the direction parallel to the base unit vector $\mathbf{e}_1$, so that $\mathbf{t}_{\mathbf{e}_1} = \mathbf{T}\mathbf{e}_1 = t\mathbf{e}_1$. By considering the stress tensor components relative to the basis $(\mathbf{e}_1, \mathbf{e}_2, \mathbf{e}_3)$, we find that the only nonvanishing component is $T_{11} = t$. From (7.19) we derive the deformation components

$$E_{11} = \frac{\lambda + \mu}{\mu(3\lambda + 2\mu)} t,$$

$$E_{22} = E_{33} = -\frac{\lambda}{2\mu(3\lambda + 2\mu)} t,$$

$$E_{12} = E_{13} = E_{23} = 0.$$

The quantities

$$E_Y = \frac{t}{E_{11}} = \frac{\mu(3\lambda + 2\mu)}{\lambda + \mu},$$

$$\sigma = -\frac{E_{22}}{E_{11}} = -\frac{E_{33}}{E_{11}} = \frac{\lambda}{2(\lambda + \mu)}, \tag{7.20}$$

are called **Young's modulus** and **Poisson's ratio**, respectively. Their meaning is as follows:

- Young's modulus indicates the ratio between the traction per unit surface area and the linear dilation produced in the same direction;

- Poisson's ratio is the ratio between the contraction, in the direction orthogonal to $\mathbf{t}$, and the dilation along $\mathbf{t}$.

More generally, when thermal phenomena are also considered, linearization of (7.11) gives

$$\mathbf{T} = \lambda(\theta_0)(\operatorname{tr} \mathbf{E})\mathbf{I} + 2\mu(\theta_0)\mathbf{E} - \beta(\theta - \theta_0)\mathbf{I},$$
$$\mathbf{h} = -K(\theta_0)\mathbf{g}, \tag{7.21}$$

where the function $K(\theta_0)$, due to (6.19), is positive:

$$K(\theta_0) > 0. \tag{7.22}$$

7.3 Perfect and Viscous Fluids

A thermoviscoelastic system S whose constitutive equation is given by $\mathbf{A}(\mathbf{F}, \dot{\mathbf{F}}, \theta, \Theta)$ is a **thermoviscous fluid** if

$$G_{C_*} = Unim \quad \forall C_*. \tag{7.23}$$

We note that (7.23) indicates that local changes of shape do not modify the response of the system, according to experimental evidence. Moreover, since the reference configuration is arbitrary, fluids do not lose their symmetry properties during deformation.

Theorem 7.3
A thermoelastic system S with constitutive equations that satisfy the principles of dissipation and material frame-indifference represents a thermoviscous fluid if and only if

$$\psi = \psi(\rho, \theta),$$
$$\eta = -\frac{\partial \psi}{\partial \theta} = \eta(\rho, \theta),$$
$$\mathbf{T}^{(e)} = -\rho^2 \frac{\partial \psi}{\partial \rho} \mathbf{I},$$
$$\mathbf{T}^{(d)} = \mathbf{f}(\rho, \mathbf{D}, \theta, \mathbf{g}),$$
$$\mathbf{h} = (K_0 \mathbf{I} + K_1 \mathbf{D} + K_2 \mathbf{D}^2)\mathbf{g}, \tag{7.24}$$

where ρ is the density, $\mathbf{f}$ is a tensor function of $\mathbf{D}$ and $\mathbf{g}$, and functions K depend on ρ, θ, and on invariants g^2, $\mathbf{g} \cdot \mathbf{Dg}$, and $\mathbf{g} \cdot \mathbf{D}^2 \mathbf{g}$.

PROOF Symmetry properties allow us to write

$$\psi(\mathbf{FH}, \theta) = \psi(\mathbf{F}, \theta) \quad \forall \mathbf{H} \in \text{Unim},$$

and this equality, taking into account $(6.14)_1$ and $(7.7)_3$, can also be written in the form

$$\psi(\mathbf{H}^T \mathbf{CH}, \theta) = \psi(\mathbf{C}, \theta) \quad \forall \mathbf{H} \in \text{Unim}. \tag{7.25}$$

Since $\det \mathbf{H} = 1$, we can put $\mathbf{H} = J\mathbf{F}^{-1}$, so that

$$\psi(J, \theta) = \psi(\mathbf{C}, \theta).$$

By reminding ourselves of the Lagrangian mass conservation, i.e., $\rho J = \rho_*$, as well as of the independence of constitutive equations of the reference configuration, we obtain

$$\psi(J, \theta) = \psi(\rho_*/\rho, \theta) = \psi(\rho, \theta),$$

so that $(7.24)_1$ is proved.

Moreover, if we put $\mathcal{I} = J^2$, then from $(6.14)_3$ it follows that

$$\mathbf{T}^{(e)} = 2\rho \mathbf{F} \frac{\partial \psi}{\partial J} \frac{d\sqrt{\mathcal{I}}}{dJ} \mathbf{F}^T = \frac{\rho}{J} \mathbf{F} \frac{\partial \psi}{\partial J} \frac{d\sqrt{\mathcal{I}}}{dJ} \mathbf{F}^T.$$

7.3. Perfect and Viscous Fluids

This condition, by recalling (3.53), can be written as

$$\mathbf{T}^{(e)} = \rho J \frac{\partial \psi}{\partial J}\mathbf{I} = \rho J \frac{\partial \psi}{\partial \rho}\frac{d\rho}{dJ}\mathbf{I} = -\rho^2 \frac{\partial \psi}{\partial \rho}\mathbf{I},$$

and $(7.24)_3$ is also proved. By using similar arguments $(7.24)_2$ follows.

To verify $(7.24)_4$, we first need to observe that the Eulerian rate of deformation tensor $\mathbf{D}$ is invariant with respect to a symmetry transformation. Then, by considering $(6.14)_4$, the symmetry property of a fluid

$$\mathbf{T}^{(d)}(\mathbf{FH}, \dot{\mathbf{F}}\mathbf{H}, \theta, \mathbf{H}^T \Theta) = \mathbf{T}^{(d)}(\mathbf{F}, \dot{\mathbf{F}}, \theta, \Theta)$$

can be put in the form

$$\mathbf{T}^{(d)}(\mathbf{H}^T \mathbf{CH}, \mathbf{H}^T \mathbf{F}^T \mathbf{DFH}, \theta, \mathbf{H}^T \Theta) = \mathbf{T}^{(d)}(\mathbf{C}, \mathbf{F}^T \mathbf{DF}, \theta, \Theta).$$

If $\mathbf{H} = J\mathbf{F}^{-1}$, it follows that

$$\mathbf{T}^{(d)}(J, J^2 \mathbf{D}, \theta, J\mathbf{g}) = \mathbf{T}^{(d)}(\mathbf{C}, \mathbf{F}^T \mathbf{DF}, \theta, \Theta)$$

and

$$\mathbf{T}^{(d)}(\mathbf{C}, \mathbf{F}^T \mathbf{DF}, \theta, \Theta) = \mathbf{T}^{(d)}(\rho, \mathbf{D}, \theta, \mathbf{g}).$$

Finally, from the expression of material frame-indifference

$$\mathbf{Q}\mathbf{T}^{(d)}(\rho, \mathbf{D}, \theta, \mathbf{g})\mathbf{Q}^T = \mathbf{T}^{(d)}(\rho, \mathbf{QDQ}^T, \theta, \mathbf{Qg}), \quad (7.26)$$

and $\mathbf{T}^{(d)}$ is proved to be an isotropic function. Finally, by considering the representation formula $(7.10)_2$, the relation $(7.24)_4$ is derived. In a similar way, $(7.24)_5$ can be verified.

The sufficient condition is rather obvious, since in (7.24) the constitutive variables are defined in the Eulerian representation so that they are invariant with respect to any unimodular transformation that preserves the mass of the reference configuration. ∎

In the absence of thermal phenomena, the constitutive equation of $\mathbf{T}^{(d)}$ assumes the form

$$\mathbf{T}^{(d)}(\rho, \mathbf{D}) = -p\mathbf{I} + k_1 \mathbf{D} + k_2 \mathbf{D}^2, \quad (7.27)$$

where p, k_1, k_2 are scalar functions of ρ and of the principal invariants I_D, II_D, III_D of $\mathbf{D}$.

If dissipation effects are also absent, then in (7.27) $k_1 = k_2 = 0$; the fluid is called **perfect** and the stress tensor has the form

$$\mathbf{T} = -\rho^2 \frac{\partial \psi}{\partial \rho}\mathbf{I} \equiv -p_0(\rho)\mathbf{I}. \quad (7.28)$$

In particular, if the components of **D** are first-order quantities and p, k_1, k_2 are analytical functions of I_D, II_D, III_D, it follows that

$$p(\rho) \simeq p_0(\rho) + \lambda(\rho)I_D,$$
$$k_1 \simeq 2\mu(\rho) + b(\rho)I_D,$$
$$k_2 \simeq 2\hat{\mu}(\rho) + c(\rho)I_D.$$

By substituting into (7.27) and neglecting higher order terms, we obtain the **Navier–Stokes behavior**

$$\mathbf{T} = -(p_0(\rho) + \lambda(\rho)I_D)\mathbf{I} + 2\mu(\rho)\mathbf{D}. \tag{7.29}$$

In this case, during the motion the fluid is subjected to shear stresses and to a pressure represented by a static component $p_0(\rho)$ and a dynamic component $\lambda(\rho)I_D$. If furthermore the fluid is *incompressible*, then $\nabla \cdot \mathbf{v} = I_D = 0$ and (7.29) becomes

$$\hat{\mathbf{T}}(\rho, \mathbf{D}) \simeq -p\mathbf{I} + 2\mu\mathbf{D}. \tag{7.30}$$

Due to the incompressibility constraint, the pressure p in (7.30) no longer depends on ρ, I_D, II_D, III_D.

By assuming that $\mathbf{Q} = -\mathbf{I}$ in (7.26) and in the isotropy condition of $\mathbf{h}$, we find that

$$\mathbf{T}^{(d)}(\rho, \mathbf{D}, \theta, \mathbf{g}) = \mathbf{T}^{(d)}(\rho, \mathbf{D}, \theta, -\mathbf{g}),$$
$$\mathbf{h}(\rho, \mathbf{D}, \theta, \mathbf{g}) = -\mathbf{h}(\rho, \mathbf{D}\theta, -\mathbf{g}),$$

and in particular

$$\mathbf{h}(\rho, \mathbf{D}, \theta, \mathbf{0}) = \mathbf{0}. \tag{7.31}$$

The relation (7.31) shows that the *heat flux vanishes with temperature gradient, for any* $\rho, \mathbf{D}$, *and* θ.

We note that this result, derived from symmetry properties of the fluid, is a more severe restriction than the one expressed by (6.20).

Finally, it can be proved [23] that the linearization in the neighborhood of an equilibrium state at constant temperature gives

$$\mathbf{T}^{(d)} = -\lambda(\rho, \theta)I_D\mathbf{I} + 2\mu\mathbf{D},$$
$$\mathbf{h} = -K(\rho, \theta)\mathbf{g}, \tag{7.32}$$

provided that terms of the order of

$$\sqrt{I_D^2 + g^2}$$

are neglected.

7.4 Anisotropic Solids

Materials such as natural crystals, porous media, fiberglass-reinforced plastics, and so on, are not isotropic. In general, they are said to be **anisotropic** because they possess some symmetry but the symmetry group does not contain the complete orthogonal group. Among anisotropic materials, crystals deserve special attention.

Let S be a hyperelastic solid and let G_{C_*} be its symmetry group, which is a *proper subgroup* of $O(3)$ in an undistorted reference configuration C_*. It can be proved,[2] starting from (7.15), that if ψ is an *analytic function* of the components of $\mathbf{C}$, then it can be expressed as a polynomial in the invariants $I_1, ..., I_n$ of $\mathbf{C}$

$$\psi = \psi(I_1, ..., I_n). \tag{7.33}$$

In the case of an isotropic material (as already seen in the previous section), the invariants $I_1, \ldots, I_n$ coincide with I_B, II_B, III_B.

From the definition (6.14) of an elastic material, it follows that

$$\mathbf{T} = 2\rho \mathbf{F} \frac{\partial \psi}{\partial I_\alpha} \frac{\partial I_\alpha}{\partial \mathbf{C}} \mathbf{F}^T \equiv 2\rho \frac{\partial \psi}{\partial I_\alpha} \mathbf{k}, \tag{7.34}$$

where the function $\mathbf{k}$ is independent of the form of the elastic energy but is determined by the material symmetries and the coordinates used to describe these symmetries.

Consequently, the constitutive equations of different elastic materials, exhibiting the same symmetry, differ only for the functions $\partial \psi / \partial I_\alpha$.

In the following discussion, the symbolic notation $\mathbf{R}_\mathbf{n}^\varphi$ is used to indicate a counterclockwise rotation of magnitude φ about the unit vector $\mathbf{n}$.

It can be shown that there are 12 groups of symmetries and 11 of them include 32 crystalline classes. The last group describes the so-called transverse isotropy (see Exercise 3), G_{C_*} being given by the identity and rotations $\mathbf{R}_\mathbf{n}^\varphi$, $0 < \varphi < 2\pi$ about a convenient unit vector $\mathbf{n}$.

In any crystalline class, in its undistorted state there are three preferred directions defined by unit vectors $\mathbf{u}_i$ ($i = 1, 2, 3$). A system of rectilinear coordinates $\mathbf{X}$ can be associated with these directions. The material symmetry is described by the orthogonal subgroup G_{C_*} which transforms these coordinates into new ones $\mathbf{X}'$, the energy form being preserved in the transformation.

To define the symmetry group G_{C_*} of a solid it is sufficient to single out a set of elements of G_{C_*}, called the generators of G_{C_*}, such that by composing and inverting all of them, the whole group G_{C_*} is obtained.

---
[2] The interested reader should refer to [24], [25].

It can be proved that the following linear transformations are present in all the generator systems of crystalline classes:

$$\mathbf{I}, \mathbf{R}_{\mathbf{u}_i}^{\pi}, \mathbf{R}_{\mathbf{u}_i}^{\pi/2}, \mathbf{R}_{\mathbf{u}_i}^{\pi/3}, \mathbf{R}_{\mathbf{u}_i}^{2\pi/3}.$$

Table 7.1 lists the generators of symmetry groups of different crystalline classes. The last column in this table gives the total number of transformations contained in the group.

Crystal Class	Generators of G_{C_*}	Order of G_{C_*}
Triclinic system *(all classes)*	$\mathbf{I}$	2
Monoclinic system *(all classes)*	$\mathbf{R}_{\mathbf{u}_3}^{\pi}$	4
Rhombic system *(all classes)*	$\mathbf{R}_{\mathbf{u}_1}^{\pi}, \mathbf{R}_{\mathbf{u}_2}^{\pi}$	8
Tetragonal system		
(3 classes)	$\mathbf{R}_{\mathbf{u}_3}^{\pi/2}$	8
(4 classes)	$\mathbf{R}_{\mathbf{u}_3}^{\pi/2}, \mathbf{R}_{\mathbf{u}_1}^{\pi}$	16
Cubic system		
(2 classes)	$\mathbf{R}_{\mathbf{u}_1}^{\pi}, \mathbf{R}_{\mathbf{u}_2}^{\pi}, \mathbf{R}_{\mathbf{p}}^{2\pi/3}$	24
(3 classes)	$\mathbf{R}_{\mathbf{u}_1}^{\pi/2}, \mathbf{R}_{\mathbf{u}_2}^{\pi/2}, \mathbf{R}_{\mathbf{u}_3}^{\pi/2}$	48
Hexagonal system		
(2 classes)	$\mathbf{R}_{\mathbf{p}}^{2\pi/3}$	6
(3 classes)	$\mathbf{R}_{\mathbf{u}_1}^{\pi}, \mathbf{R}_{\mathbf{u}_3}^{2\pi/3}$	12
(3 classes)	$\mathbf{R}_{\mathbf{u}_3}^{\pi/3}$	12
(4 classes)	$\mathbf{R}_{\mathbf{u}_1}^{\pi}, \mathbf{R}_{\mathbf{u}_3}^{\pi/3}$	24

Table 7.1

In Table 7.1, $\mathbf{p} = (\mathbf{u}_1 + \mathbf{u}_2 + \mathbf{u}_3)/\sqrt{3}$.

The complete set of polynomial bases $I_1, \ldots, I_n$ for the crystalline bases is given in [26] and [25]. In the following we consider only the first three systems.[3]

1. Triclinic system: the polynomial basis is given by all components of $\mathbf{C}$;

---

[3]Here we use Voigt's notations, according to which the indices of the components of the elasticity tensor are fixed as follows,

$$11 \to 1,\ 22 \to 2,\ 33 \to 3, 23 \to 4,\ 13 \to 5,\ 12 \to 6,$$

so that, for instance, $C_{1323} = C_{54}$.

2. Monoclinic system: the polynomial basis is given by the following 7 components:

$$C_{11}, \quad C_{22}, \quad C_{33}, \quad C_{12}^2, \quad C_{13}^2, \quad C_{23}, \quad C_{13}C_{12}.$$

3. Rhombic system: the polynomial basis is given by the following 7 components:

$$C_{11}, \quad C_{22}, \quad C_{33}, \quad C_{23}^2, \quad C_{13}^2, \quad C_{12}^2, \quad C_{12}C_{23}C_{13}.$$

In closing this section, we note the following: given a material of a crystal class with a deformation process that starts from an undistorted configuration, its free energy can be expressed as a polynomial of a convenient degree in the variables forming the polynomial basis. Finally, the stress–strain relationship can be deduced from (7.34).

7.5 Exercises

1. By using the symmetry properties, show that the strain energy function of an isotropic solid is characterized by only two independent constants.

   Hint: If the solid is isotropic, then any plane is a symmetry plane as well as any axis is a symmetry axis under any rotation, so that the following relations must hold for the nonvanishing coefficients:

   $$C_{11} = C_{22} = C_{33}$$
   $$C_{12} = C_{23} = C_{31}$$
   $$C_{44} = C_{55} = C_{66} = \frac{1}{2}(C_{11} - C_{12}),$$

   and the strain energy function assumes the form

   $$\Psi = \frac{1}{2}C_{11}\left(E_{11}^2 + E_{22}^2 + E_{33}^2\right)$$
   $$+ C_{12}\left(E_{11}E_{22} + E_{11}E_{33} + E_{22}E_{33}\right)$$
   $$+ \frac{1}{4}(C_{11} - C_{12})\left(E_{12}^2 + E_{13}^2 + E_{23}^2\right).$$

2. Using the properties of isotropic tensors, prove that only two constants are required to characterize the material response of an isotropic solid.

Isotropic tensors are such that their rectangular components are independent of any orthogonal transformation of coordinate axes.

According to this definition, all scalars (i.e., tensors of order zero) are isotropic, whereas vectors (first-order tensors) are not isotropic. The unit tensor, whose components are represented in rectangular coordinates by the Kronecker delta δ_{ij}, is isotropic. It is obvious that any scalar multiple of the unit tensor has the same property. These particular tensors exhaust the class of isotropic second-order tensors.

The most general fourth-order isotropic tensor is represented in the form
$$C_{ijhk} = \lambda \delta_{ij} \delta_{hk} + \mu \left(\delta_{ih} \delta_{jk} + \delta_{ik} \delta_{jh} \right) + \kappa \left(\delta_{ih} \delta_{jk} - \delta_{ik} \delta_{jh} \right).$$

Suppose that C_{ijhk} are the coefficients of the linear material response
$$T_{ij} = C_{ijhk} E_{hk},$$

where one or both the tensors T_{ij} and E_{hk} is symmetric. In either case, the requirement for C_{ijhk} to be symmetric with respect to ij or hk implies that $\kappa = 0$. It follows that
$$C_{ijhk} = \lambda \delta_{ij} \delta_{hk} + \mu \left(\delta_{ih} \delta_{jk} + \delta_{ik} \delta_{jh} \right)$$
is the most general expression of the coefficients of a linearized elastic isotropic solid and only two constants are needed to characterize the material properties (as found in Exercise 1).

3. Prove that the material response of a *transversely isotropic solid* is characterized by 5 independent constants.

    Hint: A transversely isotropic solid is symmetric with respect to some axis. If Ox_3 denotes this axis, then the rotation through a small angle α about Ox_3 is written as
    $$X_1^* = X_1 \cos \alpha + X_2 \sin \alpha \simeq X_1 + X_2 \alpha,$$
    $$X_2^* = X_2 \cos \alpha - X_1 \sin \alpha \simeq X_2 - X_1 \alpha,$$
    $$X_3^* = X_3.$$

    The strain energy Ψ must be invariant with respect to this transformation, so that if Ψ is expressed in terms of the strain components in the new coordinates, then all the coefficients of α must vanish.

    The result is that the number of elastic constants reduces to 5.

4. By substituting the first equality (3.53) into (7.11), prove that
    $$\mathbf{T} = f_0 \mathbf{I} + f_1 \mathbf{B} + f_{-1} \mathbf{B}^{-1},$$
    $$\mathbf{h} = (K_0 \mathbf{I} + K_{-1} \mathbf{B} + K_{-2} \mathbf{B}^{-1}) \mathbf{g},$$
    and find formulae corresponding to (7.12).

7.6 The Program LinElasticityTensor

Aim of the Program

The program `LinElasticityTensor` applies to isotropic linear materials or to anisotropic ones belonging to the triclinic, monoclinic, and rombic crystal classes. It determines the components of the elasticity tensor and the components of the linear deformation tensor **E**. Moreover, all these expressions are given following Voigt's notation.

Description of the Problem and Relative Algorithm

The program distinguishes between isotropic solids, for which $\psi = \psi(I_\mathbf{C}, II_\mathbf{C})$, and anisotropic solids, for which $\psi = \psi(I_1, \cdots, I_n)$, where $I_1, \cdots, I_n$ are the invariants of the crystal class under consideration. If the reference configuration C_* is unstressed and the state of the solid is isothermic, then the constitutive equations have the following structure:

$$\mathbf{T} = \mathbf{CE},$$

where **E** is the tensor of infinitesimal strain and C is the tensor of linear elasticity. Moreover, we know that

$$C_{ijhk} = 4\rho_* \left(\frac{\partial^2 \psi}{\partial C_{ij} \partial C_{hk}} \right),$$
$$C_{ijhk} = C_{jihk} = C_{ijkh},$$
$$C_{ijhk} = C_{hkij}.$$

On the other hand, it is possible to reduce the 81 components of C to 21 independent components.[4] If we adopt *Voigt's notation*, which uses the index transformation

$$ii \longrightarrow i, \; i = 1, 2, 3$$
$$12 \longrightarrow 6,$$
$$13 \longrightarrow 5,$$
$$23 \longrightarrow 4,$$

then the elasticity tensor can be reduced to the form $C = (C_{\alpha\beta})$, where $\alpha, \beta = 1, \cdots, 6$. If the same notation is used for **E**, then the elastic potential and the stress tensor can be written as

$$\psi = \frac{1}{2\rho_*} C_{\alpha\beta} E_\alpha E_\beta,$$
$$\mathbf{T} = C \cdot \mathbf{E}.$$

---
[4] More precisely, the first group of the above relations allows us to reduce 81 components of C to 36 independent components, and the second group reduces these to 21.

Command Line of the Program LinElasticityTensor

LinElasticityTensor[class]

Parameter List
Input Data

class   characterizes the solid and can assume the values isotropic, triclinic, monoclinic, or rombic.

Output Data

Elastic potential ψ as a function of the principal invariants of **E** or of the invariants of the crystal class to which the solid belongs;

the components of C in Voigt's notation;

the expression of ψ in Voigt's notation;

the stress tensor **T** in Voigt's notation.

Worked Examples

1. **Isotropic Solid.** To apply the program LinElasticityTensor to the class of linear and isotropic materials, we have to input the following data:

    class = isotropic;

    LinElasticityTensor[class]

    The corresponding output is

    Principal invariants in linear elasticity

    B = {I_E, II_E}

    Elastic potential as a function of the principal invariants

    $$\psi = a_{1,0}(E_{1,1} + E_{2,2} + E_{3,3}) + a_{2,0}(E_{1,1} + E_{2,2} + E_{3,3})^2$$
    $$+ a_{0,1}(-E_{1,2}^2 - E_{1,3}^2 + E_{1,1}E_{2,2} - E_{2,3}^2 + E_{1,1}E_{3,3} + E_{2,2}E_{3,3})$$

    Independent components of linear elasticity tensor
    C = (C_{ijhk}) in Voigt's notation

    $C_{1,1} = 8\rho_* a_{2,0}$

    $C_{1,2} = 4\rho_*(a_{0,1} + 2a_{2,0})$

    $C_{1,3} = 4\rho_*(a_{0,1} + 2a_{2,0})$

7.6. The Program LinElasticityTensor

$C_{2,2} = 8\rho_* a_{2,0}$

$C_{2,3} = 4\rho_* (a_{0,1} + 2a_{2,0})$

$C_{3,3} = 8\rho_* a_{2,0}$

$C_{4,4} = -8\rho_* a_{0,1}$

$C_{5,5} = -8\rho_* a_{0,1}$

$C_{6,6} = -8\rho_* a_{0,1}$

Elastic potential E_i in Voigt's notation

$\psi = -4E_4^2 a_{0,1} - 4E_5^2 a_{0,1} - 4E_6^2 a_{0,1} + 4E_1^2 a_{2,0} + 4E_2^2 a_{2,0} + 4E_3^2 a_{2,0}$
$\quad + 4E_1 E_2 (a_{0,1} + 2a_{2,0}) + 4E_1 E_3 (a_{0,1} + 2a_{2,0}) + 4E_2 E_3 (a_{0,1} + 2a_{2,0})$

Stress tensor components

$T_1 = 8\rho_* E_1 a_{2,0} + 4\rho_* E_2 (a_{0,1} + 2a_{2,0}) + 4\rho_* E_3 (a_{0,1} + 2a_{2,0})$

$T_1 = 8\rho_* E_2 a_{2,0} + 4\rho_* E_1 (a_{0,1} + 2a_{2,0}) + 4\rho_* E_3 (a_{0,1} + 2a_{2,0})$

$T_1 = 8\rho_* E_3 a_{2,0} + 4\rho_* E_1 (a_{0,1} + 2a_{2,0}) + 4\rho_* E_2 (a_{0,1} + 2a_{2,0})$

$T_1 = -8\rho_* E_4 a_{0,1}$

$T_1 = -8\rho_* E_5 a_{2,0}$

$T_1 = 8\rho_* E_6 a_{2,0}$

Exercises

Apply the program LinElasticityTensor to anisotropic linear elastic materials belonging to the following crystal classes:

1. monoclinic;

2. triclinic;

3. rombic.

Chapter 8

Wave Propagation

8.1 Introduction

The evolution of a physical system is said to have *wave behavior* if a character of *propagation* or *oscillation* or both can be attributed to it. As an example, points of a string with fixed boundary ends, oscillating in a stationary way, have an oscillatory character without propagation (see Exercise 1). On the other hand, if we consider an infinite string with a single fixed end that is subjected to an initial perturbation, then its motion has only the character of propagation (see Exercise 2), which is eventually damped if dissipation effects are present (see Exercise 3). Furthermore, when considering waves produced by a falling stone in a lake, we can distinguish the damped oscillations of material points of the surface from the propagation of the surface separating the disturbed region from the undisturbed region.

With such a premise in mind, we consider a physical phenomenon described by fields depending on position $\mathbf{x}$ and time t. Moreover, we suppose it to satisfy a system of partial differential equations. It is of interest to investigate if this system can predict the evolution of wave-type phenomena. Apparently, the answer would require that we know the solutions of the system, because their properties could allow us to identify the presence of propagation or oscillation effects. However, this approach cannot usually be pursued, owing to the complexity of the equations of mathematical physics.

If the system is *linear* and *dissipation effects* are absent, we can look for sinusoidal solutions. In fact, because of the linearity, any superposition of such solutions is still a solution of the system so that, taking advantage of the Fourier method, the most general solution can be determined. Such a procedure can no longer be applied if the system is *nonlinear*, so that we need to introduce a more general approach, due to Hadamard, which considers *wavefronts* as *singular surfaces of Cauchy's problem*. Such an approach can be applied to any system, and, for linear systems, it gives the same results as the Fourier method.

In the present chapter we first turn our attention to quasi-linear equations and then to quasi-linear systems of partial differential equations (PDEs) in order to show under which circumstances there are **characteristic surfaces** for which Cauchy's problem is ill posed. A classification of the equations and systems is given, based on the existence and the number of these surfaces. Then it is proved that such characteristic surfaces are singular surfaces for the solutions of the system. In particular, when the independent variables reduce to the usual ones, i.e., spatial and time variables, we prove the existence of a moving surface which has a speed of propagation that can be computed together with its evolution. This surface will be interpreted as a wavefront.

Finally, the last section of this chapter deals with shock waves and highlights the role of the second law of thermodynamics in the description of the phenomenon.

8.2   Cauchy's Problem for Second-Order PDEs

In a domain $\Omega \subseteq \Re^n$, the second-order **quasi-linear** partial differential equation

$$a_{ij}(\mathbf{x}, u, \nabla u)\frac{\partial^2 u}{\partial x_i \partial x_j} = h(\mathbf{x}, u, \nabla u), \quad i,j = 1, \ldots, n, \quad \mathbf{x} \in \Omega, \quad (8.1)$$

is given. In particular, if the coefficients a_{ij} are independent of u and of ∇u, then the equation is called **semilinear**; when, in addition to this condition, the function h depends linearly on u and ∇u, the equation is called **linear**.

Then, with reference to equation (8.1), the following **Cauchy's problem** is stated: If Σ_{n-1} is a regular and orientable $(n-1)$-dimensional hypersurface contained in Ω and $\mathbf{N}$ denotes the unit vector field normal to Σ_{n-1}, find a solution $u(\mathbf{x})$ of (8.1) in Ω which satisfies on Σ_{n-1} the following Cauchy's data:

$$u(\mathbf{x}) = u_0(\mathbf{x}), \quad \frac{du}{dn} \equiv \nabla u \cdot \mathbf{N} = d_0(\mathbf{x}), \quad \mathbf{x} \in \Sigma_{n-1}, \quad (8.2)$$

where $u_0(\mathbf{x})$ and $d_0(\mathbf{x})$ are assigned functions on Σ_{n-1} corresponding to the values of $u(\mathbf{x})$ and its normal derivative on this hypersurface.

A first step in discussing such a problem consists of introducing a convenient coordinate system $(\nu_1, \ldots, \nu_n)$, the **Gauss coordinates**, in the neighborhood of Σ_{n-1}, in order to simplify its formulation.

Given a Cartesian coordinate system $(O, \mathbf{u}_i)$ in $\Re^n$, let

$$x_i = r_i(\nu_1, \ldots, \nu_{n-1}), \quad i = 1, \ldots, n,$$

8.2. Cauchy's Problem for Second-Order PDEs

be the parametric equations of the surface Σ_{n-1}. In a neighborhood of Σ_{n-1} in $\Re^n$, the system of functions

$$x_i = r_i(\nu_1, \ldots, \nu_{n-1}) + \nu_n N_i \tag{8.3}$$

is considered, where ν_n is the distance of an arbitrary point of the normal from Σ_{n-1} (see Figure 8.1). Accordingly, the condition $\nu_n = 0$ again gives the equation of Σ_{n-1}.

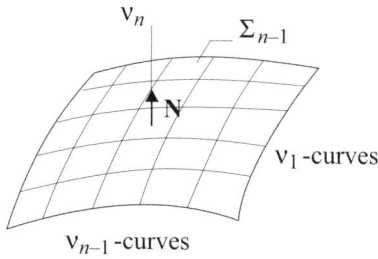

Figure 8.1

The system (8.3) defines a coordinate transformation $(\nu_1, \ldots, \nu_n) \longleftrightarrow (x_1, \ldots, x_n)$ in a neighborhood of Σ_{n-1}, and the Jacobian of (8.3), evaluated for $\nu_n = 0$, is

$$J = \begin{pmatrix} \dfrac{\partial x_1}{\partial \nu_1} & \cdots & \dfrac{\partial x_1}{\partial \nu_{n-1}} & N_1 \\ \cdots & \cdots & \cdots & \cdots \\ \dfrac{\partial x_n}{\partial \nu_1} & \cdots & \dfrac{\partial x_n}{\partial \nu_{n-1}} & N_n \end{pmatrix}.$$

We also remark that the first $n-1$ column vectors of J,

$$\mathbf{e}_i = \frac{\partial x_1}{\partial \nu_i}\mathbf{u}_1 + \cdots + \frac{\partial x_n}{\partial \nu_i}\mathbf{u}_n, \quad i = 1, \ldots, n-1,$$

are tangent to the $n-1$ coordinate curves on Σ_{n-1} and independent, since the surface is regular.

By noting that $\mathbf{N}$ is normal to Σ_{n-1}, so that it is independent of the vectors $(\mathbf{e}_i)$, we see that the determinant of J is different from zero, so that the inverse transformation

$$\nu_i = \nu_i(x_1, \ldots, x_n) \tag{8.4}$$

can be obtained from (8.3). Relative to the coordinates ν_i, the equation of the hypersurface Σ_{n-1} becomes

$$\nu_n(x_1, \ldots, x_n) \equiv f(x_1, \ldots x_n) = 0, \tag{8.5}$$

and Cauchy's data (8.2) assume the simplified form

$$u(\nu_1,\ldots,\nu_{n-1}) = u_0(\nu_1,\ldots,\nu_{n-1}), \quad \frac{\partial u}{\partial \nu_n} = d_0(\nu_1,\ldots,\nu_{n-1}). \quad (8.6)$$

Remark It is worth noting that the knowledge of the directional derivative of the function $u(\mathbf{x})$ along the normal to Σ_{n-1} gives the gradient of $u(\mathbf{x})$ in any point of Σ_{n-1}. In fact, on Σ_{n-1} it holds that

$$\nabla u = \sum_{i=1}^{n-1} \frac{\partial u}{\partial \nu_i} \mathbf{e}^i + \frac{\partial u}{\partial \nu_n} \mathbf{N},$$

where $(\mathbf{e}^i)$ are the vectors reciprocal to the $(n-1)$ tangent vectors to Σ_{n-1}.

The *Cauchy–Kovalevskaya theorem* introduced below is of basic relevance, since it asserts the *local existence* of solutions of a system of PDEs, with initial conditions on a *noncharacteristic surface*. Its proof is not straightforward and, for this reason, only particular aspects related to the wave propagation will be addressed.

Theorem 8.1 (Cauchy–Kovalevskaya theorem)
If the coefficients a_{ij}, the function u, the Cauchy data (8.2) and the implicit representation (8.5) of Σ_{n-1} are analytic functions of their arguments and, in addition, Σ_{n-1} satisfies the condition

$$a_{ij}(\mathbf{x}, u, \nabla u)\frac{\partial f}{\partial x_i}\frac{\partial f}{\partial x_j} \neq 0, \; \mathbf{x} \in \Sigma_{n-1}, \quad (8.7)$$

then there exists a unique analytic solution of the Cauchy problem (8.1)–(8.2) in a neighborhood of Σ_{n-1}.

PROOF Let us assume the coordinates $(\nu_1,\ldots,\nu_n)$ in $\Re^n$. The following expansion of u in a neighborhood of Σ_{n-1} can be written:

$$u(\mathbf{x}) = u_0(\nu_1,\ldots,\nu_{n-1},0) + \left(\frac{\partial u}{\partial \nu_n}\right)_{\mathbf{x}_0}\nu_n$$

$$+ \frac{1}{2}\left(\frac{\partial^2 u}{\partial \nu_n^2}\right)_{\mathbf{x}_0}\nu_n^2 + \cdots, \quad (8.8)$$

where $\mathbf{x}_0$ is any point of Σ_{n-1}. To prove the theorem, we need to show that

1. all the derivatives of u appearing in the above expansion can be determined by using (8.1) and Cauchy's data (8.2);

8.2. Cauchy's Problem for Second-Order PDEs

2. the series (8.8) converges uniformly towards a solution of the Cauchy problem (8.1), (8.2);

Below we only prove that, under the *assumption* (8.7), all the derivatives of the expansion (8.8) can be computed.

Let h and k be indices in the range $1, \ldots, n$; then

$$\frac{\partial u}{\partial x_i} = \frac{\partial u}{\partial \nu_h} \frac{\partial \nu_h}{\partial x_i}, \qquad (8.9)$$

$$\begin{aligned}
\frac{\partial^2 u}{\partial x_i \partial x_j} &= \frac{\partial}{\partial x_i} \left[\frac{\partial u}{\partial \nu_h} \frac{\partial \nu_h}{\partial x_j} \right] \\
&= \frac{\partial}{\partial x_i} \left[\frac{\partial u}{\partial \nu_h} \right] \frac{\partial \nu_h}{\partial x_j} + \frac{\partial u}{\partial \nu_h} \frac{\partial^2 \nu_h}{\partial x_i \partial x_j} \\
&= \frac{\partial^2 u}{\partial \nu_h \partial \nu_k} \frac{\partial \nu_h}{\partial x_j} \frac{\partial \nu_k}{\partial x_i} + \frac{\partial u}{\partial \nu_h} \frac{\partial^2 \nu_h}{\partial x_i \partial x_j}.
\end{aligned} \qquad (8.10)$$

By using the above expressions, we see that (8.1) becomes

$$\sum_{h,k=1}^n \left(a_{ij} \frac{\partial \nu_h}{\partial x_j} \frac{\partial \nu_k}{\partial x_i} \right) \frac{\partial^2 u}{\partial \nu_h \partial \nu_k} = g\left(\nu, u, \frac{\partial u}{\partial \nu}\right), \qquad (8.11)$$

where on the left-hand side we have collected the first derivatives of u with respect to ν_h and the second derivatives of ν_h with respect to x_i. We note that the latter are obtained by differentiating the known inverse transformation (8.4).

When the Cauchy data on Σ_{n-1}

$$u(\nu_1, \ldots, \nu_{n-1}) = u_0(\nu_1, \ldots, \nu_{n-1}), \qquad \frac{\partial u}{\partial \nu_n} = d_0(\nu_1, \ldots, \nu_{n-1}), \qquad (8.12)$$

are taken into account, we see that all the first derivatives of u at $\mathbf{x}_0 \in \Sigma_{n-1}$ are obtained by differentiating $(8.12)_1$ or from $(8.12)_2$. The second derivatives

$$\left(\frac{\partial^2 u}{\partial \nu_h \partial \nu_k}\right)_{\mathbf{x}_0}, \quad \left(\frac{\partial^2 u}{\partial \nu_h \partial \nu_n}\right)_{\mathbf{x}_0} = \left[\frac{\partial}{\partial \nu_h}\left(\frac{\partial u}{\partial \nu_n}\right)\right]_{\mathbf{x}_0},$$

where at least one of the indices h and k changes from 1 to $n-1$, are evaluated from Cauchy's data (8.12).

Finally, about the second derivative

$$\left(\frac{\partial^2 u}{\partial \nu_n^2}\right)_{\mathbf{x}_0},$$

we observe that the system (8.11), written at $\mathbf{x}_0 \in \Sigma_{n-1}$, when (8.5) is taken into account, allows us to obtain

$$\left[\left(a_{ij}\frac{\partial f}{\partial x_i}\frac{\partial f}{\partial x_j}\right)\frac{\partial^2 u}{\partial v_n^2}\right]_{\mathbf{x}_0} = F, \tag{8.13}$$

where F is a quantity which can be derived from Cauchy's data.

The relation (8.13) allows us to compute the desired derivative if

$$\left(a_{ij}\frac{\partial f}{\partial x_i}\frac{\partial f}{\partial x_j}\right)_{\mathbf{x}_0} \neq 0.$$

In addition, this condition allows us to compute the derivatives of higher order in $\mathbf{x}_0 \in \Sigma_{n-1}$ and to prove the remaining point (2). ∎

8.3 Characteristics and Classification of PDEs

An $(n-1)$-dimensional hypersurface Σ_{n-1}

$$f(x_1, \ldots, x_n) = 0$$

is said to be a **characteristic surface** with respect to the equation (8.1) and Cauchy's data (8.2) if $z = f(x_1, \ldots, x_n)$ is a solution of the equation

$$a_{ij}(x, u, \nabla u)\frac{\partial f}{\partial x_i}\frac{\partial f}{\partial x_j} = 0. \tag{8.14}$$

In such a case, the Cauchy problem is ill posed, in the sense that there is no uniqueness since there are several possibilities for computing the partial derivatives

$$\partial^r u/\partial v_n^r, \ r \geq 2$$

starting from the same Cauchy's data.

It is relevant to observe that, if the coefficients a_{ij} of (8.1) depend only on the coordinates x_i, equation (8.1) is linear or semilinear and the *characteristic surfaces depend on the equation but not on Cauchy's data*.

On the other hand, in the quasi-linear case (8.14) allows us to define the function $f(\mathbf{x})$ provided that the values of u and ∇u are known at any point. Since these quantities are uniquely defined from the equation and Cauchy's data, and are therefore continuous across the characteristic surface Σ_{n-1}, (8.14) can be regarded as an equation in the unknown $f(\mathbf{x})$ if the solution of (8.1) is known on at least one side of Σ_{n-1}. This is not

8.3. Characteristics and Classification of PDEs

a severe requirement, since a solution of (8.1) is often known in the form $u_0 = \text{const}$.

Accordingly, *when dealing with a quasi-linear equation, a solution u_0 from which we can compute the coefficients a_{ij} at any point $\mathbf{x}$ is supposed to be known.*

However, equation (8.14) is a first-order nonlinear partial differential equation in the unknown function $f(x_1, \ldots, x_n)$, so that its solution is not easy to find. As a consequence, it is not easy to determine the characteristic surfaces of (8.1).

Suppose that, for a fixed point $\mathbf{x} \in \Omega$, there are characteristic surfaces Σ_{n-1} containing $\mathbf{x}$. Since the components of the unit vector $\mathbf{N}$, normal to a solution Σ_{n-1} of (8.14), are $N_i = (\partial f / \partial x_i) / |\nabla f|$, equation (8.14) can be written as

$$a_{ij}(x, u_0, \nabla u_0) N_i N_j \equiv a_{ij}^0(\mathbf{x}) N_i N_j = 0, \qquad (8.15)$$

where the vectors $\mathbf{N}$ are normal to the characteristic surfaces through the point $\mathbf{x}$. The collection of the vectors verifying (8.15) give rise to a *cone* $A_{\mathbf{x}}$. In fact, $\mathbf{N} = \mathbf{0}$ is a solution of (8.15). In addition, if $\mathbf{N}$ is a solution, then so is $\mu \mathbf{N}$ provided that μ is real.

These arguments allow us to classify equations (8.1) at any $\mathbf{x} \in \mathcal{R}^n$ by considering the eigenvalues of the symmetric matrix $a_{ij}^0(\mathbf{x})$.

1. The differential equation (8.1) is called **elliptic** at $\mathbf{x}$ (or with respect to a solution u_0 if it is quasi-linear) if all the eigenvalues $\lambda_1, \ldots, \lambda_n$ of $a_{ij}^0(\mathbf{x})$ are positive or, equivalently, if the quadratic form

$$a_{ij}^0(\mathbf{x}) N_i N_j$$

is positive definite.

In this case, there exists a transformation $(x_i) \longrightarrow (\bar{x}_i)$, dependent on $\mathbf{x}$, which allows us to write *at the point $\mathbf{x}$* the quadratic form $a_{ij}^0(\mathbf{x}) N_i N_j$ in the canonical form

$$\lambda_1 \bar{N}_1^2 + \cdots + \lambda_n \bar{N}_n^2, \quad \lambda_1, \ldots, \lambda_n > 0.$$

We can also say that (8.1) is elliptic at $\mathbf{x}$ if there are no real vectors normal to the characteristic surfaces passing through $\mathbf{x}$; i.e., the cone $A_{\mathbf{x}}$ (8.15) at $\mathbf{x}$ is imaginary. So if all the eigenvalues of a_{ij} are positive, then there is no real solution of (8.14).

2. The differential equation (8.1) is called **parabolic** at $\mathbf{x}$ (or with respect to a solution u_0 if it is quasi-linear) if the matrix $a_{ij}^0(\mathbf{x})$ has at least one eigenvalue equal to zero. In this case there is a transformation $(x_i) \longrightarrow (\bar{x}_i)$, dependent on $\mathbf{x}$, which allows us to transform $a_{ij}^0(\mathbf{x}) N_i N_j$ into the canonical form

$$\lambda_1 \bar{N}_1^2 + \cdots + \lambda_m \bar{N}_m^2, \quad \lambda_1, \ldots, \lambda_m \neq 0, \quad \lambda_{m+1} = \cdots = \lambda_n = 0.$$

In this case the hyperplane $\bar{N}_1 = \cdots = \bar{N}_m = 0$ is contained in the cone $A_{\mathbf{x}}$.

3. The differential equation (8.1) is called **hyperbolic** in x (or with respect to a solution u_0 if it is quasi-linear) if all but one of the eigenvalues of the matrix $a_{ij}^0(\mathbf{x})$ have the same sign and the remaining one has the opposite sign. In this case there is a transformation $(x_i) \longrightarrow (\bar{x}_i)$, dependent on x, which allows us to transform us the quadratic form $a_{ij}^0(\mathbf{x}) N_i N_j$ into the canonical form

$$\lambda_1 \bar{N}_1^2 + \cdots + \lambda_{n-1} \bar{N}_{n-1}^2 - \lambda_n \bar{N}_n^2,$$

where the eigenvalues $\lambda_1, \ldots, \lambda_{n-1}$ have the same sign and λ_n takes the opposite sign. In this case $A_{\mathbf{x}}$ is a real cone.

8.4 Examples

1. Consider Laplace's equation

$$\Delta u \equiv \frac{\partial^2 u}{\partial x_1^2} + \frac{\partial^2 u}{\partial x_2^2} = 0$$

in the unknown function $u(x_1, x_2)$. The matrix a_{ij} is given by

$$a_{ij} = \begin{pmatrix} 1 & 0 \\ 0 & 1 \end{pmatrix}$$

with two coincident eigenvalues, equal to $\lambda = 1$. The equation is elliptic and the cone $A_{\mathbf{x}}$ is imaginary, since it is represented by the equation

$$N_1^2 + N_2^2 = 0.$$

The *characteristic curves* are given by

$$\left(\frac{\partial f}{\partial x_1} \right)^2 + \left(\frac{\partial f}{\partial x_2} \right)^2 = 0,$$

and this equation does not have real $f(x_1, x_2)$ solutions. Since Laplace's equation does not admit characteristic curves, the Cauchy problem is well posed for any curve of the x_1, x_2-plane.

2. As a second example, consider D'Alembert's equation (also called the wave equation)

$$\frac{\partial^2 u}{\partial x_1^2} - \frac{\partial^2 u}{\partial x_2^2} = 0.$$

8.4. Examples

The matrix a_{ij} is

$$a_{ij} = \begin{pmatrix} 1 & 0 \\ 0 & -1 \end{pmatrix},$$

its eigenvalues are $\lambda_1 = 1$ and $\lambda_2 = -1$, and the equation is hyperbolic. The cone $A_{\mathbf{x}}$ is written as

$$N_1^2 - N_2^2 = 0$$

and is formed by the two straight lines $N_1 - N_2 = 0$ and $N_1 + N_2 = 0$. The characteristic curves are the solutions of the equation

$$\left(\frac{\partial f}{\partial x_1}\right)^2 - \left(\frac{\partial f}{\partial x_2}\right)^2 = 0,$$

which is equivalent to the following system:

$$\left(\frac{\partial f}{\partial x_1} - \frac{\partial f}{\partial x_2}\right) = 0, \quad \left(\frac{\partial f}{\partial x_1} + \frac{\partial f}{\partial x_2}\right) = 0.$$

With the introduction of the vector fields $\mathbf{u}_1 = (1, -1)$ and $\mathbf{u}_2 = (1, 1)$, the previous system can be written as

$$\nabla f \cdot \mathbf{u}_1 = 0, \ \nabla f \cdot \mathbf{u}_2 = 0.$$

The above equations show that their solutions $f(x_1, x_2)$ are constant along straight lines parallel to the vectors $\mathbf{u}_1$ and $\mathbf{u}_2$, and, since this corresponds to the definition of characteristic curves, we conclude that the characteristic curves are represented in the plane Ox_1x_2 by families of straight lines

$$x_1 - x_2 = const, x_1 + x_2 = const.$$

In this case Cauchy problem is ill posed if Cauchy's data are assigned on these lines.

3. The heat equation

$$\frac{\partial^2 u}{\partial x_1^2} - \frac{\partial u}{\partial x_2} = 0$$

is an example of a parabolic equation, since the matrix

$$a_{ij} = \begin{pmatrix} 1 & 0 \\ 0 & 0 \end{pmatrix}$$

has eigenvalues $\lambda_1 = 1$ and $\lambda_2 = 0$. The cone $A_{\mathbf{x}}$ is defined at any point by

$$N_1^2 = 0$$

and coincides with a line parallel to the x_2 axis. The equation of the characteristic curves reduces to

$$\frac{\partial f}{\partial x_1} = 0,$$

with solutions

$$f = g(x_2) = const,$$

where g is an arbitrary function. It follows that the characteristic curves are represented by the lines

$$x_2 = const.$$

4. Finally we consider Tricomi's equation

$$\frac{\partial^2 u}{\partial x_1^2} + x_1 \frac{\partial^2 u}{\partial x_2^2} = 0.$$

Since the matrix

$$a_{ij} = \begin{pmatrix} 1 & 0 \\ 0 & x_1 \end{pmatrix}$$

has eigenvalues $\lambda_1 = 1$ and $\lambda_2 = x_1$, this equation is hyperbolic in those points of the plane Ox_1x_2 where $x_1 < 0$, parabolic if $x_1 = 0$, and elliptic if $x_1 > 0$. The equation of the cone $A_\mathbf{x}$ is given by

$$N_1^2 + x_1 N_2^2 = 0,$$

and the characteristic equation

$$\left(\frac{\partial f}{\partial x_1}\right)^2 + x_1 \left(\frac{\partial f}{\partial x_2}\right)^2 = 0$$

does not admit real solutions if $x_1 > 0$, whereas if $x_1 < 0$ it assumes the form

$$\left(\frac{\partial f}{\partial x_1} - \sqrt{|x_1|}\frac{\partial f}{\partial x_2}\right)\left(\frac{\partial f}{\partial x_1} + \sqrt{|x_1|}\frac{\partial f}{\partial x_2}\right) = 0.$$

This equation is equivalent to the following two conditions:

$$\nabla f \cdot \mathbf{v}_1 = 0, \ \nabla f \cdot \mathbf{v}_2 = 0,$$

where $\mathbf{v}_1 = (1, -\sqrt{|x_1|})$ and $\mathbf{v}_2 = (1, \sqrt{|x_1|})$, and the characteristic curves are the integrals of these fields, e.g. $x_2 = \pm\frac{2}{3}\sqrt{|x_1|^3} + c$, where c is an arbitrary constant.

8.5 Cauchy's Problem for a Quasi-Linear First-Order System

In the previous section we introduced the concept of the characteristic surface related to a second-order PDE. Here we consider the more general case of a first-order *quasi-linear* system of PDEs. It is clear that this case includes the previous one, since a second-order PDE can be reduced to a system of two first-order PDEs.

Let $\mathbf{x} = (x_1, \ldots x_n)$ be a point of a domain $\Omega \subset \Re^n$, and let $\mathbf{u} = (u_1(\mathbf{x}), \ldots, u_m(\mathbf{x}))$ be a *vector function of m components*, each depending on n variables $(x_1, \ldots, x_n)$. Assume that the vector function satisfies the m differential equations

$$\begin{cases} \left(A_{11}^1 \frac{\partial u_1}{\partial x_1} + \cdots + A_{1m}^1 \frac{\partial u_m}{\partial x_1} \right) + \cdots + \left(A_{11}^n \frac{\partial u_1}{\partial x_n} + \cdots + A_{1m}^n \frac{\partial u_m}{\partial x_n} \right) = c_1, \\ \cdots \\ \left(A_{m1}^1 \frac{\partial u_1}{\partial x_1} + \cdots + A_{mm}^1 \frac{\partial u_m}{\partial x_1} \right) + \cdots + \left(A_{m1}^n \frac{\partial u_1}{\partial x_n} + \cdots + A_{mm}^n \frac{\partial u_m}{\partial x_n} \right) = c_m, \end{cases} \tag{8.16}$$

where

$$A_{jh}^i = A_{jh}^i(\mathbf{x}, \mathbf{u}), \quad c_i = c_i(\mathbf{x}, \mathbf{u}) \tag{8.17}$$

are continuous functions of their arguments.

It is necessary to clearly specify the meaning of indices in $A_{jh}^i(\mathbf{x}, \mathbf{u})$:

$i = 1, \ldots, n$ refers to the independent variable x_i;
$j = 1, \ldots, m$ refers to the equation;
$h = 1, \ldots, m$ refers to the function u_h.

If the matrices

$$\mathbf{A}^i = \begin{pmatrix} A_{11}^i & \cdots & A_{1m}^i \\ \cdots & \cdots & \cdots \\ A_{m1}^i & \cdots & A_{mm}^i \end{pmatrix}, \quad \mathbf{c} = \begin{pmatrix} c_1 \\ \cdots \\ c_m \end{pmatrix}, \tag{8.18}$$

are introduced, system (8.16) can be written in the following concise way:

$$\mathbf{A}^1 \frac{\partial \mathbf{u}}{\partial x_1} + \cdots + \mathbf{A}^n \frac{\partial \mathbf{u}}{\partial x_n} = \mathbf{c},$$

or in the still more compact form

$$\mathbf{A}^i \frac{\partial \mathbf{u}}{\partial x_i} = \mathbf{c}. \tag{8.19}$$

The **Cauchy problem** for the system (8.19) consists of finding a solution of (8.19) which satisfies **Cauchy's data**

$$\mathbf{u}(\mathbf{x}) = \mathbf{u}_0(\mathbf{x}) \qquad \forall \mathbf{x} \in \Sigma_{n-1}, \tag{8.20}$$

where Σ_{n-1} is a regular surface of $\Re^n$.

By proceeding as in the previous section, we introduce the coordinates $(\nu_1, \ldots, \nu_n)$, so that Σ_{n-1} is represented by the equation $\nu_n = f(\mathbf{x}) = 0$. When expressed in these coordinates, system (8.19) becomes

$$\mathbf{A}^i \frac{\partial \mathbf{u}}{\partial x_i} = \mathbf{A}^i \frac{\partial \mathbf{u}}{\partial \nu_h} \frac{\partial \nu_h}{\partial x_i} = \mathbf{c}$$

or

$$\left(\mathbf{A}^i \frac{\partial \nu_h}{\partial x_i} \right) \frac{\partial \mathbf{u}}{\partial \nu_h} = \mathbf{c}, \tag{8.21}$$

whereas Cauchy's data assume the form

$$\mathbf{u}(\nu_1, \ldots, \nu_{n-1}, 0) = \tilde{\mathbf{u}}_0(\nu_1, \ldots, \nu_{n-1}). \tag{8.22}$$

According to the assumption that $A^i_{jh}(\mathbf{x}, \mathbf{u})$ and $c_i(\mathbf{x}, \mathbf{u})$ are analytic functions of their arguments, the solution can be expanded in a power series of ν_n; i.e.,

$$\mathbf{u}(\mathbf{x}) = \mathbf{u}_0(\nu_1, \ldots, \nu_{n-1}) + \left(\frac{\partial \mathbf{u}}{\partial \nu_n} \right)_{\mathbf{x}_0} \nu_n + \frac{1}{2} \left(\frac{\partial^2 \mathbf{u}}{\partial \nu_n^2} \right)_{\mathbf{x}_0} \nu_n^2 + \cdots, \tag{8.23}$$

where $\mathbf{x}_0 \in \Sigma_{n-1}$. The system (8.21) can be written in the form

$$\left(\mathbf{A}^i \frac{\partial f}{\partial x_i} \right)_{\mathbf{x}_0} \frac{\partial \mathbf{u}}{\partial \nu_n} = \mathbf{F},$$

where $\mathbf{F}$ is expressed from Cauchy's data. This form highlights the fact that $(\partial \mathbf{u}/\partial \nu_n)_{\mathbf{x}_0}$ is determined if and only if

$$\det \left(\mathbf{A}^i \frac{\partial f}{\partial x_i} \right)_{\mathbf{x}_0} \neq 0. \tag{8.24}$$

It is easy to verify that this condition allows us to determine all the derivatives $\partial^r \mathbf{u}/\partial \nu_n^r$ at any $\mathbf{x}_0 \in \Sigma_{n-1}$. Moreover, it can be proved that the series (8.23) is uniformly convergent towards a solution of the Cauchy problem (8.21), (8.22).

8.6 Classification of First-Order Systems

An $(n-1)$-dimensional hypersurface Σ_{n-1} of equation $f(\mathbf{x}) = 0$ is called a **characteristic surface** for Cauchy's problem (8.21), (8.22), if it is a solution of the equation

$$\det\left(\mathbf{A}^i \frac{\partial f}{\partial x_i}\right) = 0. \tag{8.25}$$

As we saw for a second-order PDE, if the system (8.19) is quasi-linear, then the matrices $\mathbf{A}^i$ depend on the solution $\mathbf{u}$ of the system as well as on the point $\mathbf{x}$. Therefore (8.25) allows us, at least in principle, to find the function $f(\mathbf{x})$ if a solution $\mathbf{u}_0$ is assigned. As a consequence, the classification of a quasi-linear system depends on the solution $\mathbf{u}_0$, i.e., on Cauchy's data.

Once again, the determination of characteristic surfaces requires that we solve a nonlinear first-order PDE. We remark that if a solution f of (8.25) exists, then the vector $\mathbf{N}$ normal to the characteristic surface $f = const$, has components N_i proportional to $\partial f/\partial x_i$, and the following condition holds:

$$\det(\mathbf{A}^i N_i) = 0. \tag{8.26}$$

In addition, the vectors $\mathbf{N}$ lean against a cone, since if $\mathbf{N}$ satisfies (8.26), then so does $\mu \mathbf{N}$ for any real μ.

If

$$\det \mathbf{A}^1 \neq 0; \tag{8.27}$$

then from (8.26) it follows that

$$\det(\mathbf{A}^1)^{-1}\det(\mathbf{A}^i N_i) = \det((\mathbf{A}^1)^{-1}\mathbf{A}^i N_i) = 0$$

i.e.,

$$\det\left(\mathbf{I} N_1 + \sum_{\alpha=2}^{n} \mathbf{B}^\alpha N_\alpha\right) = 0, \quad \mathbf{B}^\alpha = (\mathbf{A}^1)^{-1}\mathbf{A}^\alpha. \tag{8.28}$$

Given an arbitrary vector (N_α) of $\Re^{n-1}$, vectors normal to a characteristic surface in $\mathbf{x}$ exist if the algebraic equation of order m (8.28) admits a real solution N_1. Moreover, (8.28) represents the characteristic equation of the following eigenvalue problem:

$$\left(\sum_{\alpha=2}^{n} \mathbf{B}^\alpha N_\alpha\right)\mathbf{v} = -N_1 \mathbf{v}, \tag{8.29}$$

so that the roots of (8.28) are the opposite eigenvalues of the matrix

$$\mathbf{C} \equiv \sum_{\alpha=2}^{n} \mathbf{B}^\alpha N_\alpha, \tag{8.30}$$

which, in general, is *not symmetric*. According to all the previous considerations, system (8.19) can be classified as follows:

1. The system (8.19) is called ***elliptic*** at $\mathbf{x}$ (or for the solution $\mathbf{u}_0$ when it is quasi-linear), if, for any $\mathbf{N}$, (8.28) does not admit any real solution N_1, i.e., there are no vectors normal to characteristic surfaces passing through an arbitrary point $\mathbf{x} \in \Omega$. As a consequence, elliptic systems do not have real characteristic surfaces.

2. The system (8.19) is ***hyperbolic*** at $\mathbf{x}$ (or for the solution $\mathbf{u}_0$ when it is quasi-linear), if, for any $\mathbf{N}$, (8.28) has only real roots, some of them eventually coincident, and the eigenvectors span $\Re^m$ ([45], p. 46).

3. The system (8.19) is ***totally hyperbolic*** at $\mathbf{x}$ if, for any $\mathbf{N}$, all m roots of (8.28) are real, distinct, and the corresponding eigenvectors form a basis of $\Re^m$ ([45], p. 45).

4. The system (8.19) is ***parabolic*** at $\mathbf{x}$ if, for at least one root of (8.28), the dimension of the corresponding subset of eigenvectors is less than the algebraic multiplicity of the root itself.

We remark that in the last three cases, for any vector $\tilde{\mathbf{N}} = (N_2, \ldots, N_n) \in \Re^{n-1}$, there are several values of N_1 (with a maximum of m) which satisfy (8.28). This means that the cone $A_{\mathbf{x}}$ has several nappes (see Figure 8.2).

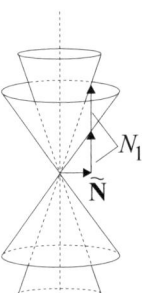

Figure 8.2

8.7 Examples

1. A second-order PDE is equivalent to a system of first-order differential equations. In order to prove that the transformation preserves

8.7. Examples

the character (elliptic, hyperbolic, or parabolic), we observe that Laplace's equation

$$\frac{\partial^2 u}{\partial x_1^2} + \frac{\partial^2 u}{\partial x_2^2} = 0,$$

with the substitution

$$\frac{\partial u}{\partial x_1} = v, \qquad \frac{\partial u}{\partial x_2} = w,$$

is equivalent to the system

$$\frac{\partial v}{\partial x_1} + \frac{\partial w}{\partial x_2} = 0,$$

$$\frac{\partial v}{\partial x_2} - \frac{\partial w}{\partial x_1} = 0,$$

where the second equation expresses that the second mixed derivatives of the function u are equal.

By comparing to (8.19), the previous system can be written as

$$\begin{pmatrix} 1 & 0 \\ 0 & -1 \end{pmatrix} \begin{pmatrix} \frac{\partial v}{\partial x_1} \\ \frac{\partial w}{\partial x_1} \end{pmatrix} + \begin{pmatrix} 0 & 1 \\ 1 & 0 \end{pmatrix} \begin{pmatrix} \frac{\partial v}{\partial x_2} \\ \frac{\partial w}{\partial x_2} \end{pmatrix} = \begin{pmatrix} 0 \\ 0 \end{pmatrix},$$

and the characteristic equation (8.25) is given by

$$\det\left(\begin{pmatrix} 1 & 0 \\ 0 & -1 \end{pmatrix} \frac{\partial f}{\partial x_1} + \begin{pmatrix} 0 & 1 \\ 1 & 0 \end{pmatrix} \frac{\partial f}{\partial x_2}\right) = 0$$

or

$$\left(\frac{\partial f}{\partial x_1}\right)^2 + \left(\frac{\partial f}{\partial x_2}\right)^2 = 0.$$

Then (8.28) becomes

$$\det\left(\begin{pmatrix} 1 & 0 \\ 0 & 1 \end{pmatrix} N_1 + \begin{pmatrix} 0 & 1 \\ -1 & 0 \end{pmatrix} N_2\right) = 0;$$

i.e.,

$$N_1^2 + N_2^2 = 0$$

and this equation admits complex roots for any real value of N_2.

By using this procedure, we ask the reader to verify that D'Alembert's equation is equivalent to a hyperbolic system.

2. The diffusion equation

$$\frac{\partial u}{\partial x_1} - \frac{\partial^2 u}{\partial x_2^2} = 0$$

is equivalent to the system

$$\frac{\partial w}{\partial x_2} = v,$$
$$\frac{\partial v}{\partial x_2} - \frac{\partial w}{\partial x_1} = 0,$$

where $\partial u/\partial x_1 = v$ and $\partial u/\partial x_2 = w$. In matrix form, this is

$$\begin{pmatrix} 0 & 0 \\ 0 & -1 \end{pmatrix} \begin{pmatrix} \frac{\partial v}{\partial x_1} \\ \frac{\partial w}{\partial x_1} \end{pmatrix} + \begin{pmatrix} 0 & 1 \\ 1 & 0 \end{pmatrix} \begin{pmatrix} \frac{\partial v}{\partial x_2} \\ \frac{\partial w}{\partial x_2} \end{pmatrix} = \begin{pmatrix} v \\ 0 \end{pmatrix},$$

and multiplying on the left by the inverse of

$$\begin{pmatrix} 0 & 1 \\ 1 & 0 \end{pmatrix}$$

gives

$$\begin{pmatrix} 0 & -1 \\ 0 & 0 \end{pmatrix} \begin{pmatrix} \frac{\partial v}{\partial x_1} \\ \frac{\partial w}{\partial x_1} \end{pmatrix} + \begin{pmatrix} 1 & 0 \\ 0 & 1 \end{pmatrix} \begin{pmatrix} \frac{\partial v}{\partial x_2} \\ \frac{\partial w}{\partial x_2} \end{pmatrix} = \begin{pmatrix} 0 \\ v \end{pmatrix}.$$

As a consequence, (8.29) becomes

$$\begin{pmatrix} 0 & -N_1 \\ 0 & 0 \end{pmatrix} \begin{pmatrix} v_1 \\ v_2 \end{pmatrix} + N_2 \begin{pmatrix} 1 & 0 \\ 0 & 1 \end{pmatrix} \begin{pmatrix} v_1 \\ v_2 \end{pmatrix} = 0.$$

It is now rather easy to verify that $N_2 = 0$ is an eigenvalue of multiplicity 2 and the corresponding eigenvector is $(v_1, 0)$, so that it spans a one-dimensional subspace. Therefore, the system is parabolic.

3. Finally, we introduce an example related to fluid mechanics (a topic addressed in Chapter 9). Let S be a perfect compressible fluid of density $\rho(t, x)$ and velocity $v(t, x)$ (oriented parallel to the axis Ox). The balance equations of mass (5.22) and momentum (5.30) of S are

$$\frac{\partial \rho}{\partial t} + v \frac{\partial \rho}{\partial x} + \rho \frac{\partial v}{\partial x} = 0,$$
$$\rho \frac{\partial v}{\partial t} + \rho v \frac{\partial v}{\partial x} + p'(\rho) \frac{\partial \rho}{\partial x} = 0,$$

where $p = p(\rho)$ is the constitutive equation.

8.7. Examples

The above system can be reduced to the form (8.19) as follows:

$$\begin{pmatrix} 1 & 0 \\ 0 & \rho \end{pmatrix} \begin{pmatrix} \frac{\partial \rho}{\partial t} \\ \frac{\partial v}{\partial t} \end{pmatrix} + \begin{pmatrix} v & \rho \\ p' & \rho v \end{pmatrix} \begin{pmatrix} \frac{\partial \rho}{\partial x} \\ \frac{\partial v}{\partial x} \end{pmatrix} = 0,$$

so that the characteristic equation (8.25) is given by

$$\det \left(\begin{pmatrix} 1 & 0 \\ 0 & \rho \end{pmatrix} \frac{\partial f}{\partial t} + \begin{pmatrix} v & \rho \\ p' & \rho v \end{pmatrix} \frac{\partial f}{\partial x} \right) = 0.$$

It can also be written in the form

$$\det \begin{pmatrix} \frac{\partial f}{\partial t} + v \frac{\partial f}{\partial x} & \rho \frac{\partial f}{\partial x} \\ p' \frac{\partial f}{\partial x} & \rho \left(\frac{\partial f}{\partial t} + v \frac{\partial f}{\partial x} \right) \end{pmatrix} = 0$$

and the expansion of the determinant gives

$$\left(\frac{\partial f}{\partial t} + v \frac{\partial f}{\partial x} \right)^2 - p' \left(\frac{\partial f}{\partial x} \right)^2 = 0.$$

Furthermore, (8.28) can be written as

$$\det \left(\begin{pmatrix} 1 & 0 \\ 0 & 1 \end{pmatrix} N_1 + \begin{pmatrix} v & \rho \\ p'/\rho & v \end{pmatrix} N_2 \right) = 0,$$

i.e.,

$$N_1^2 + 2N_2 v N_1 + N_2^2(v^2 - p') = 0.$$

This equation has two distinct roots

$$N_1 = \left(-v \pm \sqrt{p'} \right) N_2,$$

which are real if $p'(\rho) > 0$. The corresponding eigenvectors

$$\left(1, -\frac{\sqrt{p'}}{\rho} \right), \quad \left(1, \frac{\sqrt{p'}}{\rho} \right),$$

are independent, so that the system is totally hyperbolic.

8.8 Second-Order Systems

In many circumstances, the mathematical model of a physical problem reduces to a second-order quasi-linear system

$$A^{ij}_{HK} \frac{\partial^2 u_K}{\partial x_i \partial x_j} + f_H(\mathbf{x}, \mathbf{u}, \nabla \mathbf{u}) = 0, \qquad H, K = 1, \ldots, m, \tag{8.31}$$

of m equations in the unknown functions $u_1, \ldots, u_m$ depending on the variables $x_1, \ldots, x_n$.

In this case, the classification of the system can be pursued by two procedures.

The procedure discussed in the previous section can be still applied by transforming (8.31) into a first-order system of $(m + mn)$ equations by adding the mn auxiliary equations

$$\frac{\partial u_H}{\partial x_j} = v_{Hj}, \qquad H = 1, \ldots, m, \; j = 1, \ldots, n,$$

with mn auxiliary unknowns and by rewriting (8.31) in the form

$$A^{ij}_{HK} \frac{\partial v_{Kj}}{\partial x_i} + f_H(\mathbf{x}, \mathbf{u}, \mathbf{v}) = 0, \qquad H, K = 1, \ldots, m.$$

Such a procedure has the disadvantage that there is a huge increase in the number of equations: as an example, if $m = 3$ and $n = 4$, the transformed system gives rise to 15 equations.

This shortcoming calls for a different approach, so that the system (8.31) is written in the matrix form

$$\mathbf{A}^{ij} \frac{\partial^2 \mathbf{u}}{\partial x_i \partial x_j} + \mathbf{f}(\mathbf{x}, \mathbf{u}, \nabla \mathbf{u}) = 0, \tag{8.32}$$

where

$$\mathbf{A}^{ij} = \begin{pmatrix} A^{ij}_{11} & \cdots & A^{ij}_{1m} \\ \cdots & \cdots & \cdots \\ A^{ij}_{m1} & \cdots & A^{ij}_{mm} \end{pmatrix}, \qquad \mathbf{f} = \begin{pmatrix} f_1 \\ \cdots \\ f_m \end{pmatrix}.$$

We remark that the matrices $\mathbf{A}^{ij}$ are generally not symmetric. By proceeding as in the previous two sections, instead of (8.14), (8.24), we find that the characteristic surfaces are now given by the equation

$$\det\left(\mathbf{A}^{ij} \frac{\partial f}{\partial x_i} \frac{\partial f}{\partial x_j}\right) = 0. \tag{8.33}$$

8.9. Ordinary Waves

As a consequence, the vector $\mathbf{N}$ normal to the characteristic surface $f = const$, satisfies the condition

$$\det(\mathbf{A}^{ij} N_i N_j) = 0, \qquad (8.34)$$

which defines, at any $\mathbf{x}$, a cone with multiple nappes. This result allows us to extend the classification rule of first-order systems to second-order ones.

8.9 Ordinary Waves

Let C be a region of $\Re^n$. We denote by Σ_{n-1} a hypersurface dividing this region into two parts C^- and C^+, where C^+ is that part which contains the unit vector $\mathbf{N}$ normal to Σ_{n-1}. If $\mathbf{u}(\mathbf{x})$ is a C^1 function in $C - \Sigma_{n-1}$ as well as a solution of the system (8.21) in each one of the regions C^- and C^+, then the following theorem can be proved:

Theorem 8.2
The surface Σ_{n-1} is a first-order singular surface with respect to the function $\mathbf{u}(\mathbf{x})$ if and only if it is a characteristic surface for the Cauchy problem (8.21)–(8.22).

PROOF Let Σ_{n-1} be a first-order singular surface with respect to $\mathbf{u}(\mathbf{x})$. Then the following jump condition holds across Σ_{n-1} (see (2.49)):

$$\left[\!\!\left[\frac{\partial \mathbf{u}}{\partial x_i}\right]\!\!\right]_{\mathbf{x}_0} = \mathbf{a} N_i \quad \forall \mathbf{x}_0 \in \Sigma_{n-1}, \qquad (8.35)$$

where $\mathbf{a}$ is a vector field with m components. In addition, since $\mathbf{u}(\mathbf{x})$ satisfies the system (8.19) in both C^- and C^+ regions, for $\mathbf{x} \longrightarrow \mathbf{x}_0 \in \Sigma_{n-1}$, we find that

$$\mathbf{A}^i \left(\frac{\partial \mathbf{u}}{\partial x_i}\right)^{\pm}_{\mathbf{x}_0} = \mathbf{c},$$

so that

$$\mathbf{A}^i \left[\!\!\left[\frac{\partial \mathbf{u}}{\partial x_i}\right]\!\!\right]_{\mathbf{x}_0} = \mathbf{0}. \qquad (8.36)$$

Owing to (8.35), when we note that $N_i = \partial f / \partial x_i$, where $f(\mathbf{x}) = 0$ is the equation of the surface Σ_{n-1}, we conclude that

$$\left(\mathbf{A}^i \frac{\partial f}{\partial x_i}\right)_{\mathbf{x}_0} \mathbf{a} = \mathbf{0}. \qquad (8.37)$$

If the surface Σ_{n-1} is singular, then the field **a** is nonvanishing at some point $\mathbf{x}_0 \in \Sigma_{n-1}$. Equivalently, the coefficient determinant of the system (8.37) in the unknown vector **a**, has to be equal to zero. Therefore, (8.25) holds and Σ_{n-1} is a characteristic surface.

Conversely, if (8.25) holds, then (8.35) is satisfied for a nonvanishing vector **a** and Σ_{n-1} is a first-order singular surface. ∎

In its essence, the previous theorem states that the characteristic surfaces of the system (8.19) are coincident with the singular first-order surfaces of the solution $\mathbf{u}(\mathbf{x})$ of (8.19). System (8.37) is called the ***jump system*** associated with (8.19).

In order to highlight the role of the previous theorem in wave propagation, a physical phenomenon is supposed to be represented by the first-order quasi-linear system (8.21), whose unknowns are functions of the independent variables $(x_1, x_2, x_3, x_4) = (t, \mathbf{x}) \in \Re^4$, where t is the time and $\mathbf{x}$ is a spatial point. A moving surface $S(t)$, of equation $f(t, \mathbf{x}) = 0$, is supposed to subdivide a region $V \subset \Re^3$ into two parts $V^-(t)$ and $V^+(t)$, with the normal unit vector to $S(t)$ pointing towards $V^+(t)$.

If the solution $\mathbf{u}(t, \mathbf{x})$ of (8.21) exhibits a discontinuity in some of its first derivatives across $S(t)$, then $\mathbf{u}(t, \mathbf{x})$ is said to represent an ***ordinary wave***. If the function itself exhibits a discontinuity on $S(t)$, then $\mathbf{u}(t, \mathbf{x})$ represents a ***shock wave***. In both cases, $S(t)$ is the ***wavefront*** and $V^+(t)$ is the region toward which the surface $S(t)$ is moving with normal speed c_n. Accordingly, regions $V^-(t)$ and $V^+(t)$ are called the ***perturbed*** and ***undisturbed region***, respectively.

It follows that a wavefront $S(t)$ of an ordinary wave is a first-order singular surface with respect to the solution $\mathbf{u}(t, \mathbf{x})$ of the system (8.19), or, equivalently, a characteristic surface.

With these concepts in mind, we can determine the wavefront $f(t, \mathbf{x}) = 0$ by means of the theory of singular surfaces as well as by referring to characteristic surfaces. The relevant aspect relies on the fact that the system (8.19) predicts the propagation of ordinary waves if and only if its characteristics are real, i.e., the system is hyperbolic.

It could be argued that the definition of a wave introduced here does not correspond to the intuitive idea of this phenomenon. As an example, the solution of D'Alembert's equation can be expressed as a Fourier series of elementary waves. In any case, when dealing with ordinary waves of discontinuity, it is relatively easy to determine the propagation speed of the wavefront $S(t)$ and its evolution as well as the evolution of the discontinuity. Furthermore, in all those cases in which we are able to construct the solution, it can be verified a posteriori that the propagation characteristics are coincident with those derived from the theory of ordinary waves.

8.9. Ordinary Waves

If the evolution is represented by a first-order system of PDEs

$$\frac{\partial \mathbf{u}}{\partial t} + \sum_{i=1}^{3} \mathbf{B}^i(\mathbf{x}, t, \mathbf{u}) \frac{\partial \mathbf{u}}{\partial x_i} = \mathbf{b}, \qquad (8.38)$$

then the associated jump system is

$$-c_n \mathbf{a} + \sum_{i=1}^{3} \mathbf{B}^i(\mathbf{x}, t, \mathbf{u})_r n_i \mathbf{a} = \mathbf{0}, \qquad (8.39)$$

where c_n is the propagation speed of the wavefront $f(t, \mathbf{x}) = 0$, $\mathbf{n}$ is the unit vector normal to the surface, and $\mathbf{a}$ is the vector of discontinuities of first derivatives. If we introduce the $m \times m$ matrix

$$\mathbf{Q}(\mathbf{x}, t, \mathbf{u}, \mathbf{n}) = \sum_{i=1}^{3} \mathbf{B}^i(\mathbf{x}, t, \mathbf{u}) n_i, \qquad (8.40)$$

where m is the number of unknowns, i.e., the number of components of $\mathbf{u}(t, \mathbf{x})$, then system (8.39) becomes

$$\mathbf{Q}(\mathbf{x}, t, \mathbf{u}, \mathbf{n})\mathbf{a} = c_n \mathbf{a}. \qquad (8.41)$$

This equation leads to the following results (Hadamard): *Given the undisturbed state $\mathbf{u}^+(\mathbf{x}, t)$ towards which the ordinary wave propagates, the matrix $\mathbf{Q}$ is a known function of $\mathbf{r}$ and t, due to the continuity of $\mathbf{u}(\mathbf{x}, t)$ on $S(t)$. Furthermore, for a given direction of propagation $\mathbf{n}$, speeds of propagation correspond to the eigenvalues of the matrix $\mathbf{Q}$, and discontinuities of first derivatives are the eigenvectors of $\mathbf{Q}$.*

The existence of waves requires that eigenvalues and eigenvectors be real.

Once the propagation speed $c_n(\mathbf{x}, t, \mathbf{u}^+, \mathbf{n})$ has been determined, the evolution of the wavefront can be derived by referring to the speed of propagation of $S(t)$ (see Chapter 4):

$$\frac{\partial f}{\partial t} = -c_n(\mathbf{x}, t, \mathbf{u}^+, \mathbf{n}) |\nabla f|. \qquad (8.42)$$

Since $S(t)$ is a moving surface, $\partial f/\partial t \neq 0$; furthermore, $\mathbf{n} = \nabla f/|\nabla f|$ and c_n is a homogeneous function of zero order with respect to $\partial f/\partial x_i$. Therefore, (8.42) reduces to the **eikonal equation**:

$$\frac{\partial f}{\partial t} + c_n\left(\mathbf{x}, t, \mathbf{u}^+, \frac{\partial f}{\partial x_i}\right) = 0. \qquad (8.43)$$

Finding the solution of this equation is far from being a simple task.

Here again it may happen that the physical problem can be represented by m second-order PDEs with m unknowns depending on $x_1, \ldots, x_n$. This system could be transformed into a new first-order system of mn equations with mn unknowns, to which the previous results apply, by introducing the new unknowns

$$\frac{\partial u_i}{\partial x_j} = v_{ij},$$

and the additional equations

$$\frac{\partial v_{ij}}{\partial x_h} = \frac{\partial v_{ih}}{\partial x_j},$$

expressing the invertibility of second derivatives of functions u_i. Because such a procedure greatly increases the number of equations, the extension of the previous procedure to second-order systems appears to be more convenient. From this perspective, if the evolution is represented by the system

$$\frac{\partial^2 \mathbf{u}}{\partial t^2} + \mathbf{B}_{ij}(t, \mathbf{x}, \mathbf{u}, \nabla \mathbf{u}) \frac{\partial^2 \mathbf{u}}{\partial x_i \partial x_j} + \mathbf{f}(t, \mathbf{x}, \mathbf{u}, \nabla \mathbf{u}) = \mathbf{0}, \qquad (8.44)$$

then an ordinary wave is defined as a second-order singular surface $S(t)$ of the solution $\mathbf{u}(t, \mathbf{x})$, i.e., as a surface across which the second-order derivatives (some or all of them) are discontinuous; similarly, $S(t)$ is a shock wave if the function or its first derivatives are discontinuous across it.

If the system (8.44) is written in the regions V^- and V^+, then by considering the limit of $\mathbf{x}$ towards a point $\mathbf{r} \in S(t)$ and by subtracting the results, we obtain the jump system associated with (8.44):

$$\left[\left[\frac{\partial^2 \mathbf{u}}{\partial t^2}\right]\right]_\mathbf{r} + \mathbf{B}_{ij}(\mathbf{x}, t, \mathbf{u}, \nabla \mathbf{u})_\mathbf{r} \left[\left[\frac{\partial^2 \mathbf{u}}{\partial x_i \partial x_j}\right]\right]_\mathbf{r} = \mathbf{0}. \qquad (8.45)$$

By recalling jump expressions derived in Section 4.5, this becomes

$$\mathbf{Q}(\mathbf{x}, t, \mathbf{u}, \nabla \mathbf{u}, \mathbf{n})\mathbf{a} \equiv \mathbf{B}_{ij}(\mathbf{x}, t, \mathbf{u}, \nabla \mathbf{u})_\mathbf{r} n_i n_j \mathbf{a} = c_n^2 \mathbf{a}, \qquad (8.46)$$

where $\mathbf{n} = (n_i)$ is the unit vector normal to $S(t)$ and $\mathbf{a}$ is the discontinuity vector.

It has been proved that (Hadamard): *Given an undisturbed state $\mathbf{u}^+(\mathbf{x}, t)$ towards which the ordinary wave $S(t)$ propagates, then due to the continuity of $u(x,t)$ across $S(t)$, the matrix Q is a known function of t and r. Furthermore, given a propagation direction n, the speeds of propagation are the roots of eigenvalues of the matrix*

$$\mathbf{B}_{ij}(t, \mathbf{x}, \mathbf{u}, \nabla \mathbf{u})_\mathbf{r} n_i n_j,$$

and the discontinuity vectors are its eigenvectors.

We give the following examples:

1. When we consider D'Alembert's equation

$$\frac{\partial^2 u}{\partial t^2} = \Delta u,$$

the jump system is written as

$$(\mathbf{n} \otimes \mathbf{n})\mathbf{a} = c_n^2 \mathbf{a}.$$

In a reference frame $Oxyz$ in which Ox is parallel to $\mathbf{n}$, where $\mathbf{n} = (1,0,0)$, the eigenvalues of the tensor $\mathbf{n} \otimes \mathbf{n}$ are 0 (with multiplicity 2) and 1 and the eigenvectors have components $(0, a_2, a_3)$ and $(a_1, 0, 0)$. It follows that the speeds of propagation along $\mathbf{n}$ are ± 1.

2. The system of example 3, Section 8.3, can be written in the form (8.38)

$$\begin{pmatrix} \dfrac{\partial \rho}{\partial t} \\ \dfrac{\partial v}{\partial t} \end{pmatrix} + \begin{pmatrix} v & \rho \\ \dfrac{p'}{\rho} & v \end{pmatrix} \begin{pmatrix} \dfrac{\partial \rho}{\partial x} \\ \dfrac{\partial v}{\partial x} \end{pmatrix} = 0$$

and, since $\mathbf{n} = (1, 0, 0)$, it follows that

$$\mathbf{Q} = \begin{pmatrix} v & \rho \\ \dfrac{p'}{\rho} & v \end{pmatrix}.$$

The eigenvalues of this matrix are

$$c_n = v \pm \sqrt{p'(\rho)},$$

so that the result previously obtained is recovered.

8.10 Linearized Theory and Waves

We have already seen how far from straightforward it is to obtain a solution of the quasi-linear system (8.38). But it can also be observed that, under the assumption $\mathbf{b} = \mathbf{0}$, any arbitrary *constant* vector function $\mathbf{u} = \mathbf{u}_0$ is a solution of the system. More generally, suppose that a solution $\mathbf{u}_0$ of (8.38) is already known, so that we can ask under which circumstances the function

$$\mathbf{u} = \mathbf{u}_0 + \mathbf{v}, \tag{8.47}$$

obtained by adding a *small perturbation* $\mathbf{v}$ to the undisturbed state $\mathbf{u}_0$, is still a solution of the system.

We say that there is a small perturbation if $|\mathbf{v}|$ and $|\nabla\mathbf{v}|$ are first-order quantities, so that their products and power can be neglected. It will be shown in the following discussion that such an assumption implies that $\mathbf{v}$ is a solution of the *linearized system* (8.38), so that we can apply many techniques in order to find an analytical expression of $\mathbf{v}$. In addition, we also note that the assumption of small perturbation, to be consistent, requires that we ascertain the stability of the undisturbed state. This observation is related to the nonlinear nature of the system (8.38), so that it could happen that a solution corresponding to initial data $\mathbf{u}_0 + \mathbf{v}_0$, with small $\mathbf{v}_0$,

- it does not remain over time of the order of $\mathbf{v}_0$; or
- only exists in a finite time interval.

In both cases the solution of the linearized problem does not represent an approximate solution of the nonlinear problem.

Within this framework, we introduce the following examples to clarify the previous considerations.

1. Consider the nonlinear equation

$$\frac{\partial u}{\partial t} + u\frac{\partial u}{\partial x} = 0, \tag{8.48}$$

which admits the solution $u_0 = 0$. The integral curves of the vector field $\mathbf{V} \equiv (1, u)$ are the solution of the system

$$\frac{dt}{ds} = 1,$$
$$\frac{dx}{ds} = u,$$

which corresponds to the equation

$$\frac{dx}{dt} = u(t, x(t)). \tag{8.49}$$

On these curves, (8.48) gives

$$\frac{du}{ds} = 0,$$

i.e., $u(t, x(t)) = const = u(0, x(0)) \equiv u_0(\xi)$, where $\xi = x(0)$. From (8.49) it follows that along the lines

$$x - u_0(\xi)t = \xi, \tag{8.50}$$

the solution $u(t, x)$ of (8.48) is constant, so that at (t, x) it assumes the same values that it has at the point ξ:

$$u = u_0(x - u_0(\xi)t),$$

8.10. Linearized Theory and Waves

showing a character of wave solution. We can conclude that, given the values $u_0(\xi)$ of the function u on the axis Ox, the solution of (8.48) is determined. If the initial datum $u_0(\xi)$ is not constant, then the straight lines (8.50) are no longer parallel and they intersect at some point of the plane t, x. This means that the wave is a shock (Figure 8.3).

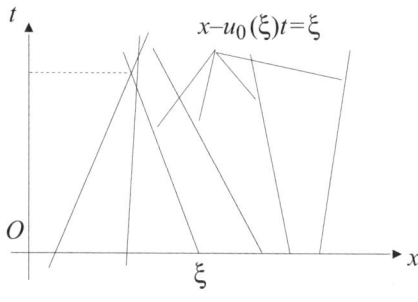

Figure 8.3

On the other hand, consider the Cauchy problem related to (8.48), with
$$u_0(\xi) = 1 + v_0(\xi),$$
where $v_0(\xi)$ is a first-order quantity. If (8.48) is linearized in the neighborhood of $u = 1$, then we obtain for the perturbation v the equation
$$\frac{\partial v}{\partial t} + \frac{\partial v}{\partial x} = 0,$$
to be solved with the Cauchy datum $v_0(\xi)$. By using the previous procedure, we conclude that the function $v(t, x)$ is constant along the parallel lines $x - t = \xi$ and that it assumes the form
$$v(t, x) = v_0(x - t) = v_0(\xi).$$
In this case the linear solution is defined at any time t and represents a wave moving with constant speed, equal to 1.

2. As a second example we consider the equation
$$\frac{\partial u}{\partial t} + \frac{\partial u}{\partial x} - u^2 = 0, \tag{8.51}$$
which admits the solution $u = 0$. If we introduce the vector field $\mathbf{V} \equiv (1, 1)$, whose integral curves are the parallel lines
$$x - t = \xi,$$

then the previous equation along these lines gives
$$\frac{du}{ds} = u^2.$$

The general solution of (8.51) then is
$$u = \frac{u_0(\xi)}{1 - u_0(\xi)t} = \frac{u_0(x-t)}{1 - u_0(x-t)t}, \qquad (8.52)$$

where $u_0(\xi) = u_0(0, x)$. If $u_0(\xi) > 0$, then the solution along the straight line $x - t = \xi$ is infinity at the time $t_* > 0$, such that $u_0(\xi)t_* = 1$.

By linearizing in the neighborhood of $u = 0$, we obtain the following equation for the perturbation v:
$$\frac{\partial v}{\partial t} + \frac{\partial v}{\partial x} = 0,$$

which has the wave solution
$$v = v_0(\xi) = v_0(x - t),$$

defined at any time t.

When we can apply the linearizing procedure, presuming that the undisturbed state is a stable one, then by substituting the solution (8.47) into (8.38) and by neglecting nonlinear terms of $|\mathbf{v}|$ and $|\nabla \mathbf{v}|$, we obtain that the perturbation $\mathbf{v}$ must be a solution of the linear system
$$\frac{\partial \mathbf{v}}{\partial t} + \sum_{i=1}^{3} \mathbf{B}^i(\mathbf{x}, t, \mathbf{u}_0) \frac{\partial \mathbf{v}}{\partial x_i} = 0. \qquad (8.53)$$

In this case we can try to obtain the solution as a Fourier series of elementary waves; i.e., we can try to determine whether a sinusoidal wave of amplitude $\mathbf{A}$, wavelength λ, and propagating in the direction of the unit vector $\mathbf{n}$ with speed U, written as
$$\mathbf{v}(t, \mathbf{x}) = \mathbf{A} \sin \frac{2\pi}{\lambda}(\mathbf{n} \cdot \mathbf{x} - Ut), \qquad (8.54)$$

can be a solution of (8.53). Since
$$\frac{\partial \mathbf{v}}{\partial t} = -\frac{2\pi}{\lambda} U \mathbf{A} \cos \frac{2\pi}{\lambda}(\mathbf{n} \cdot \mathbf{x} - Ut),$$
$$\frac{\partial \mathbf{v}}{\partial x_i} = \frac{2\pi}{\lambda} n_i \mathbf{A} \cos \frac{2\pi}{\lambda}(\mathbf{n} \cdot \mathbf{x} - Ut),$$

8.11. Shock Waves

then by substituting into (8.53) we find that

$$-U\mathbf{A} + \sum_{i=1}^{3} \mathbf{B}^i(\mathbf{x}, t, \mathbf{u}_0) n_i \mathbf{A} = \mathbf{0},$$

which is the same as (8.39) if we assume that $\mathbf{u} = \mathbf{u}_0$. We reach the conclusion that *a plane sinusoidal wave is a solution of the linear system* (8.53) *if and only if the speed of propagation is an eigenvalue of*

$$\mathbf{Q}^i(\mathbf{x}, t, \mathbf{u}_0, \mathbf{n}) \equiv \sum_{i=1}^{3} \mathbf{B}^i(\mathbf{x}, t, \mathbf{u}_0) n_i$$

and the amplitude is the corresponding eigenvector.

8.11 Shock Waves

As we mentioned in Section 8.5, a singular surface $\Sigma(t)$ is a shock wavefront if the function $\mathbf{u}$ exhibits a discontinuity across it. Usually, we meet with shock waves when we are dealing with systems of PDEs related to *conservation laws*

$$\frac{\partial g_i(\mathbf{u})}{\partial t} + \frac{\partial}{\partial x_j} \Psi_{ij}(t, \mathbf{x}, \mathbf{u}) = b_i(t, \mathbf{x}, \mathbf{u}), \tag{8.55}$$

$i = 1, \ldots, m$ and $j = 1, \ldots, n$, i.e. to those systems which are equivalent to integral balance laws (see (5.2) and (5.3))

$$\frac{d}{dt} \int_V g_i(\mathbf{u}) \, dc = - \int_{\partial V} \Psi_{ij} N_j \, d\sigma + \int_V b_i \, dc,$$

where V is an arbitrary volume of $\Re^3$ and $\mathbf{N}$ is the unit vector normal to ∂V.

As discussed in Chapter 5 (see (5.3)), on the discontinuity surface $\Sigma(t)$ the following **Rankine–Hugoniot jump conditions** hold:

$$[[g_i(\mathbf{u}) c_n - \Psi_{ij}(t, \mathbf{x}, \mathbf{u}) n_j]] = 0, \tag{8.56}$$

where c_n is the normal component of the speed of propagation of $\Sigma(t)$ and $\mathbf{n}$ is the unit vector normal to $\Sigma(t)$.

Let us now assume that a given phenomenon is represented by a system of n PDEs such as (8.55), with the unknowns $\mathbf{u} = (u_1, \ldots, u_m)$. When the constitutive law $\boldsymbol{\Psi}(t, \mathbf{x}, \mathbf{u})$ has been defined, the system (8.56) gives a finite relation between the two limits $\mathbf{u}^-$ and $\mathbf{u}^+$ of the field $\mathbf{u}$ on the two sides of the wavefront, where $\mathbf{u}^+$ are the values relative to the undisturbed region.

The aim is now to analyze the system (8.56) with reference to the following three cases:

- *linear shock*;
- *small intensity shock*;
- *finite shock*.

Let us start by assuming that both functions g_i and Ψ_{ij} are linearly dependent on the unknown functions u_i:

$$g_i = K_{ij}(t,\mathbf{x})u_j, \qquad \Psi_{ij} = H_{ijk}(t,\mathbf{x})u_k,$$

i.e., the system (8.55) is linear or semi-linear. Accordingly, the Rankine–Hugoniot conditions are

$$(K_{ij}c_n - H_{ikj}n_k)[[u_j]] = 0,$$

and they constitute an *algebraic system* for the jumps $[[u_j]]$. In this case, it is then apparent that our problem reduces to the usual eigenvalue problem, as we saw for ordinary waves.

We assume that the jumps are of small intensity, then we can write

$$g_i = K_{ij}u_j + \frac{1}{2}K^{(1)}_{ijh}u_j u_h + \cdots,$$

$$\Psi_{ij} = H_{ijk}u_k + \frac{1}{2}H^{(1)}_{ijhk}u_h u_k + \cdots.$$

By substituting into the Rankine–Hugoniot conditions (8.56), as well as by neglecting terms higher than first order and taking into account the symmetry conditions

$$K^{(1)}_{ijh} = K^{(1)}_{ihj}, \qquad H^{(1)}_{ijhk} = H^{(1)}_{ijkh},$$

we get

$$(K_{ij}c_n - H_{ijk}n_k)[[u_j]] + \frac{1}{2}(K^{(1)}_{ijh}c_n - H^{(1)}_{ikjh}n_k)[[u_j u_h]] = 0.$$

By also observing that

$$[[AB]] = A^+B^+ - A^-B^- = A^+B^+ - A^-B^+ + A^-B^+ - A^-B^-$$
$$= [[A]]B^+ + [[B]]A^- = [[A]]B^+ + [[B]](A^+ - [[A]])$$
$$\simeq [[A]]B^+ + [[B]]A^+,$$

8.11. Shock Waves

we find that the previous relation assumes the form

$$(K_{ij}c_n - H_{ikj}n_k)[[u_j]]$$
$$+ \frac{1}{2}(K^{(1)}_{ijh}c_n - H^{(1)}_{ikjh}n_k)([[u_j]]u_h^+ + [[u_h]]u_j^+) = 0$$

and it follows that

$$(L_{ij}c_n - M_{ikj}n_k)[[u_j]] = 0, \tag{8.57}$$

where

$$L_{ij} = K_{ij} + K^{(1)}_{ijh}u_h^+,$$
$$M_{ikj} = H_{ikj} + H^{(1)}_{ikjh}u_h^+.$$

In the case of small intensity shocks, we conclude that the problem of determining the speed of propagation and the jumps is reduced to a linear algebraic problem.

In the most general case, the Rankine–Hugoniot conditions correlate the values of $\mathbf{u}^-$, corresponding to the perturbed region, with those of $\mathbf{u}^+$, corresponding to the undisturbed region, when *the speed of propagation c_n is known*. In fact, equations (8.56) do not allow us to solve the problem, because, due to the presence of c_n, there are fewer equations than unknowns. All we can do is express the values of $\mathbf{u}^-$ as functions of c_n, called the **shock intensity**.

We remark that even if we know the value of c_n, the problem is still difficult to solve. First, the system (8.56) could have no real solutions $u_1^-, \ldots, u_m^-$ in the disturbed region; i.e., the state $u_1^+, \ldots, u_m^+$ could be incompatible with the propagation of shock waves.

It could also happens that the system (8.56), which is nonlinear in the general case, could admit more than one solution compatible with the choice $u_1^+, \ldots, u_m^+$, which is equivalent to having several modes of propagation of shock waves. As this doesn't match physical reality, we need to introduce a rule for selecting the one meaningful solution.

But before introducing such a rule, we also need to remark that the indeterminacy arises from the definition of shock waves, according to which discontinuities of zero order are assumed. In reality, around the wavefront there exists a boundary layer where the fields $\mathbf{u}$ change rapidly but with continuity from values pertinent to the perturbed state to those of the undisturbed state. This boundary layer can also be predicted on a theoretical basis by introducing more sophisticated constitutive laws. These laws generate a system of PDEs of higher order, with terms of higher order being multiplied by a small parameter, which creates the boundary layer (see Chapter 9).

In order to avoid this difficulty, an alternative approach consists of adding to (8.56) a scalar condition which reflects the second law of thermodynamics (see also Chapter 9).

One choice consists of assuming that the entropy jump across the shock wavefront is positive. Another possibility is represented by the **Lax condition** [44]:

$$v_n(\mathbf{u}^+) < c_n < v_n(\mathbf{u}^-), \tag{8.58}$$

where $v_n(\mathbf{u}^+)$ and $v_n(\mathbf{u}^-)$ denote the advancing speeds of the ordinary discontinuity waves in the states $\mathbf{u}^+$ and $\mathbf{u}^-$, respectively. In other words, the Lax condition states that, before the arrival of the shock wave, the undisturbed region must be in a supersonic state, while the disturbed region is in a subsonic state. This also means that an observer who moves with the velocity of the perturbed region sees the shock waves moving toward him with a supersonic speed. On the other hand, an observer moving with the velocity of the undisturbed region sees the shock wave going away with subsonic speed.

8.12 Exercises

1. Find a stationary solution for a string with fixed ends and length l.

   We must find a solution $u(x,t)$ to the problem

   $$\frac{\partial u}{\partial t^2} = v^2 \frac{\partial u}{\partial x^2},$$

   $$u(0,t) = u(l,t) = 0,$$

   which exhibits, at any point, a character of oscillation but not of propagation. To this end, consider the function

   $$u(x,t) = A(t)B(x),$$

   where $B(0) = B(l) = 0$ and $A(t)$ is periodic. It can be verified that this function is the required solution if and only if it has the form

   $$u(x,t) = A \sin\left(\frac{2\pi}{\lambda} vt + \varphi\right) \sin\left(\frac{2\pi}{\lambda} x + \varphi\right),$$

   where A and φ are arbitrary constants and $\lambda = l/k$, with k a positive integer.

2. Find a sinusoidal solution representing a wave propagating on a string of infinite length.

   In this case we need to verify that the function

   $$u = A \sin \frac{2\pi}{\lambda}(x - vt)$$

   is a solution of D'Alembert's equation.

8.12. Exercises

3. Verify that the function

$$u(x,t) = Ae^{-ht} \sin \frac{2\pi}{\lambda}(x - vt)$$

is a solution of D'Alembert's equation in presence of a dissipation term

$$\frac{\partial^2 u}{\partial t^2} - v^2 \frac{\partial^2 u}{\partial x^2} = -h\frac{\partial u}{\partial t}, \quad h > 0.$$

4. Classify the following PDEs by determining the corresponding characteristic curves:

$$x^2 y \frac{\partial^2 u}{\partial x^2} + xy \frac{\partial^2 u}{\partial x \partial y} - y^2 \frac{\partial^2 u}{\partial y^2} = 0, \quad (8.59)$$

$$(x \log y) \frac{\partial^2 u}{\partial x^2} + 4y \frac{\partial^2 u}{\partial y^2} = 0,$$

$$\frac{\partial^2 u}{\partial x \partial y} - xy \frac{\partial^2 u}{\partial y^2} = 0,$$

$$x^2 y \frac{\partial^2 u}{\partial x^2} - xy^2 \frac{\partial^2 u}{\partial y^2} = 0.$$

Derive values of k for which the equation

$$\frac{\partial^2 u}{\partial x^2} - kx \frac{\partial^2 u}{\partial x \partial y} + 4x^2 \frac{\partial^2 u}{\partial y^2} = 0$$

is elliptic, parabolic, or hyperbolic.

5. Verify that the quasi-linear equation

$$u^2 \frac{\partial^2 u}{\partial x^2} + 3 \frac{\partial u}{\partial x} \frac{\partial u}{\partial y} \frac{\partial^2 u}{\partial x \partial y} - u^2 \frac{\partial^2 u}{\partial y^2} = 0$$

is hyperbolic for any solution u.

6. Verify that the quasi-linear equation

$$\left[1 - \left(\frac{\partial u}{\partial x}\right)^2\right] \frac{\partial^2 u}{\partial x^2} - 2 \frac{\partial u}{\partial x} \frac{\partial u}{\partial y} \frac{\partial^2 u}{\partial x \partial y} + \left[1 - \left(\frac{\partial u}{\partial y}\right)^2\right] \frac{\partial^2 u}{\partial y^2} = 0$$

is hyperbolic for those solutions for which $|\nabla u| > 1$ and it is elliptic for those solutions for which $|\nabla u| < 1$.

7. Classify the linear system

$$\frac{\partial u}{\partial x} + 2\frac{\partial u}{\partial t} + \frac{\partial v}{\partial x} + 3\frac{\partial v}{\partial t} - u + v = 0,$$

$$3\frac{\partial u}{\partial x} + \frac{\partial u}{\partial t} - 2\frac{\partial v}{\partial x} - \frac{\partial v}{\partial t} - 2v = 0.$$

8. Classify the quasi-linear system

$$\rho \frac{\partial u}{\partial x} + u \frac{\partial \rho}{\partial x} + \frac{\partial \rho}{\partial t} = -2\rho \frac{u}{x},$$
$$\rho u \frac{\partial u}{\partial x} + \rho \frac{\partial u}{\partial t} + c^2 \frac{\partial \rho}{\partial x} = 0.$$

8.13 The Program PdeEqClass

Aim of the Program PdeEqClass

The program `PdeEqClass` classifies the quasi-linear second-order PDEs.

Description of the Problem and Relative Algorithm

Given a quasi-linear second-order PDE in a domain $\Omega \subseteq \Re^n$

$$a_{ij}(\mathbf{x}, u, \nabla u) \frac{\partial^2 u}{\partial x_i \partial x_j} = h(\mathbf{x}, u, \nabla u), \qquad i,j = 1, \cdots, n, \qquad \mathbf{x} \in \Omega,$$

a point $\mathbf{x}_0$ and a solution $u_0(\mathbf{x})$, the program `PdeEqClass` determines the coefficient matrix $\mathbf{A} = (a_{i,j})$ and the matrix $\mathbf{A}_0$, which is obtained by evaluating $\mathbf{A}$ at $\mathbf{x}_0$ with respect to the known solution u_0. Moreover, it calculates the eigenvalues of $\mathbf{A}$ if `option=symbolic` is chosen or those of $\mathbf{A}_0$ if `option= numeric` is chosen. In the latter case the PDE is classified.

We recall that, for *quasi-linear* PDEs, the characteristic surfaces depend both on the equation and the Cauchy data. Consequently, the classification is relative to a point $\mathbf{x}_0$ for the given solution $u_0(\mathbf{x})$.[1] On the other hand, for *semi-linear* and *linear*[2] PDEs the characteristic surfaces depend only on the equation and therefore the classification is relative to any point $\mathbf{x}_0$.

Command Line of the Program PdeEqClass

`PdeEqClass[eq, var, unk, point, unk0, option]`

Parameter List

Input Data

eq = second-order quasi-linear PDE, where all the terms containing the second derivative appear on the left-hand side, while all the other terms appear on the right-hand side;

---

[1] It is clear that this procedure is applicable when a solution is known, for instance when $u_0(\mathbf{x}) = const$.

[2] The equation is *semi-linear* if it has the form

$$a_{ij}(\mathbf{x}) \frac{\partial^2 u}{\partial x_i \partial x_j} = h(\mathbf{x}, u, \nabla u), \qquad i,j = 1, \cdots, n, \qquad \mathbf{x} \in \Omega,$$

and *linear* if $h(\mathbf{x}, u, \nabla u)$ is a linear function of u and ∇u.

> var = list of independent variables;
>
> unk = unknown function;
>
> point = coordinates of the point $\mathbf{x}_0$ at which the PDE has to be classified;
>
> unk0 = known solution u_0 of PDE when it is quasi-linear;
>
> option = option for the symbolic or numeric calculation of the eigenvalues of $\mathbf{A}$ and $\mathbf{A}_0$, respectively.

Output Data

> the coefficient matrix $\mathbf{A}$;
>
> the matrix $\mathbf{A}_0$ obtained by evaluating $\mathbf{A}$ at point and with respect to the solution unk0;
>
> the eigenvalues of $\mathbf{A}$ if option = symbolic;
>
> the eigenvalues of $\mathbf{A}_0$ if option = numeric;
>
> the elliptic, hyperbolic, or parabolic character of the PDE if option = numeric.

Use Instructions

The equation has to be given in such a way that on the left-hand side the second derivatives appear, while the right-hand side includes all the terms involving the unknown function and its first derivatives. The PDE we wish to examine must to be given according with the syntax both of Mathematica and the program, i.e., in the form

$$\text{eq} = \mathtt{a_{ij}}(\mathbf{x}, \mathtt{u}, \nabla \mathtt{u})\mathtt{u}_{\mathtt{x_i},\mathtt{x_j}} == \mathtt{h}(\mathbf{x}, \mathtt{u}, \nabla \mathtt{u}),$$

where the two sides of the equation are separated by a double equal sign and the symbol $\dfrac{\partial^2 \mathtt{u}}{\partial \mathtt{x_i} \partial \mathtt{x_j}}$ is written as $\mathtt{u}_{\mathtt{x_i},\mathtt{x_j}}$.

If the equation is semi-linear or linear, then only the input datum point has to be given, not unk0.

Worked Examples

1. Laplace's Equation: Consider Laplace's equation

$$v_{x,x} + v_{y,y} + v_{z,z} = 0.$$

8.13. The Program PdeEqClass

To apply the program `PdeEqClass`, it is sufficient to input the following data:

eq = v$_{x,x}$ + v$_{y,y}$ + v$_{z,z}$ == 0;
var = {x, y, z};
unk = v;
point = {x$_0$, y$_0$, z$_0$};
unk0 = v$_0$;
option = numeric;
PdeEqClass[eq, var, unk, point, unk0, option]

In output we obtain

Coefficient matrix

$$A = \begin{pmatrix} 1 & 0 & 0 \\ 0 & 1 & 0 \\ 0 & 0 & 1 \end{pmatrix}$$

Eigenvalues of A
$\lambda_1 = 1$

The Pde is elliptic.

2. Classify the following equation:

$$u_{x,x} + 2k u_{t,x} + k^2 u_{t,t} = 0.$$

To apply the program `PdeEqClass`, it is sufficient to input the following data:

u$_{x,x}$ + 2k u$_{t,x}$ + k^2 u$_{t,t}$ == 0;
var = {t, x};
unk = u;
point = {t$_0$, x$_0$};
unk0 = u$_0$;
option = numeric;
PdeEqClass[eq, var, unk, point, unk0, option]

In output we obtain

Coefficient matrix

$$A = \begin{pmatrix} k^2 & k \\ k & 1 \end{pmatrix}$$

Eigenvalues of A

$\lambda_1 = 0$

$\lambda_2 = 1 + k^2$

The Pde is parabolic.

3. To verify that the quasi-linear equation

$$u^2 u_{x,x} + 3u_x u_y u_{x,y} - u^2 u_{y,y} = 0$$

is hyperbolic for any nonvanishing solution $u = u(x, y)$, we can start by applying the program `PdeEqClass` with the following input data:

$u^2 u_{x,x} + 3u_x u_y u_{x,y} - u^2 u_{y,y} == 0;$

var = $\{x, y\}$;

unk = u;

point = $\{x_0, y_0\}$;

unk0 = u_0;

option = numeric;

PdeEqClass[eq, var, unk, point, unk0, option]

The output is

Coefficient matrix

$$A = \begin{pmatrix} u^2 & \dfrac{3u_x u_y}{2} \\ \dfrac{3u_x u_y}{2} & -u^2 \end{pmatrix}$$

Matrix A_0 obtained by evaluating A at point $\{x_0, y_0\}$ and with respect to the solution $u_0[x, y]$

$$A_0 = \begin{pmatrix} u_0[x,y]^2 & \dfrac{3}{2} u_0^{(0,1)}[x,y] u_0^{(1,0)}[x,y] \\ \dfrac{3}{2} u_0^{(0,1)}[x,y] u_0^{(1,0)}[x,y] & -u_0[x,y]^2 \end{pmatrix}$$

Eigenvalues of A_0

$$\lambda_1 = -\sqrt{u_0[x,y]^4 + \frac{9}{4} u_0^{(0,1)}[x,y]^2 u_0^{(1,0)}[x,y]^2}$$

$$\lambda_2 = \sqrt{u_0[x,y]^4 + \frac{9}{4} u_0^{(0,1)}[x,y]^2 u_0^{(1,0)}[x,y]^2}$$

The equation cannot be classified since not all the eigenvalues have a definite sign!

8.13. The Program PdeEqClass

As one can see, the output supplies only some help to the user in classifying the equation. In fact, the program cannot decide the sign of the eigenvalues as long as the input data are not given in the numerical form. That is due to the fact that, when the input data are assigned in symbolic form, Mathematica is not able to distinguish between real and complex quantities.

Exercises

Apply the program `PdeEqClass` to the following differential equations.

1. D'Alembert's equation

$$\frac{\partial^2 u}{\partial x^2} - \frac{1}{v^2}\frac{\partial^2 u}{\partial t^2} = 0.$$

2. Heat equation

$$\frac{\partial^2 u}{\partial x^2} - \frac{\partial u}{\partial y} = 0.$$

3. Tricomi's equation

$$\frac{\partial^2 u}{\partial x^2} + x^2 \frac{\partial^2 u}{\partial y^2} = 0.$$

4. Classify the PDE

$$x\frac{\partial^2 u}{\partial x^2} - 4\frac{\partial^2 u}{\partial xy} = 0$$

in the region $x > 0$.

5. Show that the equation

$$yz\frac{\partial^2 u}{\partial x^2} + zx\frac{\partial^2 u}{\partial y^2} + xy\frac{\partial^2 u}{\partial z^2} = 0$$

is elliptic in the regions $x > 0, y > 0, z > 0$ and $x < 0, y < 0, z < 0$, and is hyperbolic almost everywhere else.

6.

$$x \log y \frac{\partial^2 u}{\partial x^2} + 4y\frac{\partial^2 u}{\partial y^2} = 0.$$

7.

$$x^2 y \frac{\partial^2 u}{\partial x^2} - xy^2 \frac{\partial^2 u}{\partial y^2} = 0.$$

8.
$$y^2 \frac{\partial^2 u}{\partial x^2} + x \frac{\partial^2 u}{\partial y^2} = 0.$$

9.
$$-(y^2 + z^2)\frac{\partial^2 u}{\partial x^2} + (z^2 + x^2)\frac{\partial^2 u}{\partial y^2} + (x^2 + y^2)\frac{\partial^2 u}{\partial z^2} = 0.$$

10.
$$(y^2 - z^2)\frac{\partial^2 u}{\partial x^2} + (z^2 - x^2)\frac{\partial^2 u}{\partial y^2} + (x^2 - y^2)\frac{\partial^2 u}{\partial z^2} = 0.$$

11.
$$(y - x)\frac{\partial^2 u}{\partial x^2} + x^2 \frac{\partial^2 u}{\partial y^2} = 0.$$

12. Verify that the quasi-linear equation

$$\left[1 - \left(\frac{\partial u}{\partial x}\right)^2\right]\frac{\partial^2 u}{\partial x^2} - 2\frac{\partial u}{\partial x}\frac{\partial u}{\partial y}\frac{\partial^2 u}{\partial x \partial y} + \left[1 - \left(\frac{\partial u}{\partial y}\right)^2\right]\frac{\partial^2 u}{\partial y^2} = 0$$

is hyperbolic for those solutions for which $|\nabla u| > 1$ and it is elliptic for those solutions for which $|\nabla u| < 1$.

8.14 The Program PdeSysClass

Aim of the Program PdeSysClass

The program `PdeSysClass` classifies the quasi-linear first-order systems of PDEs.

Description of the Problem and Relative Algorithm

Consider the quasi-linear first-order systems of PDEs of m equations in the m unknown functions $(u_1, \cdots, u_m)$ depending on n variables $(x_1 \equiv t, \cdots, x_n)$

$$\left(A_{11}^1 \frac{\partial u_1}{\partial x_1} + \cdots + A_{1m}^1 \frac{\partial u_m}{\partial x_1}\right) + \cdots + \left(A_{11}^n \frac{\partial u_1}{\partial x_n} + \cdots + A_{1m}^n \frac{\partial u_m}{\partial x_n}\right) = c_1,$$

$$\cdots$$

$$\left(A_{m1}^1 \frac{\partial u_1}{\partial x_1} + \cdots + A_{mm}^1 \frac{\partial u_m}{\partial x_1}\right) + \cdots + \left(A_{m1}^n \frac{\partial u_1}{\partial x_n} + \cdots + A_{mm}^n \frac{\partial u_m}{\partial x_n}\right) = c_m,$$

$$(8.60)$$

where $A_{ij}^k(\mathbf{x}, \mathbf{u})$ and $c_i(\mathbf{x}, \mathbf{u})$ are continuous functions of their arguments.

By introducing the matrices

$$\mathbf{A}^i = \begin{pmatrix} A_{11}^i & \cdots & A_{1m}^i \\ \cdots & \cdots & \cdots \\ A_{m1}^i & \cdots & A_{mm}^i \end{pmatrix}, \quad \mathbf{c} = \begin{pmatrix} c_1 \\ \cdots \\ c_m \end{pmatrix},$$

the system (8.60) can be put in the form (see Chapter 8)

$$\mathbf{A}^i \frac{\partial \mathbf{u}}{\partial x_i} = \mathbf{c}. \tag{8.61}$$

The program `PdeSysClass` determines the characteristic equation associated with the system (8.61)

$$\det\left(\mathbf{I}N_1 + \sum_{\alpha=2}^n \mathbf{B}^\alpha N_\alpha\right) = 0, \tag{8.62}$$

where $\mathbf{B}^\alpha = (\mathbf{A}^1)^{-1}\mathbf{A}^\alpha$ and $\mathbf{N} \equiv (N_\alpha)$ represent the components of an arbitrary vector of $\Re^{n-1}$. It is well known that the roots N_1 of the characteristic equation (8.62) coincide with the opposite of the eigenvalues of the matrix

$$\mathbf{B} = \sum_{\alpha=2}^n \mathbf{B}^\alpha N_\alpha.$$

Then, if option = numeric, the program calculates the eigenvalues of **B** with their respective algebraic and geometric multiplicities. Finally, it classifies the system (8.60) at any point x_0 (but for the given solution u_0 if the system is quasi-linear).

Command Line of the Program PdeSysClass

PdeSysClass[sys, var, unk, point, unk0, option]

Parameter List

Input Data

> sys = quasi-linear first-order system of PDEs, where the terms containing the first derivatives appear on the left-hand side while the right-hand side contains the functions and the known terms;
>
> var = list of independent variables;
>
> unk = list of unknown functions;
>
> point = coordinates of the point x_0 at which the system of PDEs has to be classified;
>
> unk0 = known solution u_0 of the system of PDEs when it is quasi-linear;
>
> option = option for the symbolic or numeric calculations of the solutions of the associated characteristic equation.

Output Data

> the matrices A_i of the coefficients of the system of PDEs;
>
> the matrices A_i^0 obtained by evaluating A_i at point and with respect to the solution unk0;
>
> the solutions of the characteristic equation (8.62);
>
> the eigenvalues of $\mathbf{B} = \sum_{\alpha=2}^{n} B^\alpha N_\alpha$ if option = numeric;
>
> the algebraic and geometric multiplicities of the distinct eigenvalues of **B** if option = numeric;
>
> the elliptic, hyperbolic, totally hyperbolic, or parabolic character of the system of PDEs if option = numeric.

Use Instructions

The system of PDEs has to be given in such a way that on the left-hand side the first derivatives appear, while the right-hand side includes all the terms involving the unknown functions and the known terms. The system of PDEs we wish to examine must to be given according to the syntax both of Mathematica and the program, i.e., in the form

$$\text{eq1} = (A_{11}^1 u_{1x_1} + \cdots + A_{1m}^1 u_{mx_1}) + \cdots + (A_{11}^n u_{1x_n} + \cdots + A_{1m}^n u_{mx_n}) == c_1,$$
...
$$\text{eqm} = (A_{m1}^1 u_{1x_1} + \cdots + A_{mm}^1 u_{mx_1}) + \cdots + (A_{m1}^n u_{1x_n} + \cdots + A_{mm}^n u_{mx_n}) == c_m,$$

where the two sides of the any equation are separated by a double equal sign and the symbol $\dfrac{\partial u}{\partial x_i}$ is written as u_{x_i}.

For a semi-linear or linear system, only the input datum `point` has to be given, not `unk0`.

We remark that to classify the system of PDEs when `option = numeric`, the arbitrary unit vector $\mathbf{N} \equiv (N_\alpha)$ of $\Re^{n-1}$ is chosen.

Worked Examples

1. Laplace's equation: Consider the system in $\Re^2$

$$\begin{cases} v_x + w_y = 0, \\ v_y - w_x = 0, \end{cases}$$

which is equivalent to Laplace's equation

$$u_{x,x} + u_{y,y} = 0.$$

To apply the program `PdeSysClass`, it is sufficient to input the following data:

eq1 = v$_x$ + w$_y$ == 0;
eq2 = v$_x$ − w$_y$ == 0;
sys = {eq1, eq2};
var = {x, y};
unk = {v, w};
point = {x$_0$, y$_0$};
unk0 = {v$_0$, w$_0$};
option = numeric;
PdeSysClass[sys, var, unk, point, unk0, option]

In output we obtain

Coefficient matrices

$$A_1 = \begin{pmatrix} 1 & 0 \\ 0 & -1 \end{pmatrix}$$

$$A_2 = \begin{pmatrix} 0 & 1 \\ 1 & 0 \end{pmatrix}$$

Solutions of the characteristic equation

$$\det(IN_1 + \sum_{\alpha=2}^{n} B^\alpha N_\alpha) = 0$$

$N_1 = -iN_2$

$N_1 = iN_2$

Eigenvalues of $B = \sum_{\alpha=2}^{n} B^\alpha N_\alpha = \begin{pmatrix} 0 & 1 \\ -1 & 0 \end{pmatrix}$

$\lambda_1 = -i$

$\lambda_2 = i$

Algebraic and geometric multiplicities of the distinct eigenvalues of B

$\lambda_1 = -i$: AlgMult $= 1$ GeoMult $= 1$

$\lambda_2 = i$: AlgMult $= 1$ GeoMult $= 1$

The system of PDEs is elliptic.

2. Classify the linear system in $\Re^2$

$$\begin{cases} u_x + 2u_y + v_x + 3v_y = u - v, \\ 3u_x + u_y - 2v_x - v_y = 2v. \end{cases}$$

To apply the program PdeSysClass, it is sufficient to input the following data:

eq1 $= u_x + 2u_y + v_x + 3v_y == u - v$;

eq2 $= 3u_x + u_y - 2v_x - v_y == 2v$;

sys $= \{eq1, eq2\}$;

var $= \{x, y\}$;

unk $= \{u, v\}$;

point $= \{x_0, y_0\}$;

unk0 $= \{u_0, v_0\}$;

option $=$ numeric;

PdeSysClass[sys, var, unk, point, unk0, option]

8.14. The Program PdeSysClass

In output we obtain

Coefficient matrices

$$A_1 = \begin{pmatrix} 1 & 1 \\ 3 & -2 \end{pmatrix}$$

$$A_2 = \begin{pmatrix} 2 & 3 \\ 1 & -1 \end{pmatrix}$$

Solutions of the characteristic equation
$$\det(\text{IN} +_1 \sum_{\alpha=2}^{n} B^\alpha N_\alpha) = 0$$

$$N_1 = -\frac{1}{2}(3 + \sqrt{5})N_2$$

$$N_1 = \frac{1}{2}(-3 + \sqrt{5})N_2$$

Eigenvalues of $B = \sum_{\alpha=2}^{n} B^\alpha N_\alpha = \begin{pmatrix} 1 & 1 \\ 1 & 2 \end{pmatrix}$

$$\lambda_1 = \frac{1}{2}(3 - \sqrt{5})$$

$$\lambda_2 = \frac{1}{2}(3 + \sqrt{5})$$

Algebraic and geometric multiplicities of the distinct eigenvalues of B

$\lambda_1 = \frac{1}{2}(3 - \sqrt{5})$: AlgMult $= 1$ GeoMult $= 1$

$\lambda_2 = \frac{1}{2}(3 + \sqrt{5})$: AlgMult $= 1$ GeoMult $= 1$

The system of PDEs is totally hyperbolic.

Exercises

Apply the program PdeSysClass to the following systems.

1. D'Alembert's equation

$$\begin{cases} v_x - w_t = 0, \\ w_x - a^2 v_t = 0, \end{cases} \quad a \in \Re.$$

2. Heat equation

$$\begin{cases} w_x - v_y = 0, \\ v_x = -w. \end{cases}$$

3. Equations of 2D fluid dynamics

$$\begin{cases} \rho_t + v\rho_x + \rho v_x = 0, \\ \rho v_t + \rho v v_x + p'(\rho)\rho_x = 0. \end{cases}$$

4. Equations of 3D fluid dynamics

$$\begin{cases} \rho_t + u\rho_x + v\rho_y + \rho u_x + \rho v_y = 0, \\ \rho u_t + \rho u u_x + \rho v u_y + p'(\rho)\rho_x = 0, \\ \rho v_t + \rho u v_x + \rho v v_y + p'(\rho)\rho_y = 0. \end{cases}$$

5. Classify the system

$$\begin{cases} u_x + xu_t + v_x + 3v_t = u - v, \\ 3u_x + u_t - 2v_x - v_t = 2v, \end{cases}$$

along the straight line $x_0 = 2$, i.e. at the arbitrary point $(t_0, x_0 = 2)$.

6. Classify the system

$$\begin{cases} u_x + 2u_t + v_x + 3v_t = u - v, \\ 3u_x + u_t - 2v_x - v_t = 2v. \end{cases}$$

7. Classify the system

$$\begin{cases} \rho u_x + u\rho_x + \rho_t = -\dfrac{2\rho u}{x}, \\ \rho u u_x + \rho u_t + c^2 \rho_x = 0, \end{cases}$$

at any point with respect to the solution $(\rho = \rho_0, u = 0)$.

8. Tricomi's equation: Classify the system

$$\begin{cases} v_y - w_x = 0, \\ v_x + xw_y = 0, \end{cases}$$

at any point $(x = x_0, y)$ for any x_0 in $\Re$.

8.15 The Program WavesI

Aim of the Program WavesI

The program WavesI determines the characteristic equation of a quasi-linear first-order system of PDEs as well as the advancing speeds of the characteristic surfaces.

Description of the Problem and Relative Algorithm

Consider a quasi-linear first-order system of PDEs of m equations in the m unknown functions $(u_1, \cdots, u_m)$ depending on the n variables $(x_1 \equiv t, \cdots, x_n)$ in the form

$$A_{ij}^k(\mathbf{x}, \mathbf{u}) \frac{\partial u_j}{\partial x_k} = F_i(\mathbf{x}, \mathbf{u}), \qquad (8.63)$$

where $A_{ij}^k(\mathbf{x}, \mathbf{u})$ and $F_i(\mathbf{x}, \mathbf{u})$ are continuous functions of their arguments.

The hypersurface Σ_{n-1} of equation $f(\mathbf{x}) = 0$ is a characteristic surface for the Cauchy problem

$$\begin{cases} \mathbf{A}^i \dfrac{\partial \mathbf{u}}{\partial x_i} = \mathbf{F}, \\ \mathbf{u}(\mathbf{x}) = \mathbf{u}_0(\mathbf{x}) \qquad \forall \mathbf{x} \in \Sigma_{n-1}, \end{cases}$$

where

$$\mathbf{A}^i = \begin{pmatrix} A_{11}^i & \cdots & A_{1m}^i \\ \cdots & \cdots & \cdots \\ A_{m1}^i & \cdots & A_{mm}^i \end{pmatrix}$$

if it is a singular first-order surface of the function $\mathbf{u}(\mathbf{x})$, i.e., if

$$\det\left(\mathbf{A}^i \frac{\partial f}{\partial x_i}\right)_{\mathbf{x}_0} = 0. \qquad (8.64)$$

Moreover, the following jump conditions hold: [3]

$$\left[\!\left[\frac{\partial \mathbf{u}}{\partial t}\right]\!\right]_{\mathbf{x}_0} = -c_n \mathbf{a},$$

$$\left[\!\left[\frac{\partial \mathbf{u}}{\partial x_i}\right]\!\right]_{\mathbf{x}_0} = n_{i-1} \mathbf{a} \qquad \forall i = 2, \cdots, n, \quad \mathbf{x}_0 \in \Sigma_{n-1},$$

---

[3] Remember that the time occupies the first place, i.e., $x_1 \equiv t$, in the list of the independent variables.

where **n** is the unit vector normal to Σ_{n-1} at $\mathbf{x}_0$, c_n is the normal component of the advancing speed, and **a** is the discontinuity field of the same order as **u**. These last relations, together with (8.64), allow us to calculate the normal components of the advancing speeds of Σ_{n-1}.

Command Line of the Program WavesI

WavesI[sys, unk, var]

Parameter List

Input Data

sys = quasi-linear first-order system of PDEs, where the terms containing the first derivatives appear on the left-hand side, and the right-hand side contains the functions and the known terms;

unk = list of the unknown functions;

var = list of the independent variables; the first one is the time.

Output Data

the canonical form of the system;

the characteristic equation;

the normal speed of Σ_{n-1}.

Use Instructions

The system of PDEs has to be given in such a way that on the left-hand side the first derivatives appear and the right-hand side includes all the terms involving the unknown functions and the known terms. Moreover, the time t occupies the first place in the list of the independent variables.

The system of PDEs we wish to examine must to be given according to the syntax both of Mathematica and the program; i.e., any equation has to be written in the form

$$eq_i = A_{ij}^k \left(u_j\right)_{x_k} == F_i,$$

where the two sides are separated by a double equal sign and the symbol $\dfrac{\partial u_j}{\partial x_k}$ is written as $\left(u_j\right)_{x_k}$.

8.15. The Program WavesI

Worked Examples

1. Hydrodynamic equations: In the absence of body forces, consider the hydrodynamic equations of a *compressible perfect fluid*

$$\begin{cases} \rho \dot{\mathbf{v}} + p'(\rho)\nabla\rho = 0, \\ \dot{\rho} + \rho \nabla \cdot \mathbf{v} = 0, \end{cases}$$

where the pressure p is a function of the density ρ. To verify the existence of *acoustic waves* in the fluid, it is possible to use the program WavesI with the following data:[4]

eq1 $= \rho(v_1)_t + \rho v_1 (v_1)_{x1} + \rho v_2 (v_1)_{x2} + p'[\rho](\rho)_{x1} == 0$;
eq2 $= \rho(v_2)_t + \rho v_1 (v_2)_{x1} + \rho v_2 (v_2)_{x2} + p'[\rho](\rho)_{x2} == 0$;
eq3 $= (\rho)_t + v_1(\rho)_{x1} + v_2(\rho)_{x2} + \rho(v_1)_{x1} + \rho(v_2)_{x2} == 0$;
sys $= \{\text{eq1}, \text{eq2}, \text{eq3}\}$;
unk $= \{v_1, v_2, \rho\}$;
var $= \{t, x1, x2\}$;
WavesI[sys, unk, var]

The output is

Canonical form of the system

$$\begin{pmatrix} \rho & 0 & 0 \\ 0 & \rho & 0 \\ 0 & 0 & 1 \end{pmatrix} \begin{pmatrix} (v_1)_t \\ (v_2)_t \\ \rho_t \end{pmatrix} + \begin{pmatrix} \rho v_1 & 0 & p'[\rho] \\ 0 & \rho v_1 & 0 \\ \rho & 0 & v_1 \end{pmatrix} \begin{pmatrix} (v_1)_{x_1} \\ (v_2)_{x_1} \\ \rho_{x_1} \end{pmatrix}$$
$$+ \begin{pmatrix} \rho v_2 & 0 & 0 \\ 0 & \rho v_2 & p'[\rho] \\ 0 & \rho & v_2 \end{pmatrix} \begin{pmatrix} (v_1)_{x_2} \\ (v_2)_{x_2} \\ \rho_{x_2} \end{pmatrix} = \begin{pmatrix} 0 \\ 0 \\ 0 \end{pmatrix}$$

Characteristic equation in a matrix form

$$\text{Det}\left[\begin{pmatrix} \rho & 0 & 0 \\ 0 & \rho & 0 \\ 0 & 0 & 1 \end{pmatrix} f_t + \begin{pmatrix} \rho v_1 & 0 & p'[\rho] \\ 0 & \rho v_1 & 0 \\ \rho & 0 & v_1 \end{pmatrix} f_{x_1} + \begin{pmatrix} \rho v_2 & 0 & 0 \\ 0 & \rho v_2 & p'[\rho] \\ 0 & \rho & v_2 \end{pmatrix} f_{x_2} \right] = 0$$

The explicit characteristic equation

$$\rho^2 \left(f_t + f_{x_1} v_1 + f_{x_2} v_2\right)$$
$$\times \left(f_t^2 + 2 f_{x_1} f_{x_2} v_1 v_2 + 2 f_t \left(f_{x_1} v_1 + f_{x_2} v_2\right) + f_{x_1}^2 \left(v_1^2 - p'[\rho]\right) \right.$$
$$\left. + f_{x_2}^2 \left(v_2^2 - p'[\rho]\right)\right) = 0$$

---

[4] We remark that in Mathematica the functional dependence is expressed by brackets. Therefore, instead of $p = p(\rho)$ we have to write $p = p[\rho]$.

Normal speed of $\Sigma(t)$

$c_{n,1} = n_1 v_1 + n_2 v_2$

$c_{n,2} = n_1 v_1 + n_2 v_2 - \sqrt{p'[\rho]}$

$c_{n,3} = n_1 v_1 + n_2 v_2 + \sqrt{p'[\rho]}$

Exercises

Verify the existence of characteristic surfaces for the following first-order quasi-linear systems of PDEs.

1. Laplace's equation
$$\triangle u = 0.$$

In order to check that there is no real characteristic surface for this equation by the program WavesI, we need to transform the previous equation into an equivalent first-order system. But it is well known that this transformation can be done in infinite ways, without modifying either the mathematical or the physical properties. However, in this process it is essential to recall that the program WavesI determines the characteristic surfaces as discontinuity surfaces for the *first derivatives* of the involved unknown functions. In other words, only the transformations for which the new unknowns coincide with the first derivatives of the function u are allowed. In particular, we can achieve this by introducing the functions

$$u_1 = u_x, \quad u_2 = u_y,$$

and reducing the Laplace equation to the system

$$\begin{cases} (u_1)_y - (u_2)_x = 0, \\ (u_1)_x + (u_2)_y = 0, \end{cases}$$

where the first equation expresses Schwarz's condition for the second derivative of u.

To make this clearer, we note that the transformation

$$\begin{cases} u_x = u_1, \\ u_y = u_2, \\ (u_1)_x + (u_2)_y = 0, \end{cases}$$

cannot be used in the program WavesI. In fact, with these new unknowns u_1, u_2, and u, the program attempts to determine the characteristic surfaces as discontinuity surfaces for the first derivatives of u too.

8.15. The Program WavesI

2. D'Alembert's equation

$$\frac{\partial^2 u}{\partial x^2} - \frac{1}{v^2}\frac{\partial^2 u}{\partial t^2} = 0.$$

Similarly to the previous exercise, by introducing the notations $u_1 = u_t$ and $u_2 = u_x$, the wave equation becomes equivalent to the following system:

$$\begin{cases} (u_1)_x - (u_2)_t = 0, \\ (u_2)_x - \dfrac{1}{v^2}(u_1)_t = 0. \end{cases}$$

3. Heat equation

$$\frac{\partial u}{\partial t} - \frac{\partial^2 u}{\partial x^2} = 0 \equiv \begin{cases} (u_1)_x - (u_2)_t = 0, \\ u_1 - (u_2)_x = 0. \end{cases}$$

4. Tricomi's equation

$$\frac{\partial^2 u}{\partial t^2} - t\frac{\partial^2 u}{\partial x^2} = 0 \equiv \begin{cases} (u_1)_x - (u_2)_t = 0, \\ (u_1)_t - t(u_2)_x = 0. \end{cases}$$

5. Generalized Tricomi's equation

$$\frac{\partial^2 u}{\partial t^2} - g(t)\frac{\partial^2 u}{\partial x^2} = 0 \equiv \begin{cases} (u_1)_x - (u_2)_t = 0, \\ (u_1)_t - g(t)(u_2)_x = 0. \end{cases}$$

6. Maxwell's equations

$$\begin{cases} \nabla \times \mathbf{E} = -\mu\dfrac{\partial \mathbf{H}}{\partial t}, \\ \nabla \times \mathbf{H} = \epsilon\dfrac{\partial \mathbf{E}}{\partial t}, \end{cases}$$

where $\mathbf{E}$ and $\mathbf{H}$ are the electric and magnetic fields, and μ and ϵ are the dielectric and magnetic permeability constants, respectively.

Maxwell's equations are equivalent to the following first-order system:

$$\begin{cases} (E_z)_y - (E_y)_z + \mu(H_x)_t = 0, \\ (E_x)_z - (E_z)_x + \mu(H_y)_t = 0, \\ (E_y)_x - (E_x)_y + \mu(H_z)_t = 0, \\ (H_z)_y - (H_y)_z - \epsilon(E_x)_t = 0, \\ (H_x)_z - (H_z)_x - \epsilon(E_y)_t = 0, \\ (H_y)_x - (H_x)_y - \epsilon(E_z)_t = 0, \end{cases}$$

where $E_x, E_y, E_z, H_x, H_y,$ and H_z denote the components of **E** and **H** along the axes. The existence of *electromagnetic waves* can be easily verified with the program WavesI.

7. 2D-linear elasticity equation

$$\rho^* \frac{\partial^2 \mathbf{u}}{\partial t^2} = (\lambda + \mu)\nabla\nabla \cdot \mathbf{u} + \mu \triangle \mathbf{u},$$

where ρ^* is the mass density in the reference configuration, **u** is the displacement field, and λ and μ are the Lamé coefficients. By introducing the new unknowns

$$\mathbf{u}_1 = \frac{\partial \mathbf{u}}{\partial x}, \quad \mathbf{u}_2 = \frac{\partial \mathbf{u}}{\partial t}, \quad \mathbf{u}_3 = \frac{\partial \mathbf{u}}{\partial y},$$

the above equation becomes equivalent to the following first-order system composed by six equations in the unknowns $\mathbf{u}_1, \mathbf{u}_2$ and $\mathbf{u}_3$:

$$\begin{cases} (u_{1,x})_t - (u_{2,x})_x = 0, \\ (u_{1,y})_t - (u_{2,y})_x = 0, \\ (u_{2,x})_y - (u_{3,x})_t = 0, \\ (u_{2,y})_y - (u_{3,y})_t = 0, \\ \rho(u_{2,x})_t - (\lambda + \mu)\left[(u_{1,x})_x + (u_{1,y})_x\right] - \mu\left[(u_{1,x})_x + (u_{3,x})_y\right] = 0, \\ \rho(u_{2,y})_t - (\lambda + \mu)\left[(u_{3,x})_y + (u_{3,y})_x\right] - \mu\left[(u_{2,y})_x + (u_{3,y})_y\right] = 0. \end{cases}$$

The use of the program WavesI shows that the longitudinal waves travel with a greater velocity than the transverse waves.

8.
$$\begin{cases} u_t - q_x = 0, \\ q_t - a(u)q_x = 0. \end{cases}$$

9.
$$\begin{cases} u_t - q_x = 0, \\ q_t - a(u)q_x - b(u)v_x = 0, \\ q_x - v = 0. \end{cases}$$

8.16 The Program WavesII

Aim of the Program WavesII

The program WavesII evaluates the characteristic equation of a system of second-order quasi-linear PDEs as well as the advancing speeds of the characteristic surfaces and the second derivative jumps. Moreover, in the 2D case, it distinguishes between transverse and longitudinal waves.

Description of the Problem and Relative Algorithm

Consider a second-order quasi-linear system of m equations in m unknowns $(u_1, \cdots, u_m)$ depending on n variables $(x_1 \equiv t, \cdots, x_n)$:

$$A_{ij}^{hk}(\mathbf{x}, \mathbf{u}, \nabla \mathbf{u}) \frac{\partial^2 u_j}{\partial x_h \partial x_k} = F_i(\mathbf{x}, \mathbf{u}, \nabla \mathbf{u}), \tag{8.65}$$

where $A_{ij}^k(\mathbf{x}, \mathbf{u}, \nabla \mathbf{u})$ and $F_i(\mathbf{x}, \mathbf{u}, \nabla \mathbf{u})$ are continuous functions of their arguments.

A hypersurface Σ_{n-1} of equation $f(\mathbf{x}) = 0$ is a characteristic surface relative to the Cauchy problem

$$\begin{cases} \mathbf{A}^{ij} \dfrac{\partial^2 \mathbf{u}}{\partial x_i \partial x_j} = \mathbf{F}, \\ \mathbf{u}(\mathbf{x}) = \mathbf{u}_0(\mathbf{x}) & \forall \mathbf{x} \in \Sigma_{n-1}, \\ \dfrac{d\mathbf{u}}{d\mathbf{n}} \equiv \nabla \mathbf{u} \cdot \mathbf{n} = \mathbf{d}_0(\mathbf{x}) & \forall \mathbf{x} \in \Sigma_{n-1}, \end{cases}$$

where

$$\mathbf{A}^{ij} = \begin{pmatrix} A_{11}^{ij} & \cdots & A_{1m}^{ij} \\ \cdots & \cdots & \cdots \\ A_{m1}^{ij} & \cdots & A_{mm}^{ij} \end{pmatrix}$$

if it is a second-order singular surface for the function $\mathbf{u}(\mathbf{x})$, i.e., if

$$\det \left(\mathbf{A}^{ij} \frac{\partial f}{\partial x_i} \frac{\partial f}{\partial x_j} \right)_{\mathbf{x}_0} = 0. \tag{8.66}$$

Moreover, the following jump conditions hold:[5]

$$\left[\left[\frac{\partial^2 \mathbf{u}}{\partial t^2} \right]\right]_{\mathbf{x}_0} = c_n^2 \mathbf{a},$$

---
[5] We recall that the time always occupies the first position in the list of the variables, i.e., $x_1 \equiv t$.

$$\left[\left[\frac{\partial^2 \mathbf{u}}{\partial x_i \partial x_j}\right]\right]_{\mathbf{x}_0} = n_{i-1}n_{j-1}\mathbf{a} \qquad \forall i,j = 2,\cdots,n,$$

$$\left[\left[\frac{\partial^2 \mathbf{u}}{\partial x_i \partial t}\right]\right]_{\mathbf{x}_0} = -c_n n_{i-1}\mathbf{a} \qquad \forall i = 2,\cdots,n, \qquad \mathbf{x}_0 \in \Sigma_{n-1},$$

where $\mathbf{n}$ is the unit vector normal to Σ_{n-1} at $\mathbf{x}_0$, c_n is the advancing velocity of Σ_{n-1}, and $\mathbf{a}$ is the discontinuity field of the same order as $\mathbf{u}$. These last relations, together with (8.66), allow us to evaluate the advancing velocity of Σ_{n-1} and the jump fields.

In the planar case, the program distinguishes between transverse and longitudinal waves. In particular, the velocity c_n refers to a *transverse wave* if the second derivative jumps on the characteristic surface are orthogonal to $\mathbf{n}$, that is, the following relations hold:

$$\begin{cases} \mathbf{a} \cdot \mathbf{n} = 0, \\ \mathbf{a} \times \mathbf{n} \neq 0; \end{cases}$$

similarly, c_n refers to a *longitudinal wave* when the jumps are parallel to $\mathbf{n}$:

$$\begin{cases} \mathbf{a} \cdot \mathbf{n} \neq 0, \\ \mathbf{a} \times \mathbf{n} = 0. \end{cases}$$

Command Line of the Program WavesII

WavesII[sys, unk, var]

Parameter List

Input Data

sys = second-order quasi-linear system of PDEs, where all the terms containing the second derivative appear on the left-hand side, and all the other terms appear on the right-hand side;

unk = list of the unknowns;

var = list of the independent variables where the first one is the time.

Output Data

characteristic equation associated with the system;

the normal speed of Σ_{n-1};

jump vectors;

in the 2D case the distinction between transverse and longitudinal waves.

8.16. The Program WavesII

Use Instuctions

The system has to be given in such a way that on the left-hand sides the second derivatives appear while the right-hand sides include all the terms involving the unknown functions and their first derivatives. The time t occupies the first position of the list of the independent variables. Each equation of the system must to be written with the syntax both of Mathematica and the program, i.e., in the form

$$eq_i = A_{ij}^{hk}(u_j)_{x_h,x_k} == F_i,$$

where the two sides of the equation are separated by a double equal sign and the symbol $\dfrac{\partial^2 u_j}{\partial x_h \partial x_k}$ is written as $(u_j)_{x_h,x_k}$.

Worked Examples

1. Consider the 2D linear elasticity equation

$$\rho^* \frac{\partial^2 \mathbf{u}}{\partial t^2} = (\lambda + \mu)\nabla\nabla \cdot \mathbf{u} + \mu \triangle \mathbf{u},$$

where ρ^* is the mass density, $\mathbf{u} \equiv (u_x, u_y)$ is the displacement field, and λ and μ are the Lamé coefficients.

To verify the existence of longitudinal and transverse waves, we can use the program WavesII by inputting the following input data:

eq1 = $\rho(u_x)_{t,t} - (\lambda+\mu)((u_x)_{x,x} + (u_y)_{x,y}) - \mu((u_x)_{x,x} + (u_x)_{y,y}) == 0$;
eq2 = $\rho(u_y)_{t,t} - (\lambda+\mu)((u_x)_{y,x} + (u_y)_{y,y}) - \mu((u_y)_{x,x} + (u_y)_{y,y}) == 0$;
sys = $\{eq1, eq2\}$;
unk = $\{u_x, u_y\}$;
var = $\{t, x, y\}$;
WavesII[sys, unk, var]

The corresponding output is

Characteristic equation
$(-\lambda - 2\mu + \rho c_n^2)(-\mu + \rho c_n^2) = 0$

Normal speed of $\Sigma(t)$

$c_{n,1} = -\dfrac{\sqrt{\mu}}{\sqrt{\rho}}$

$c_{n,2} = \dfrac{\sqrt{\mu}}{\sqrt{\rho}}$

$$c_{n,3} = -\frac{\sqrt{\lambda + 2\mu}}{\sqrt{\rho}}$$

$$c_{n,4} = \frac{\sqrt{\lambda + 2\mu}}{\sqrt{\rho}}$$

Jump vectors

$$a_1 = \frac{a_2 n_1}{n_2}$$

$$a_1 = -\frac{a_2 n_2}{n_1}$$

The velocity $c_{n,1} = -\frac{\sqrt{\mu}}{\sqrt{\rho}}$ refers to a transverse wave.

The velocity $c_{n,2} = \frac{\sqrt{\mu}}{\sqrt{\rho}}$ refers to a transverse wave.

The velocity $c_{n,3} = -\frac{\sqrt{\lambda + 2\mu}}{\sqrt{\rho}}$ refers to a longitudinal wave.

The velocity $c_{n,4} = \frac{\sqrt{\lambda + 2\mu}}{\sqrt{\rho}}$ refers to a longitudinal wave.

Exercises

Apply the program WavesII to the following systems.

1. Navier–Stokes equations

$$\begin{cases} \rho \dot{\mathbf{v}} = (\lambda + \mu)\nabla\nabla \cdot \mathbf{v} + \mu \triangle \mathbf{v} - p'(\rho)\nabla\rho, \\ \dot{\rho} + \rho \nabla \cdot \mathbf{v} = 0. \end{cases}$$

The program WavesII shows that, due to the viscosity, there is no propagation in the fluid.

2. Laplace's equation

$$\triangle u = 0.$$

3. D'Alembert's equation

$$\frac{\partial^2 u}{\partial x^2} - \frac{1}{v^2}\frac{\partial^2 u}{\partial t^2} = 0.$$

4. Heat equation

$$\frac{\partial u}{\partial t} - \frac{\partial^2 u}{\partial x^2} = 0.$$

5. Tricomi's equation
$$\frac{\partial^2 u}{\partial t^2} - t\frac{\partial^2 u}{\partial x^2} = 0.$$

6. Generalized Tricomi's equation
$$\frac{\partial^2 u}{\partial t^2} - g(t)\frac{\partial^2 u}{\partial x^2} = 0.$$

Chapter 9

Fluid Mechanics

9.1 Perfect Fluid

A *perfect fluid* or *Euler fluid* was defined (see (7.28)) to be a fluid S whose stress tensor is expressed by the constitutive equation

$$\mathbf{T} = -p\mathbf{I}, \qquad p > 0, \tag{9.1}$$

where the positive scalar p is called the *pressure*. If the pressure

$$p = p(\rho) \tag{9.2}$$

is a given function of the mass density ρ, then the fluid is called *compressible*. On the other hand, if the fluid is density preserving, so that the pressure does not depend on ρ, then the fluid is said to be *incompressible*. In this case the pressure p is no longer defined by a constitutive equation; it is a function of $\mathbf{x}$ to be deduced from the momentum balance. By assuming (9.1), it follows that the stress acting on any unit surface with normal $\mathbf{N}$ is given by

$$\mathbf{t} = \mathbf{T}\mathbf{N} = -p\mathbf{N}, \tag{9.3}$$

so that $\mathbf{t}$ is normal to the surface, is directed towards the interior, and its intensity does not depend on $\mathbf{N}$ (*Pascal's principle*). When dealing with *static conditions* ($\mathbf{v} = \mathbf{0}$) and conservative body forces, i.e. $\mathbf{b} = -\nabla U$, the momentum balance equation (5.30) is

$$\nabla p = -\rho \nabla U. \tag{9.4}$$

Experimental evidence leads us to assume that $dp/d\rho > 0$ and the function $p = p(\rho)$ can be inverted. If we introduce the notation

$$h(p) = \int \frac{dp}{\rho(p)}, \tag{9.5}$$

245

then
$$\nabla h(p) = \frac{dh}{dp}\nabla p = \frac{1}{\rho}\nabla p,$$
and (9.4) becomes
$$\nabla(h(p) + U) = 0. \tag{9.6}$$
The equation (9.6) shows that in the whole volume occupied by the fluid we have
$$h(p) + U = const. \tag{9.7}$$
In particular, if the fluid is incompressible ($\rho = const$), then instead of (9.7), we have
$$\frac{p}{\rho} + U(\mathbf{x}) = const. \tag{9.8}$$
Both (9.7) and (9.8) allow us to say that potential surfaces coincide with the isobars.

9.2 Stevino's Law and Archimedes' Principle

It is relevant to observe how many conclusions can be inferred from (9.7) and (9.8).

1. Let S be an incompressible fluid on the boundary of which acts a uniform pressure p_0 (atmospheric pressure). If the vertical axis z is oriented downward, then the relation $U(\mathbf{x}) = -gz + const$ implies that the *free surface is a horizontal plane*.

2. Let S be an incompressible fluid subjected to a uniform pressure p_0 on its boundary. If the cylinder which contains the fluid is rotating around a vertical axis a with uniform angular velocity ω, then its *free surface is a paraboloid of rotation around a*. In fact, the force acting on the arbitrary particle P is given by $\mathbf{g} + \omega^2 \overrightarrow{PQ}$, where Q is the projection of P on a. It follows that

$$U = -gz - \frac{\omega^2}{2}(x^2 + y^2),$$

and then (9.8) proves that the free surface is a paraboloid (see Figure 9.1).

9.2. Stevino's Law and Archimedes' Principle

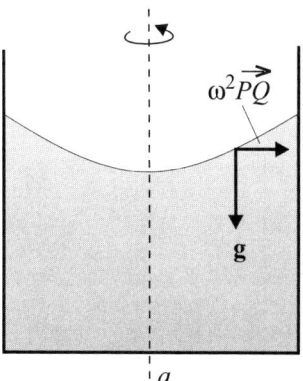

Figure 9.1

3. Assume that the arbitrary constant in the expression of U is such that if $z = 0$, then $U(0) = 0$. From (9.8) it follows that

$$\frac{p(z)}{\rho} - gz = \frac{p_0}{\rho},$$

i.e.,

$$p(z) = p_0 + \rho gz, \quad z > 0, \qquad (9.9)$$

and **Stevino's law** is obtained: *The pressure linearly increases with depth by an amount equal to the weight of the fluid column acting on the unit surface.*

4. The simplest constitutive equation (9.2) is given by a linear function relating the pressure to the mass density. By assuming that the proportionality factor is a linear function of the absolute temperature θ, the corresponding constitutive equation defines a **perfect gas**

$$p = R\rho\theta, \qquad (9.10)$$

where R is the universal gas constant. Let the gas be at equilibrium at constant and uniform temperature when subjected to its weight action. Assuming the axis Oz is oriented upward, we get $U(z) = gz + const$ and, taking into account (9.10), equation (9.7) becomes

$$R\theta \int_{p_0}^{p} \frac{dp}{p} = -gz,$$

so that

$$p(z) = p_0 \exp\left(-\frac{gz}{R\theta}\right). \qquad (9.11)$$

5. ***Archimedes' principle*** is a further consequence of (9.8): *The buoyant force on a body submerged in a liquid is equal to the weight of the liquid displaced by the body.*

   To prove this statement, consider a body S submerged in a liquid, as shown in Figure 9.2.

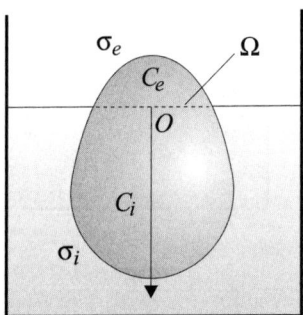

Figure 9.2

If p_0 is the atmospheric pressure, then the force acting on S is given by

$$\mathbf{F} = -\int_{\sigma_e} p_0 \mathbf{N}\, d\sigma - \int_{\sigma_i} (p_0 + \rho g z) \mathbf{N}\, d\sigma, \qquad (9.12)$$

where $\mathbf{N}$ is the outward unit vector normal to the body surface σ, σ_i is the submerged portion of σ, and σ_e is the portion above the waterline. By adding and subtracting on the right-hand side of (9.12) the integral of $-p_0\mathbf{N}$ over Ω (see Figure 9.2), (9.12) becomes

$$\mathbf{F} = -\int_{\partial C_e} p_0 \mathbf{N}\, d\sigma - \int_{\partial C_i} (p_0 + \rho g z) \mathbf{N}\, d\sigma, \qquad (9.13)$$

so that the knowledge of $\mathbf{F}$ requires the computation of the integrals in (9.13). To do this, we define a virtual pressure field (continuous on Ω) in the interior of the body:

$$p = p_0 \quad \text{on } C_e,$$
$$p = p_0 + \rho g z \quad \text{on } C_i.$$

Applying Gauss's theorem to the integrals in (9.13), we obtain

$$\int_{\partial C_e} p_0 \mathbf{N}\, d\sigma = \int_{C_e} \nabla p_0\, dV = 0, \quad \int_{\partial C_i} p_0 \mathbf{N}\, d\sigma = \int_{C_i} \nabla p_0\, dV = 0,$$

$$\int_{\partial C_i} \rho g z \mathbf{N}\, d\sigma = \rho g \int_{C_i} \nabla z\, dV.$$

9.2. Stevino's Law and Archimedes' Principle

Finally,
$$\mathbf{F} = -\rho g V_i \mathbf{k}, \tag{9.14}$$
where $\mathbf{k}$ the unit vector associated with Oz.

The equation (9.14) gives the resultant force acting on the body S. To complete the equilibrium analysis, the momentum $\mathbf{M}_O$ of pressure forces with respect to an arbitrary pole O has to be explored. If $\mathbf{r} = \overrightarrow{OP}$, then

$$\mathbf{M}_O = -p_0 \int_{\sigma_e} \mathbf{r} \times \mathbf{N} \, d\sigma - \int_{\sigma_i} \mathbf{r} \times (p_0 + \rho g z) \mathbf{N} \, d\sigma. \tag{9.15}$$

Again, by adding and subtracting on the right-hand side of (9.15) the integral of $-p_0 \mathbf{r} \times \mathbf{N}$ over Ω, (9.15) becomes

$$\mathbf{M}_O = -p_0 \int_{\partial C_e} \mathbf{r} \times \mathbf{N} \, d\sigma - \int_{\partial C_i} \mathbf{r} \times (p_0 + \rho g z) \mathbf{N} \, d\sigma.$$

By applying Gauss's theorem, we get

$$-p_0 \int_{C_e} \epsilon_{ijl} \frac{\partial x_j}{\partial x_l} \, dV = -p_0 \int_{C_e} \epsilon_{ijl} \delta_{jl} \, dV = 0,$$

$$\int_{C_i} \frac{\partial}{\partial x_l} [\epsilon_{ijl} x_j (p_0 + \rho g z)] \, dV$$

$$= -\int_{C_i} [\epsilon_{ijl} \delta_{jl} (p_0 + \rho g z) + \epsilon_{ijl} x_j \rho g \delta_{3l}] \, dV$$

$$= -\rho g \int_{C_i} \epsilon_{ij3} x_j \, dV.$$

Finally,
$$\mathbf{M}_0 = -\rho g \left[x_{2C} \mathbf{i} - x_{1C} \mathbf{j} \right] V_i, \tag{9.16}$$
where $\mathbf{i}$ and $\mathbf{j}$ are orthonormal base vectors on the horizontal plane and

$$x_{1C} V_i = \int_{C_i} x_1 \, dV, \qquad x_{2C} V_i = \int_{C_i} x_2 \, dV. \tag{9.17}$$

Expression (9.16) shows that the momentum of the pressure forces vanishes if the line of action of the buoyant force passes through the centroid of the body. The centroid of the displaced liquid volume is called **center of buoyancy**.

In summary: *A body floating in a liquid is at equilibrium if the buoyant force is equal to its weight and the line of action of the buoyant force passes through the centroid of the body.* It can be proved that the equilibrium is stable if the center of buoyancy is above the centroid and it is unstable if the center of buoyancy is below the centroid.

An extension of Archimedes' principle to perfect gases is discussed in the Exercise 3.

9.3 Fundamental Theorems of Fluid Dynamics

The momentum balance equation (5.30), when applied to a perfect fluid subjected to conservative body forces, is written as

$$\rho \dot{\mathbf{v}} = -\nabla p - \rho \nabla U; \tag{9.18}$$

with the additional introduction of (9.5), it holds that

$$\dot{\mathbf{v}} = -\nabla (h(p) + U). \tag{9.19}$$

By recalling the above definitions, the following theorems can be proved.

Theorem 9.1 (W. Thomson, Lord Kelvin)
In a barotropic flow under conservative body forces, the circulation around any closed material curve γ is preserved; i.e., it is independent of time:

$$\frac{d}{dt} \int_\gamma \mathbf{v} \cdot d\mathbf{s} = 0. \tag{9.20}$$

PROOF If γ is a material closed curve, then there exists a closed curve γ^* in C^* such that γ is the image of γ^* under the motion equation $\gamma = \mathbf{x}(\gamma^*, t)$. It follows that

$$\frac{d}{dt} \int_\gamma v_i \, dx_i = \frac{d}{dt} \int_{\gamma^*} v_i \frac{\partial x_i}{\partial X_j} dX_j = \int_{\gamma^*} \frac{d}{dt} \left(v_i \frac{\partial x_i}{\partial X_j} dX_j \right)$$

$$= \int_{\gamma^*} \left(\dot{v}_i \frac{\partial x_i}{\partial X_j} + v_i \frac{\partial \dot{x}_i}{\partial X_j} \right) dX_j = \int_\gamma \left(\dot{v}_i + v_i \frac{\partial v_i}{\partial x_j} \right) dx_j,$$

and, taking into account (9.19), it is proved that

$$\frac{d}{dt} \int_\gamma \mathbf{v} \cdot d\mathbf{s} = \int_\gamma \left(\dot{\mathbf{v}} + \frac{1}{2} \nabla v^2 \right) \cdot d\mathbf{s} = -\int_\gamma \nabla \left(h(p) + U - \frac{1}{2} v^2 \right) \cdot d\mathbf{s} = 0. \tag{9.21}$$

∎

Theorem 9.2 (Lagrange)
If at a given instant t_0 the motion is irrotational, then it continues to be irrotational at any $t > t_0$, or equivalently, vortices cannot form.

PROOF This can be regarded as a special case of Thomson's theorem. Suppose that in the region C_0 occupied by the fluid at the instant t_0 the

9.3. Fundamental Theorems of Fluid Dynamics

condition $\boldsymbol{\omega} = \mathbf{0}$ holds. Stokes's theorem requires that

$$\Gamma_0 = \int_{\gamma_0} \mathbf{v} \cdot d\mathbf{s} = 0$$

for any material closed curve γ_0. But Thomson's theorem states that $\Gamma = 0$ for all $t > t_0$, so that also $\boldsymbol{\omega}(t) = \mathbf{0}$ holds at any instant. ∎

Theorem 9.3 (Bernoulli)
In a steady flow, along any particle path, i.e., along the trajectory of an individual element of fluid, the quantity

$$H = \frac{1}{2}v^2 + h(p) + U, \tag{9.22}$$

is constant. In general, the constant H changes from one streamline to another, but if the motion is irrotational, then H is constant in time and over the whole space of the flow field.

PROOF By recalling (4.18) and the time independence of the flow, (9.19) can be written as

$$\dot{\mathbf{v}} = (\nabla \times \mathbf{v}) \times \mathbf{v} + \frac{1}{2}\nabla v^2 = -\nabla(h(p) + U). \tag{9.23}$$

A scalar multiplication by $\mathbf{v}$ gives the relation

$$\dot{\mathbf{v}} \cdot \nabla \left(\frac{1}{2}v^2 + h(p) + U \right) = 0,$$

which proves that (9.22) is constant along any particle path. If the steady flow is irrotational, then (9.23) implies $H = const$ through the flow field at any time. ∎

In particular, if the fluid is incompressible, then Bernoulli's theorem states that in a steady flow along any particle path (or through the flow field if the flow is irrotational) the quantity H is preserved; i.e.,

$$H = \frac{1}{2}v^2 + \frac{p}{\rho} + U = const. \tag{9.24}$$

The Bernoulli equation is often used in another form, obtained by dividing (9.24) by the gravitational acceleration

$$h_z + h_p + h_v = const,$$

where $h_z = U/g$ is the **gravity head** or **potential head**, $h_p = p/\rho g$ is the **pressure head** and $h_v = v^2/2g$ is the **velocity head**.

In a steady flow, a ***stream tube*** is a tubular region Σ within the fluid bounded by streamlines. We note that streamlines cannot intersect each other. Because $\partial \rho / \partial t = 0$, the balance equation (5.22) gives $\nabla \cdot (\rho \mathbf{v}) = 0$, and by integrating over a volume V defined by the sections σ_1 and σ_2 of a stream tube (see Figure 9.3), we obtain

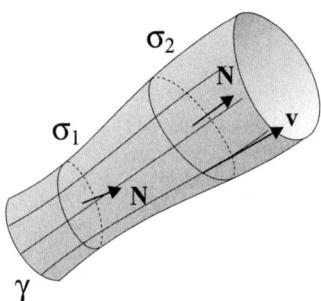

Figure 9.3

$$Q = \int_{\sigma_1} \rho \mathbf{v} \cdot \mathbf{n}\, d\sigma = \int_{\sigma_2} \rho \mathbf{v} \cdot \mathbf{n}\, d\sigma. \qquad (9.25)$$

This relation proves that the ***flux*** is constant across any section of the stream tube. If the fluid is incompressible, then (9.25) reduces to

$$\int_{\sigma_1} \mathbf{v} \cdot \mathbf{n}\, d\sigma = \int_{\sigma_2} \mathbf{v} \cdot \mathbf{n}\, d\sigma. \qquad (9.26)$$

The local angular speed ω is also called the ***vortex vector*** and the related integral curves are ***vortex lines***; furthermore, a ***vortex tube*** is a surface represented by all vortex lines passing through the points of a (nonvortex) closed curve. By recalling that a vector field $\mathbf{w}$ satisfying the condition $\nabla \cdot \mathbf{w} = 0$ is termed *solenoidal*, and that $2\nabla \cdot \omega = \nabla \cdot \nabla \times \mathbf{v} = 0$, we conclude that the field ω is solenoidal. Therefore vortex lines are closed if they are limited, and they are open if unconfined. We observe that Figure 9.3 can also be used to represent a vortex tube if the vector $\mathbf{v}$ is replaced by ω.

The following examples illustrate some relevant applications of Bernoulli's equation.

1. Consider an open vessel with an orifice at depth h from the free surface of the fluid. Suppose that fluid is added on the top, in order to keep constant the height h. Under these circumstances, it can be proved that the *velocity of the fluid leaving the vessel through the orifice is equal to that of a body falling from the elevation h with initial velocity*

9.3. Fundamental Theorems of Fluid Dynamics

equal to zero (this result is known as **Torricelli's theorem**, because it was found long before Bernoulli's work). Assuming that at the free surface we have $\mathbf{v} = \mathbf{0}$ and $z = 0$, it follows that $H = p_0/\rho$ and, by applying (9.24), we derive the relation

$$H = \frac{p_0}{\rho} = \frac{v^2}{2} - gh + \frac{p_0}{\rho},$$

so that $v = \sqrt{2gh}$.

2. *In a horizontal pipe of variable cross section, the pressure of an incompressible fluid in steady motion decreases in the converging section.*

First, the mass balance equation (9.26) requires that

$$v_1 \sigma_1 = v_2 \sigma_2, \qquad (9.27)$$

so that *the fluid velocity increases in the converging section and decreases in the diverging section.*

Furthermore, since $U = \rho z = const$ along the stream tube, (9.24) implies that

$$\frac{v_1^2}{2} + \frac{p_1}{\rho} = \frac{v_2^2}{2} + \frac{p_2}{\rho},$$

and this proves that the pressure decreases in a converging section. This result is applied in **Venturi's tube**, where a converging section acts as a nozzle, by increasing the fluid velocity and decreasing its pressure.

Theorem 9.4 (First Helmholtz's theorem)
The flux of the vortex vector across any section of a vortex tube is constant.

PROOF Let σ_1 and σ_2 be two sections of a vortex tube T and consider the closed surface Σ defined by σ_1, σ_2 and the lateral surface of T. By applying Gauss's theorem, we have

$$\int_\Sigma \boldsymbol{\omega} \cdot \mathbf{N} \, d\sigma = \int_V \nabla \cdot \boldsymbol{\omega} \, dV = \frac{1}{2} \int_V \nabla \cdot (\nabla \times \mathbf{v}) \, dV = 0,$$

where $\mathbf{N}$ is the unit outward vector normal to Σ. The definition of a vortex tube implies that $\boldsymbol{\omega}$ is tangent to Σ at any point, so that the theorem is proved since

$$\int_\Sigma \boldsymbol{\omega} \cdot \mathbf{N} \, d\sigma = \int_{\sigma_1} \boldsymbol{\omega} \cdot \mathbf{N}_1 \, d\sigma = \int_{\sigma_2} \boldsymbol{\omega} \cdot \mathbf{N}_2 \, d\sigma = 0,$$

where $\mathbf{N}_1$ is the unit vector normal to σ_1, pointing towards the interior of the tube, and $\mathbf{N}_2$ is the outward unit vector normal to σ_2. ∎

From this theorem it also follows that the particle vorticity increases if the vortex curves are converging.

Theorem 9.5 (Second Helmholtz's theorem)
Vortex lines are material lines.

PROOF At the instant $t_0 = 0$, the vector w is supposed to be tangent to the surface σ_0. Denote by $\sigma(t)$ the material surface defined by the particles lying upon σ_0 at the instant t_0. We have to prove that $\sigma(t)$ is a vortex surface at any arbitrary instant. First, we verify that the circulation Γ along any closed line γ_0 on σ_0 vanishes. In fact, if A is the portion of σ_0 contained in γ_0, it holds that

$$\Gamma = \int_{\gamma_0} \mathbf{v} \cdot d\mathbf{s} = \int_A \nabla \times \mathbf{v} \cdot \mathbf{N} \, d\sigma = 2 \int_A w \cdot \mathbf{N} \, d\sigma = 0,$$

since w is tangent to A. According to Thomson's theorem, the circulation is preserved along any material curve, so that, if $\gamma(t)$ is the image of γ_0, it follows that

$$\int_{\gamma(t)} \mathbf{v} \cdot d\mathbf{s} = \int_{A(t)} \nabla \times \mathbf{v} \cdot \mathbf{N} d\sigma = 2 \int_{A(t)} w \cdot \mathbf{N} d\sigma = 0.$$

Since $A(t)$ is arbitrary, $w \cdot \mathbf{N} = 0$ and the theorem is proved. ∎

The theorem can also be stated by saying that the vortex lines are constituted by the same fluid particles and are transported during the motion. Examples include the smoke rings, whirlwinds and so on.

9.4 Boundary Value Problems for a Perfect Fluid

The motion of a perfect compressible fluid S subjected to body forces $\mathbf{b}$ is governed by the momentum equation (see (9.18))

$$\dot{\mathbf{v}} = -\frac{1}{\rho}\nabla p(\rho) + \mathbf{b} \tag{9.28}$$

and the mass conservation

$$\dot{\rho} + \rho \nabla \cdot \mathbf{v} = 0. \tag{9.29}$$

9.4. Boundary Value Problems for a Perfect Fluid

Equations (9.28) and (9.29) are a first-order system for the unknowns $\mathbf{v}(\mathbf{x}, t)$ and $\rho(\mathbf{x}, t)$ and, to find a unique solution, both initial and boundary conditions have to be specified.

If we consider the motion in a fixed and compact region C of the space, (e.g., a liquid in a container with rigid walls), the initial conditions are

$$\mathbf{v}(\mathbf{x}, 0) = \mathbf{v}_0(\mathbf{x}), \qquad \rho(\mathbf{x}, 0) = \rho_0(\mathbf{x}) \qquad \forall \mathbf{x} \in C, \tag{9.30}$$

and the boundary condition is

$$\mathbf{v} \cdot \mathbf{N} = 0 \qquad \forall \mathbf{x} \in \partial C, \quad t > 0. \tag{9.31}$$

This boundary condition states that the fluid can perform any tangential motion on the fixed surface, whose unit normal is $\mathbf{N}$.

The problem is then to find in $C \times [0, t]$ the fields $\mathbf{v}(\mathbf{x}, t)$ and $\rho(\mathbf{x}, t)$ that satisfy the balance equations (9.28), (9.29), the initial conditions (9.30), and the boundary condition (9.31).

If the fluid is incompressible ($\rho = const$), then the equation (9.28) becomes

$$\dot{\mathbf{v}} = -\frac{1}{\rho}\nabla p + \mathbf{b}, \tag{9.32}$$

while the mass conservation (9.29) leads us to the condition

$$\nabla \cdot \mathbf{v} = 0. \tag{9.33}$$

The unknowns of the system (9.32) and (9.33) are given by the fields $\mathbf{v}(\mathbf{x}, t)$ and $p(\mathbf{x}, t)$, and the appropriate initial and boundary conditions are

$$\begin{aligned} \mathbf{v}(\mathbf{x}, 0) &= \mathbf{v}_0(\mathbf{x}) & \forall \mathbf{x} \in C, \\ \mathbf{v} \cdot \mathbf{N} &= 0 & \forall \mathbf{x} \in \partial C, \, t > 0. \end{aligned} \tag{9.34}$$

A more complex problem arises when a part of the boundary is represented by a *moving* or *free surface* $f(\mathbf{x}, t) = 0$. In this case, finding the function f is a part of the boundary value problem. The moving boundary $\partial C'$, represented by $f(\mathbf{x}, t) = 0$, is a *material surface*, since a material particle located on it has to remain on this surface during the motion. This means that its velocity c_N along the unit normal $\mathbf{N}$ to the free surface is equal to $\mathbf{v} \cdot \mathbf{N}$; that is, $f(\mathbf{x}, t)$ has to satisfy the condition (see (4.33))

$$\frac{\partial}{\partial t} f(\mathbf{x}, t) + \mathbf{v}(\mathbf{x}, t) \cdot \nabla f(\mathbf{x}, t) = 0.$$

In addition, on the free surface it is possible to prescribe the value of the pressure, so that the *dynamic boundary conditions* are

$$\begin{aligned} \frac{\partial}{\partial t} f(\mathbf{x}, t) + \mathbf{v}(\mathbf{x}, t) \cdot \nabla f(\mathbf{x}, t) &= 0, \\ p &= p_e \qquad \forall \mathbf{x} \in \partial C', \quad t > 0, \end{aligned} \tag{9.35}$$

where p_e is the prescribed external pressure.

On the fixed boundary part, the previous impenetrability condition (9.32) applies, and if the boundary C extends to infinity, then conditions related to the asymptotic behavior of the solution at infinity have to be added.

9.5 2D Steady Flow of a Perfect Fluid

The following two conditions define an irrotational steady motion of an incompressible fluid S:

$$\nabla \times \mathbf{v} = \mathbf{0}, \qquad \nabla \cdot \mathbf{v} = 0, \qquad (9.36)$$

where $\mathbf{v} = \mathbf{v}(\mathbf{x})$. The first condition allows us to deduce the existence of a **velocity** or **kinetic potential** $\varphi(\mathbf{x})$ such that

$$\mathbf{v} = \nabla \varphi, \qquad (9.37)$$

where φ is a single- or a multiple-valued function, depending on whether the motion region C is connected or not.[1]

In addition, taking into account $(9.36)_2$, it holds that

$$\Delta \varphi = \nabla \cdot \nabla \varphi = 0. \qquad (9.38)$$

The equation (9.38) is known as **Laplace's equation** and its solution is a **harmonic function**.

Finally, it is worthwhile to note that, in dealing with a *two-dimensional* (2D) flow, the velocity vector $\mathbf{v}$ at any point is parallel to a plane π and it is independent of the coordinate normal to this plane. In this case, if a system $Oxyz$ is introduced, where the axes x and y are parallel to π and the z axis is normal to this plane, then we have

$$\mathbf{v} = u(x,y)\mathbf{i} + v(x,y)\mathbf{j},$$

where u, v are the components of $\mathbf{v}$ on x and y, and $\mathbf{i}, \mathbf{j}$ are the unit vectors of these axes.

If now C is a simply connected region of the plane Oxy, conditions (9.36) become

$$-\frac{\partial u}{\partial y} + \frac{\partial v}{\partial x} = 0, \qquad \frac{\partial u}{\partial x} + \frac{\partial v}{\partial y} = 0. \qquad (9.39)$$

---

[1] If C is not a simply connected region, then the condition $\nabla \times \mathbf{v} = \mathbf{0}$ does not imply that $\int_\gamma \mathbf{v} \cdot d\mathbf{s} = 0$ on any closed curve γ, since this curve could not be the boundary of a surface contained in C. In this case, Stokes's theorem cannot be applied.

9.5. 2D Steady Flow of a Perfect Fluid

These conditions allow us to state that the two differential forms $\omega_1 = u\,dx + v\,dy$ and $\omega_2 = -v\,dx + u\,dy$ are integrable, i.e. there is a function φ, called the **velocity potential** or the **kinetic potential**, and a function ψ, called the **stream potential** or the **Stokes potential**, such that

$$d\varphi = u\,dx + v\,dy, \qquad d\psi = -v\,dx + u\,dy. \tag{9.40}$$

From $(9.36)_2$ it follows that the curves $\varphi = const$ are at any point normal to the velocity field. Furthermore, since $\nabla\varphi \cdot \nabla\psi = 0$, the curves $\psi = const$ are flow lines.

It is relevant to observe that (9.40) suggest that the functions φ and ψ satisfy the Cauchy–Riemann conditions

$$\frac{\partial\varphi}{\partial x} = \frac{\partial\psi}{\partial y}, \qquad \frac{\partial\varphi}{\partial y} = -\frac{\partial\psi}{\partial x}, \tag{9.41}$$

so that the complex function

$$F(z) = \varphi(x,y) + i\psi(x,y) \tag{9.42}$$

is **holomorphic** and represents a **complex potential**. Then the complex potential can be defined as the holomorphic function whose real and imaginary parts are the velocity potential φ and the stream potential ψ, respectively. The two functions φ and ψ are harmonic and the derivative of $F(z)$,

$$V \equiv F'(z) = \frac{\partial\varphi}{\partial x} + i\frac{\partial\psi}{\partial x} = u - iv = |V|e^{-i\theta}, \tag{9.43}$$

represents the **complex velocity**, with $|V|$ being the modulus of the velocity vector and θ the angle that this vector makes with the x axis.

Within the context of considerations developed in the following discussion, it is relevant to remember that the line integral of a holomorphic function vanishes around any arbitrary closed path in a simply connected region, since the Cauchy–Riemann equations are necessary and sufficient conditions for the integral to be independent of the path (and therefore it vanishes for a closed path).

The above remarks lead to the conclusion that a 2D irrotational flow of an incompressible fluid is completely defined if a harmonic function $\varphi(x,y)$ or a complex potential $F(z)$ is prescribed, as it is shown in the examples below.

Example 9.1
Uniform motion. Given the complex potential

$$F(z) = U_0(x + iy) = U_0 z, \tag{9.44}$$

it follows that $V = U_0$, and the 2D motion

$$\mathbf{v} = U_0 \mathbf{i} \qquad (9.45)$$

is defined, where $\mathbf{i}$ is the unit vector of the axis Ox. The kinetic and Stokes potentials are $\varphi = U_0 x$, $\psi = U_0 y$, and the curves $\varphi = const$ and $\psi = const$ are parallel to Oy and Ox (see Figure 9.4), respectively. This example shows that the complex potential (9.44) can be introduced in order to describe a 2D uniform flow, parallel to the wall $y = 0$.

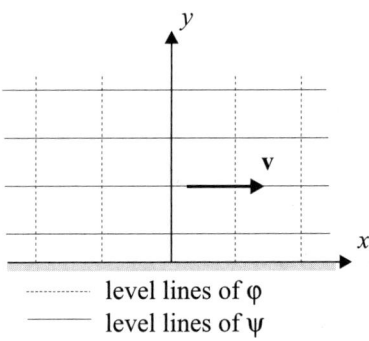

level lines of φ
level lines of ψ

Figure 9.4: Uniform motion

Example 9.2
Vortex potential. Let a 2D flow be defined by the complex potential

$$F(z) = -i\frac{\Gamma}{2\pi} \ln z = -i\frac{\Gamma}{2\pi} \ln r e^{i\theta} = \frac{\Gamma}{2\pi}\theta - i\frac{\Gamma}{2\pi} \ln r, \qquad (9.46)$$

where r, θ are polar coordinates. It follows that

$$\varphi = \frac{\Gamma}{2\pi}\theta = \frac{\Gamma}{2\pi}\arctan\frac{y}{x},$$
$$\psi = -i\frac{\Gamma}{2\pi}\ln r = -\frac{\Gamma}{2\pi}\ln(x^2 + y^2),$$

and the curves $\varphi = const$ are straight lines through the origin, while the curves $\psi = const$ are circles whose center is the origin (see Figure 9.5).

Accordingly, the velocity components become

$$u = \frac{\partial \varphi}{\partial x} = -\frac{\Gamma}{2\pi}\frac{y}{x^2 + y^2} = -\frac{\Gamma}{2\pi}\frac{\sin\theta}{r},$$
$$v = \frac{\partial \varphi}{\partial y} = \frac{\Gamma}{2\pi}\frac{x}{x^2 + y^2} = \frac{\Gamma}{2\pi}\frac{\cos\theta}{r}.$$

9.5. 2D Steady Flow of a Perfect Fluid

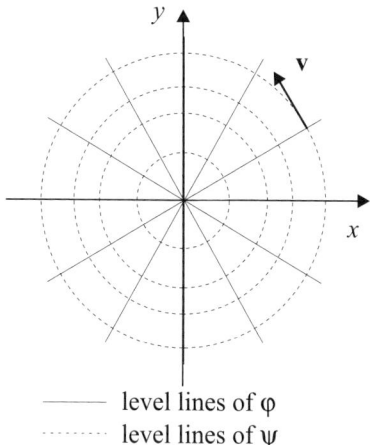

— level lines of φ
······ level lines of ψ

Figure 9.5: Vortex potential

It is relevant to observe that the circulation around a path γ bordering the origin is given by
$$\int_\gamma \mathbf{v} \cdot d\mathbf{s} = \Gamma,$$
so that it does not vanish if $\Gamma \neq 0$. This is not in contradiction with the condition $\nabla \times \mathbf{v} = \mathbf{0}$ since the plane without the origin is no longer a simply connected region. In Section 9.7, further arguments will be addressed in order to explain why the circulation around an obstacle does not vanish in a 2D flow.

The complex potential (9.46) can then be used with advantage to describe the uniform 2D flow of particles rotating around the axis through the origin and normal to the plane Oxy.

Example 9.3
Sources and sinks. For pure radial flow in the horizontal plane, the complex potential is taken:
$$F(z) = \frac{Q}{2\pi} \ln z = \frac{Q}{2\pi} \ln re^{i\theta} = \frac{Q}{2\pi}(\ln r + i\theta), \qquad Q > 0.$$
so that
$$\varphi = \frac{Q}{2\pi} \ln r = \frac{Q}{2\pi} \log(x^2 + y^2), \qquad \psi = \frac{Q}{2\pi}\theta = \frac{Q}{2\pi} \arctan \frac{y}{x}.$$
The curves $\varphi = const$ are circles and the curves $\psi = const$ are straight lines through the origin. The velocity field is given by
$$\mathbf{v} = \frac{Q}{2\pi r} \mathbf{e}_r,$$

where $\mathbf{e}_r$ is the unit radial vector (see Figure 9.6). Given an arbitrary closed path γ around the origin, the radial flow pattern associated with the above field is said to be either a **source**, if

$$\int_\gamma \mathbf{v} \cdot \mathbf{n}\, d\sigma = Q > 0, \qquad (9.47)$$

or a **sink**, if $Q < 0$. The quantity $Q/2\pi$ is called the **strength** of the source or sink.

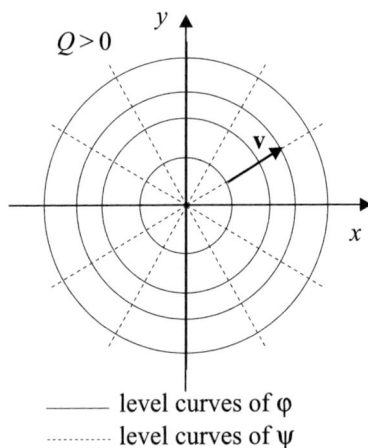

——— level curves of φ
·········· level curves of ψ

Figure 9.6: Sources and Sinks

Example 9.4
Doublet. A **doublet** is the singularity obtained by taking to zero the distance between a source and a sink having the same strength. More precisely, consider a source and a sink of equal strength, the source placed at the point A, with $z_1 = -ae^{i\alpha}$, and the sink placed at B, with $z_2 = ae^{i\alpha}$. The complex potential for the combined flow is then

$$F(z) = \frac{Q}{2\pi} \ln(z + ae^{i\alpha}) - \frac{Q}{2\pi} \ln(z - ae^{i\alpha}) \equiv f(z, a), \qquad (9.48)$$

The reader is invited to find the kinetic and Stokes potential (see Figure 9.7).

If points A and B are very near to each other, i.e. $a \simeq 0$, it follows that

$$F(z) = f(z, 0) + f'(z, 0) a = \frac{Q}{2\pi} \frac{1}{z} 2ae^{i\alpha}.$$

9.5. 2D Steady Flow of a Perfect Fluid

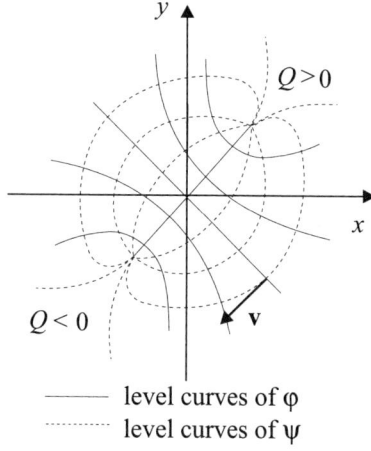

Figure 9.7: Doublet

In particular, if $a \to 0$, so that $Qa/\pi \to m$, then we obtain the potential of a doublet (source–sink):
$$F(z) = \frac{m}{z} e^{i\alpha}. \tag{9.49}$$

Furthermore, if $\alpha = 0$, e.g., if the source and the sink are on the x axis, then
$$F(z) = \frac{m}{z} = \frac{my}{x^2 + y^2} - i \frac{mx}{x^2 + y^2} \equiv \varphi + i\psi \tag{9.50}$$

and the flow lines $\psi = const$ are circles through the origin.

Example 9.5
Consider the 2D flow with complex potential
$$F(z) = U_0 z + \frac{U_0 a^2}{z} \tag{9.51}$$

i.e., the flow is obtained as the superposition of a uniform flow parallel to the x axis (in the positive direction) and the flow due to a dipole (source–sink) system of strength $m = U_0 a^2$. Assuming $z = x + iy$, then from (9.51) it follows that

$$\varphi = U_0 x \left(1 + \frac{a^2}{x^2 + y^2}\right), \qquad \psi = U_0 y \left(1 - \frac{a^2}{x^2 + y^2}\right).$$

The flow lines $\psi = const$ are represented in Figure 9.8 (for $U_0 = a = 1$).
Note that the condition $x^2 + y^2 = a^2$ corresponds to the line $\psi = 0$. Furthermore, if the region internal to this circle is substituted with the cross section of a cylinder, (9.51) can be assumed to be the complex potential of

the velocity field around a cylinder. The components of the velocity field are given by

$$u = \frac{\partial \varphi}{\partial x} = U_0\left(1 - a^2\frac{x^2 - y^2}{r^4}\right) = U_0\left(1 - \frac{a^2}{r^2}\cos 2\theta\right),$$

$$v = \frac{\partial \varphi}{\partial y} = U_0\frac{a^2}{r^2}\sin 2\theta.$$

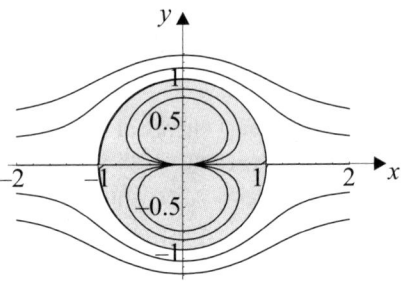

Figure 9.8

It should be noted that, although singular points associated with a doublet do not actually occur in real fluids, they are interesting because the flow pattern associated with a doublet is a useful approximation far from singular points, and it can be combined to advantage with other nonsingular complex potentials.

Example 9.6
All the previous examples, with the exception of Example 9.2, refer to 2D irrotational flows whose circulation along an arbitrary closed path is vanishing. We now consider the complex potential

$$F(z) = U_0 z + \frac{U_0 a^2}{z} + i\frac{\Gamma}{2\pi}\ln\frac{z}{a}. \tag{9.52}$$

This represents a flow around a cylinder of radius a obtained by superposing a uniform flow, a flow generated by a dipole, and the flow due to a vortex. The dipole and the vortex are supposed to be located on the center of the section. In this case, according to (9.47), the vorticity is Γ, and the vortex can be supposed to be produced by a rotation of the cylinder around its axis. Flow lines are given by the complex velocity

$$V = \frac{dF}{dz} = U_0\left(1 - \frac{a^2}{z^2}\right) + i\frac{\Gamma}{2\pi z}.$$

9.5. 2D Steady Flow of a Perfect Fluid

In order to find **stagnation points** $(V = 0)$, we must solve the equation $dF/dz = 0$. Its roots are given by

$$z_{v=0} = \frac{1}{4U_0\pi}\left(-i\Gamma \pm \sqrt{(16U_0^2\pi^2 a^2 - \Gamma^2)}\right). \tag{9.53}$$

Three cases can be distinguished: $\Gamma^2 < 16\pi^2 a^2 U^2$, $\Gamma^2 = 16\pi^2 a^2 U^2$, and $\Gamma^2 > 16\pi^2 a^2 U^2$, whose corresponding patterns are shown in Figures 9.9, 9.10, and 9.11 (assuming $U_0 = a = 1$).

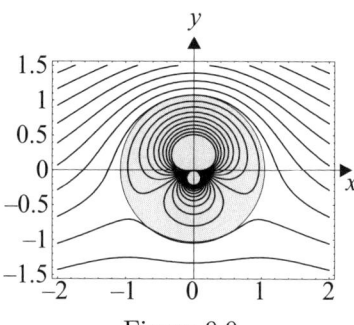

Figure 9.9

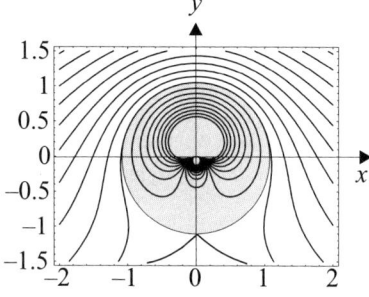

Figure 9.10

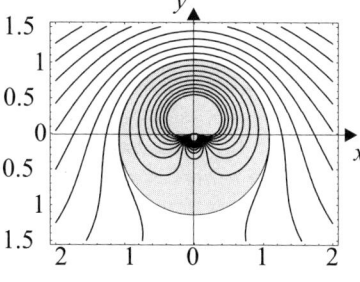

Figure 9.11

At this stage the reader should be aware that these examples, although of practical relevance, are based on equations of steady-state hydrodynamics, which in some occasions give rise to inaccurate or paradoxical results. As an example, in the next section we will analyze D'Alembert's paradox, according to which the drag of a fluid on an obstacle is zero. In order to remove this paradox, it will be necessary to introduce a *boundary layer* within which viscosity cannot be disregarded.

9.6 D'Alembert's Paradox and the Kutta–Joukowsky Theorem

In spite of the attractive results obtained in the previous section, the assumption of a perfect fluid cannot be used to explain some important phenomena occurring in fluid dynamics.

In particular, the condition $\mathbf{v} \cdot \mathbf{n} = 0$ assumed on the boundary ∂C means that the fluid is free to move with respect to the wall boundary, while there is actually no relative movement and the appropriate boundary condition on ∂C should be $\mathbf{v} = \mathbf{0}$. It has also been shown that, when considering the irrotational steady flow of a perfect fluid, the motion is described by a velocity potential φ that is the solution of a Neumann's problem for Laplace's equation. It follows that the boundary condition $\mathbf{v} = \nabla \varphi = \mathbf{0}$ on ∂C, appropriate for a real fluid, is a redundant condition, not admissible for Laplace's equation.

This example is just one of the contradictions arising when dealing with inviscid fluids. The aim of this section is to discuss the most relevant of these.

Let V be a fixed (nonmaterial) volume. Then the momentum balance equation is written as

$$\frac{d}{dt} \int_V \rho \mathbf{v} \, dc = \int_{\partial V} (\mathbf{T} - \rho \mathbf{v} \otimes \mathbf{v}) \cdot \mathbf{N} \, d\sigma + \int_V \rho \mathbf{b} \, dc. \qquad (9.54)$$

If a fixed solid S of volume C is placed in a steady flow of an incompressible fluid, then equation (9.54) allows us to find the drag on S, assuming that $\mathbf{b} = \mathbf{0}$. Since the flow is steady, the left-hand side of (9.54) vanishes; furthermore, if Δ is the region bounded by the surface ∂C of the solid and an arbitrary surface Σ that contains C (see Figure 9.12), then by recalling that $\mathbf{v} \cdot \mathbf{n} = 0$ on ∂C, we find that

$$\int_{\partial C} \mathbf{T} \cdot \mathbf{N} \, d\sigma + \int_\Sigma (\mathbf{T} - \rho \mathbf{v} \otimes \mathbf{v}) \cdot \mathbf{N} \, d\sigma = 0.$$

9.6. D'Alembert's Paradox and the Kutta–Joukowsky Theorem

The first integral is just the opposite of the force $\mathbf{F}$ acting on S, so that the previous relation becomes

$$\mathbf{F} = \int_{\Sigma} (\mathbf{T} - \rho \mathbf{v} \otimes \mathbf{v}) \cdot \mathbf{N} \, d\sigma. \tag{9.55}$$

From similar arguments it can be proved that the torque of the force acting on S with respect to a pole O can be obtained from (5.30), so that

$$\mathbf{M}_O = \int_{\Sigma} \mathbf{r} \times (\mathbf{T} - \rho \mathbf{v} \otimes \mathbf{v}) \cdot \mathbf{N} \, d\sigma. \tag{9.56}$$

At this stage we recall the following theorem of the potential theory, without a proof.

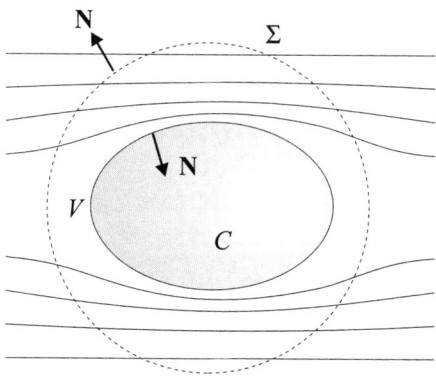

Figure 9.12

Theorem 9.6
If in the region surrounding a solid C an irrotational flow of an incompressible fluid satisfies the condition

$$\lim_{r \to \infty} \mathbf{v} = \mathbf{V},$$

where $\mathbf{V}$ is a constant vector (representing the undisturbed motion at infinity) and r is the distance of any point from an arbitrary origin, then the velocity field assumes the asymptotic behavior

$$\mathbf{v} = \mathbf{V} + O(r^{-3}). \tag{9.57}$$

In a 2D motion, (9.57) can be replaced by the following relation:

$$\mathbf{v} = V\mathbf{i} + \frac{\Gamma}{2\pi r^2}(-y\mathbf{i} + x\mathbf{j}) + O(r^{-2}), \tag{9.58}$$

where **i** and **j** are the unit vectors of the axes Ox and Oy, **i** is parallel to **V**, and the circulation

$$\Gamma = \oint \mathbf{v} \cdot d\mathbf{s} \tag{9.59}$$

refers to any closed path surrounding the solid body.

Note that in two dimensions the condition $\nabla \times \mathbf{v} = \mathbf{0}$ does not imply that $\Gamma = 0$, since the region surrounding the obstacle is not simply connected. However, by applying Stokes's theorem, it can be proved that Γ assumes a value which is independent of the path.

By taking into account the asymptotic behavior (9.57) and Bernoulli's theorem (9.24), where $U = 0$ in the absence of body forces, we find that

$$p = p_0 + \frac{1}{2}\rho(V^2 - v^2) = p_0 + O(r^{-3}), \tag{9.60}$$

where p_0 is the pressure at infinity. Substituting $\mathbf{T} \cdot \mathbf{N} = [-p_0 + 0(r^{-3})]\mathbf{N}$ into (9.55) gives

$$\mathbf{F} = -\int_\Sigma (p_0\mathbf{I} - \rho\mathbf{V} \otimes \mathbf{V})\mathbf{N}\,d\sigma + \int_\Sigma O(r^{-3})\,d\sigma,$$

so that, assuming that the arbitrary surface Σ is a sphere of radius R, we have

$$\mathbf{F} = -(p_0\mathbf{I} - \rho\mathbf{V} \otimes \mathbf{V})\int_\Sigma \mathbf{N}\,d\sigma + O(R^{-1}). \tag{9.61}$$

Finally, by applying Gauss's theorem to the right-hand side, we find that

$$\mathbf{F} = \mathbf{0},$$

i.e., *the irrotational flow of a perfect fluid gives zero drag on any obstacle placed in a fluid stream* (**D'Alembert's paradox**).

In a 2D motion, from (9.58), we have

$$v^2 = V^2 - V\frac{\Gamma}{\pi r^2}y + O(r^{-2}).$$

Substituting into Bernoulli's theorem instead of (9.60), we get

$$p = p_0 + \frac{1}{2}\rho\frac{\Gamma}{\pi r^2}yV + O(r^{-2}). \tag{9.62}$$

By combining with (9.55), the force acting on the obstacle assumes the expression

$$\mathbf{F} = -(p_0 + \rho V^2)\,\mathbf{i}\int_\Sigma N_x\,dl - p_0\mathbf{j}\int_\Sigma N_y\,dl$$
$$- \frac{\Gamma}{2\pi R^2}V\mathbf{j}\int_\Sigma (xN_x + yN_y)\,dl$$
$$- \frac{\Gamma}{2\pi R^2}V\mathbf{i}\int_\Sigma (yN_x - xN_y)\,dl + \int_\Sigma O(r^{-2})\,dl,$$

where Σ is now a circle of radius R surrounding the body C. Since integrals

$$\int_\Sigma N_x\,dl, \qquad \int_\Sigma N_y\,dl,$$

vanish, the previous expression becomes

$$\mathbf{F} = -\frac{\Gamma}{2\pi R^2}V\mathbf{j}\int_\Sigma (xN_x + yN_y)\,dl - \frac{\Gamma}{2\pi R^2}V\mathbf{i}\int_\Sigma (yN_x - xN_y)\,dl + O(r^{-1}),$$

In addition we have $(xN_x + yN_y) = \mathbf{R}\cdot\mathbf{N} = R$ and $(yN_x - xN_y) = 0$, so that the **Blasius formula**

$$\mathbf{F} = -\rho\Gamma V\mathbf{j} \tag{9.63}$$

is obtained.

This formula shows that *although a steady flow of an inviscid fluid predicts no drag on an obstacle in the direction of the relative velocity in the unperturbed region, it can predict a force normal to this direction*. This is a result obtained independently by W.M. Kutta in 1902 and N.E. Joukowski (sometimes referred as Zoukowskii) in 1906, known as the **Kutta–Joukowski theorem**. Such a force is called **lift**, and it is important for understanding why an airplane can fly.

Before closing this section, we observe that our inability to predict the drag for an inviscid fluid in the direction of relative velocity does not means we should abandon the perfect fluid model. Viscosity plays an important role around an obstacle, but, far from the obstacle, the motion can still be conveniently described according to the assumption of an inviscid fluid.

9.7 Lift and Airfoils

A *lifting wing* having the form of an infinite cylinder of appropriate cross section normal to rulings, is usually called an **airfoil**.

Airplane wings are obviously cylinders of finite length, and the effects of this finite length have an important role in the theory of lift. Nevertheless, even considering an infinitely long cylinder, essential aspects of lift can be identified.

First, we must justify a nonvanishing circulation around a wing. To do this, consider a wing L of section S and the fluid motion around it, described by the complex potential (9.51) at time t_0. Let γ be any material curve surrounding the wing (see Figure 9.13).

In the region $R^2 - S$, the motion is assumed to be irrotational with the exception of a small portion $\Delta \cup d$, near the wing, whose presence is justified by the following arguments: viscosity acts in a boundary layer Δ just around the wing and it produces a vortex of area d at the point A. This vortex grows away from the wing and disappears into the fluid mass, but it is continuously generated. The *material curve* γ is assumed to be large enough to contain S at any instant t. Since in the region $R^2 - S \cup \Delta \cup d$, which is not a simply connected region, the motion is irrotational, Thomson's theorem implies that the circulation on γ is zero at any time. Furthermore, by applying Stokes's theorem to the region bounded by the oriented curves γ, γ_1, and γ_2, we see that the circulation on γ_1 must be equal and opposite to the circulation on γ_2.

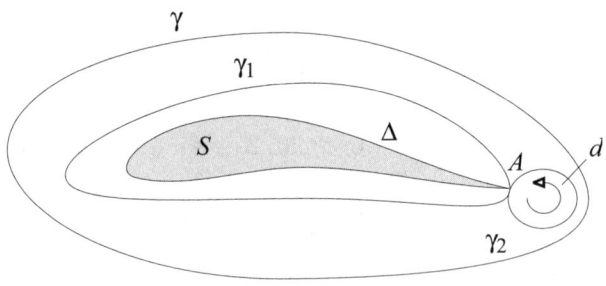

Figure 9.13

At this stage, it becomes useful to transform the flow field we are dealing with into another flow field which is easier to determine. For this reason, we recall the definition of *conformal mapping* together with some other properties. If a complex variable w is an analytic function of z, i.e. $w = \Phi(z)$, then there is a connection between the shape of a curve in the z plane and the shape of the curve in the w plane, as a consequence of the properties of an analytic function. In fact, the value of the derivative is independent of the path the increments dx and dy follow in going to zero. Since it can be proved that this transformation preserves angles and their orientation, it is called conformal mapping.

As already shown in Section 9.5, the flow around a given profile can be described through a convenient complex potential. The simplest wing profile was suggested by Joukowski. It is obtained from the complex potential $F(\zeta)$ of the motion around a circular cylinder (Example 5) in the plane $\zeta = \xi + i\eta$, by means of the conformal mapping $z = x + iy = \Phi(\zeta)$.

Note that lift can occur if there is an asymmetry due either to the asymmetry of the body or to a misalignment between the body and the approaching flow. The angle of misalignment is called the **angle of attack**. The angle of attack to the cylinder of the velocity vector is denoted by φ,

9.7. Lift and Airfoils

as shown in Figure 9.14, and the motion is supposed to be described by the complex potential

$$\hat{F}(\zeta) = U_0 \left[\left(\zeta - be^{i\theta}\right) e^{-i\varphi} + \frac{a^2}{\left(\zeta - be^{i\theta}\right) e^{-i\varphi}} \right]. \tag{9.64}$$

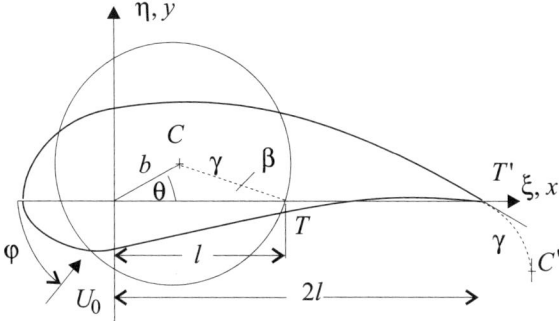

Figure 9.14

It is obtained from the potential introduced in Section 9.5, Example 5, i.e., $F(\zeta') = U_0(\zeta' + a^2/\zeta')$, by applying the transformations $\zeta' \to \zeta'' = \zeta' e^{i\varphi} \to \zeta = \zeta'' + be^{i\theta}$, corresponding to a rigid rotation φ of the axes and a rigid translation of the origin into $be^{i\theta}$.

Joukowsky's transformation is defined as

$$z = \Phi(\zeta) = \zeta + \frac{l^2}{\zeta}, \tag{9.65}$$

which is conformal everywhere, with the exception of the origin, which is mapped to infinity. The inverse transformation cannot be carried out globally, since from (9.65) it follows that

$$\zeta = \frac{z \pm \sqrt{z^2 - 4l^2}}{2}.$$

It can also be proved that the region external to the circle γ of radius a is in one-to-one correspondence with the region external to the curve given by the image Γ of the circle γ.

As a consequence, Joukowsky's conformal mapping allows us to define, in the region external to Γ, a new complex potential

$$F(z) = \hat{F}(\Phi^{-1}(z))$$

which describes the motion around the profile Γ. Such a profile, provided that the parameter l is conveniently selected, is the wing shape in Figure

9.14. In particular, the point $T(l,0)$ in the plane (ξ,η), corresponding to the sharp trailing edge, is mapped into $T'(2l,0)$ in the plane (x,y). This point is singular for the derivative $d\hat{F}/d\zeta$ of (9.64), since

$$\left(\frac{dF}{dz}\right)_{z=(2l,0)} = \left(\frac{d\hat{F}}{d\zeta}\right)_{\zeta=(l,0)} \left(\frac{d\zeta}{dz}\right)_{z=(2l,0)}$$

$$= \left(\frac{d\hat{F}}{d\zeta}\right)_{\zeta=(l,0)} \frac{1}{(dz/d\zeta)_{\zeta=(l,0)}}$$

$$= \left(\frac{d\hat{F}}{d\zeta}\right)_{\zeta=(l,0)} \frac{1}{\left(1-\frac{l^2}{\zeta^2}\right)_{\zeta=(l,0)}}.$$

A nonvanishing circulation around the wing is justified by Joukowsky assuming that the *velocity has a finite value at any point*.

In order to satisfy this condition, Joukowsky introduces the complex potential due to the circulation (see Example 6 in Section 9.5), so that the velocity is zero at T'. The complex potential $F(z)$, due to the superposition of different contributions, is then given by

$$F(z) = U_0 \left[\left(\zeta - be^{i\theta}\right)e^{-i\varphi} + \frac{a^2}{\left(\zeta - be^{i\theta}\right)e^{-i\varphi}}\right] + i\frac{\Gamma}{2\pi}\ln\frac{\left(\zeta - be^{i\theta}\right)e^{-i\varphi}}{a}, \tag{9.66}$$

where ζ and z correspond to each other through (9.65). Figure 9.15 shows streamlines around the cylinder obtained by considering the real part of the right-hand side of (9.66), i.e., the kinetic potential $\varphi(\xi,\eta)$. Figure 9.16 shows the wing profile and the related streamlines.

The complex velocity at $T'(2l,0)$ in the z plane is

$$\left(\frac{dF}{dz}\right)_{z=(2l,0)} = \left(\frac{d\hat{F}}{d\zeta}\right)_{\zeta=(l,0)} \frac{d\zeta}{dz}.$$

Since $\left(\frac{d\zeta}{dz}\right)_{(2l,0)} = 1/(1 - l^2/\zeta^2)_{\zeta=(l,0)} = \infty$, to obtain a finite value of $(dF/dz)_{(2l,0)}$ the derivative $(dF/d\zeta)_{(l,0)}$ needs to vanish. This condition is satisfied by observing that

$$\frac{d\hat{F}}{d\zeta} = U_0 \left[e^{-i\varphi} - \frac{a^2 e^{i\varphi}}{\left(\zeta - be^{i\theta}\right)^2}\right] + i\frac{\Gamma}{2\pi}\frac{1}{\zeta - be^{i\theta}}, \tag{9.67}$$

and from Figure 9.14 it also follows that $\zeta - be^{i\theta} = \zeta - l + ae^{i(\pi-\beta)}$; therefore, when $\zeta = l$, from (9.67) we deduce that

$$U_0\left(1 - e^{-2i(\pi-\beta)+2i\varphi}\right) + i\frac{\Gamma}{2\pi a}e^{-i(\pi-\beta)+i\varphi} = 0.$$

9.7. Lift and Airfoils

Equating to zero both the real and imaginary parts, we get the value of Γ:

$$\Gamma = -4\pi a U_0 \sin(\pi - \beta - \varphi) = 4\pi a U_0 \sin(\beta + \varphi), \qquad (9.68)$$

which depends on the velocity of the undisturbed stream, on the apparent attack angle φ, and on β. The last two parameters define the dimension and curvature of the wing profile, given a and U_0.

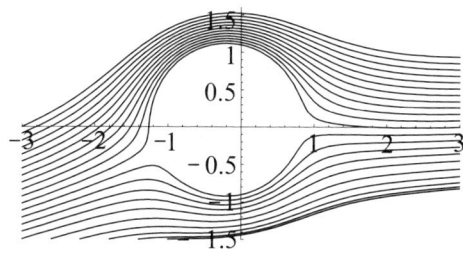

Figure 9.15

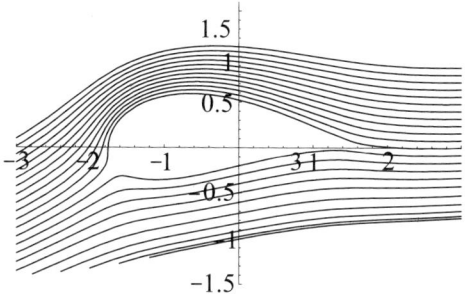

Figure 9.16

Increasing the angle of attack of an airfoil increases the flow asymmetry, resulting in greater lift. Experimental evidence shows that if the total angle of attack $\varphi + \beta$ reaches a critical value, the airfoil *stalls*, and the lift drops dramatically.

There are three programs attached to this chapter: Wing, Potential, and Joukowsky. The program Wing gives the curve Γ in the plane z corresponding to the circle of unit radius through Joukowsky's map.

The program Potential provides a representation of the streamlines corresponding to a given complex potential $F(z)$.

Finally, the program Joukowski gives the streamlines around a wing, allowing changes of the angle of attack as well as the coordinates of the center of the cylinder.

9.8 Newtonian Fluids

The simplest assumption, that the difference between the stress in a moving fluid and the stress at equilibrium is linearly related to the rate of deformation tensor, is due to Newton (1687). The 3D case was studied by Navier (1821) for an incompressible fluid, and later by Poisson (1831) for the general case.

Due to these contributions, we have that the constitutive equations of a linear compressible or incompressible viscous fluid (see Section 7.3), respectively are

$$\mathbf{T} = [-p(\rho) + \lambda(\rho) I_D] \mathbf{I} + 2\mu(\rho) \mathbf{D}, \tag{9.69}$$
$$\mathbf{T} = -p\mathbf{I} + 2\mu\mathbf{D}. \tag{9.70}$$

If (9.69) is introduced into (5.30)$_1$, the **Navier–Stokes equation**

$$\rho\dot{\mathbf{v}} = \rho\mathbf{b} - \nabla p + \nabla(\lambda \nabla \cdot \mathbf{v}) + \nabla \cdot (2\mu\mathbf{D}) \tag{9.71}$$

is obtained. Consequently, for a *compressible viscous fluid* we must find the fields $\mathbf{v}(\mathbf{x}, t)$ and $\rho(\mathbf{x}, t)$ which satisfy the system

$$\begin{aligned}\rho\dot{\mathbf{v}} &= \rho\mathbf{b} - \nabla p + \nabla(\lambda \nabla \cdot \mathbf{v}) + \nabla \cdot (2\mu\mathbf{D}), \\ \dot{\rho} + \rho \nabla \cdot \mathbf{v} &= 0,\end{aligned} \tag{9.72}$$

in the domain C occupied by the fluid, as well as the boundary condition on a fixed wall

$$\mathbf{v} = \mathbf{0} \qquad \text{on } \partial C \tag{9.73}$$

and the initial conditions

$$\rho(\mathbf{x}, 0) = \rho_0(\mathbf{x}), \quad \mathbf{v}(\mathbf{x}, 0) = \mathbf{v}_0(\mathbf{x}) \qquad \forall \mathbf{x} \in C. \tag{9.74}$$

For the case of an *incompressible fluid*, for which $\rho = \text{const}$, $\mu = \text{const}$, $\nabla \cdot \mathbf{v} = 0$, and

$$2\nabla \cdot \mathbf{D} = \Delta v + \nabla(\nabla \cdot \mathbf{v}) = \Delta v,$$

equation (9.71) becomes

$$\dot{\mathbf{v}} = \mathbf{b} - \frac{1}{\rho}\nabla p + \nu\Delta v,$$

where $\nu = \mu/\rho$ is the **coefficient of kinematic viscosity**.

Then the problem reduces to finding the fields $\mathbf{v}(\mathbf{x}, t)$ and $p(\mathbf{x}, t)$ that satisfy the system

$$\begin{aligned}\dot{\mathbf{v}} &= \mathbf{b} - \frac{1}{\rho}\nabla p + \frac{\mu}{\rho}\Delta v \\ \nabla \cdot \mathbf{v} &= 0,\end{aligned} \tag{9.75}$$

the boundary condition (9.73), and the initial condition (9.74).

9.9 Applications of the Navier–Stokes Equation

In this section we consider a steady flow of a linear viscous fluid characterized by a velocity field parallel to the axis Ox in the absence of body forces.

If Oy and Oz are two other axes, which together with Ox form an orthogonal frame of reference, then

$$\mathbf{v} = (x, y, z)\mathbf{i}, \qquad (9.76)$$

where $\mathbf{i}$ is the unit vector associated with the Ox axis. The introduction of (9.76) into (9.75) gives

$$\frac{\partial v}{\partial x} v + \frac{1}{\rho}\frac{\partial p}{\partial x} = \frac{\mu}{\rho}\Delta v,$$

$$\frac{\partial p}{\partial y} = \frac{\partial p}{\partial z} = 0,$$

$$\frac{\partial v}{\partial x} = 0.$$

These equations imply that

$$p = p(x), \qquad v = v(y, z),$$

so that the pressure and the velocity field depend only on x and on y, z, respectively. It follows that they are both equal to the same constant A:

$$\frac{\partial p}{\partial x} = \rho A, \qquad \frac{\mu}{\rho}\left(\frac{\partial^2 v}{\partial y^2} + \frac{\partial^2 v}{\partial z^2}\right) = A. \qquad (9.77)$$

The first equation tells us that p is a linear function of x, so that if p_0 and p_1 are the values at $x = 0$ and $x = l$, it follows that

$$p = \frac{p_1 - p_0}{l} x + p_0. \qquad (9.78)$$

Therefore, $(9.77)_2$ becomes

$$\frac{\partial^2 v}{\partial y^2} + \frac{\partial^2 v}{\partial z^2} = \frac{p_1 - p_0}{\mu l}. \qquad (9.79)$$

We integrate (9.79) when $v = v(y)$ and the fluid is confined between the two plates $y = 0$ and $y = h$ of infinite dimension, defined by $y = 0$ and $y = h$. Moreover, the second plate is supposed to move with uniform velocity V along Ox and p_0, p_1 are supposed to be equal. We have

$$v = ay + b,$$

where a and b are constant. If there is no relative slip at $y = 0$ and $y = h$, i.e.,
$$v(0) = 0, \qquad v(h) = V,$$
we get
$$v = \frac{V}{h} y. \tag{9.80}$$

Now, a flow in a cylinder with axis Ox is taken into account. In cylindrical coordinates (r, φ, x), defined by
$$r = \sqrt{y^2 + z^2}, \quad \varphi = \arctan \frac{z}{y}, \quad x = x,$$
provided that $v(y, z) = v(r)$, equation (9.79) can be written as
$$\frac{d}{dr}\left(r \frac{dv}{dr} \right) = \frac{p_1 - p_0}{\mu l} r, \tag{9.81}$$
and the integration gives
$$v = \frac{p_1 - p_0}{4\mu l} r^2 + A \log r + B, \tag{9.82}$$
where A and B are arbitrary constants. The condition that v has a finite value at $r = 0$ requires $A = 0$. Furthermore, the adhesion condition $v(a) = 0$, where a is the radius of the cylinder, leads to
$$B = -\frac{p_1 - p_0}{4\mu l} a^2,$$
so that (9.82) becomes
$$v = \frac{p_1 - p_0}{4\mu l} (r^2 - a^2). \tag{9.83}$$

The solution (9.83) allows us to compute the flux Q across the tube:
$$Q = \rho \int_0^a v 2\pi r \, dr = \pi \rho a^4 \frac{p_0 - p_1}{8\mu l}. \tag{9.84}$$

Formula (9.84) is used to determine the viscosity μ, by measuring Q, $p_1 - p_0$, and the mass density of the fluid.

9.10 Dimensional Analysis and the Navier–Stokes Equation

It is well known that *dimensional analysis* is a very useful technique in modeling a physical phenomenon. By using such a procedure, it can be

9.10. Dimensional Analysis and the Navier–Stokes Equation

proved that some terms of the Navier–Stokes equation can be neglected with respect to other terms, so that the complexity of the equation is reduced.

For instance, the term $\nabla p/\rho$ in equation (9.75) cannot be compared with $\nu \Delta \mathbf{v}$, since their order of magnitude is not known a-priori. To make this comparison possible we must introduce suitable reference quantities. If, for sake of simplicity, our attention is restricted to liquids, i.e. to equations (9.75), a characteristic length L and a characteristic velocity U are necessary. As an example, for a solid body placed in a moving liquid, L can be identified with a characteristic dimension of the body[2] and U with the uniform velocity of the liquid particles that are very far from the body. In any case, both L and U have to be selected in such a way that the dimensionless quantities

$$\mathbf{v}^* = \frac{\mathbf{v}}{U}, \quad \mathbf{r}^* = \frac{\mathbf{r}}{L}, \quad \tau = \frac{t}{L/U} \tag{9.85}$$

have the order of magnitude of unity. It follows that

$$\dot{\mathbf{v}} = \frac{\partial \mathbf{v}}{\partial t} + \mathbf{v} \cdot \nabla \mathbf{v} = \frac{U}{L/U} \frac{\partial \mathbf{v}^*}{\partial \tau} + \frac{U^2}{L} \mathbf{v}^* \cdot \nabla^* \mathbf{v}^*,$$

$$\frac{1}{\rho} \nabla p = \frac{1}{\rho L} \nabla^* p,$$

$$\nu \Delta \mathbf{v} = \nu \Delta^* \mathbf{v}^*,$$

$$\nabla \cdot \mathbf{v} = \frac{U}{L} \nabla^* \cdot \mathbf{v}^*,$$

where the derivatives appearing in the operators ∇^*, $\nabla^* \cdot$, and Δ^* are relative to the variables $(x^*, y^*, z^*) = \mathbf{r}^*$. The substitution of these relations into (9.75), after dividing by U^2/L, leads to

$$\dot{\mathbf{v}}^* = -\frac{1}{\rho U^2} \nabla^* p + \frac{\nu}{UL} \Delta^* \mathbf{v}^*,$$

$$\nabla^* \cdot \mathbf{v}^* = 0. \tag{9.86}$$

If the quantity ρU^2 is assumed to be a reference pressure, then the introduction of the dimensionless pressure

$$p^* = \frac{p}{\rho U^2} \tag{9.87}$$

allows us to write the system (9.86) in the form

$$\dot{\mathbf{v}}^* = -\nabla^* p^* + \frac{1}{R} \Delta^* \mathbf{v}^*,$$

$$\nabla^* \cdot \mathbf{v}^* = 0, \tag{9.88}$$

---

[2] In the 2D case, L can be a characteristic dimension of the cross section.

where the dimensionless quantity

$$R = \frac{UL}{\nu} \tag{9.89}$$

is known as the **Reynolds number**.

We are now in a position to note that the quantities $\dot{\mathbf{v}}^*$, $\nabla^* p$, and $\Delta^* \mathbf{v}^*$ in (9.88) are of the order of unity, whereas the term which multiplies the Laplacian on the right-hand side can be neglected for large values of R, i.e., for low viscosity.

In addition, it can be noted that the solution of (9.88) is the same for problems involving the same Reynolds number. This conclusion allows us to build physical models of the real problem in the laboratory by conveniently choosing parameters U, L, and v in order to preserve the Reynolds number (*scaling*).

If $R \to \infty$, then equations (9.88) reduce to dimensionless equations of a perfect liquid. But care must be used in stating that (9.88) can be applied at any point of the domain. In fact, when $R \to \infty$, equations (9.88) are not compatible with the boundary condition $\mathbf{v}^* = \mathbf{0}$, so that they can be used in the whole domain, with the exception of a narrow region around the body. In this region, which is called the **boundary layer**, (9.88) still applies, because of the influence of viscosity.

9.11 Boundary Layer

In the previous section, the boundary layer has been defined as *the thin layer in which the effect of viscosity is important*, independent of the value of the Reynolds number. This definition derives from the hypothesis, introduced by Prandtl (1905), that viscosity effects are important (that is, comparable with convection and inertia terms) in layers attached to solid boundaries whose thickness approaches zero as the Reynolds number goes to infinity, while the same effects can be neglected outside these layers.

There are two specific aspects related to Prandtl's hypothesis to be noted: first, the fact that the boundary layer is thin compared with other dimensions allows us to introduce some approximations in the motion equations. Second, in the boundary layer the liquid velocity has a transition from the zero value at the boundary (according to the no-slip condition) to a finite value outside the layer.

Accordingly, with these remarks, it is appropriate to search for the solution of (9.75) in terms of a power series in the small parameter $\varepsilon = 1/R$, i.e.,

$$\mathbf{v} = \mathbf{v}_0 + \mathbf{v}_1 \epsilon + \mathbf{v}_2 \epsilon^2 + \cdots, \tag{9.90}$$

9.11. Boundary Layer

where $\mathbf{v_0}$ is the solution for perfect liquid. However, by substituting (9.90) into (9.88), we see how this approach does not allow us to determine terms such as $\mathbf{v}_1, \mathbf{v}_2, \ldots$ since the parameter ϵ multiplies the highest derivatives. Therefore, the need arises for a new perturbation method, called the method of *singular perturbations*.

Before going into the details of the problem, we recall the basic ideas of both *regular* and *singular perturbation methods*.

In modeling reality, it often happens that the effect to be described is produced by a main cause and a secondary one, called a perturbation of the main cause. For instance, the motion of a planet (effect) is primarily caused by its gravitational attraction to the sun; however, the action of the heavier planets cannot be completely neglected. In this case, when writing the equations mathematically describing the phenomenon in nondimensional form, the term related to the secondary cause is multiplied by a small dimensionless parameter ε.

The *regular perturbation method* assumes that the solution of the *perturbed problem* can be represented by a power series of ε. Referring to (9.90), the term $\mathbf{v}_o$ is called the *leading-order term*, and, if the method works properly, it is the solution of the *unperturbed problem* (in our example, it is the solution for the perfect liquid).

By substituting the power series expansion into the differential equations and auxiliary conditions, we obtain a set of equations which allow us to determine $\mathbf{v}_1, \mathbf{v}_2$, etc.

This procedure does not always give an approximate solution and its failure can be expected when the small parameter ε multiplies the highest derivatives appearing in the equations. In fact, in this case, setting ε equal to zero changes the mathematical character of the problem. Related to this aspect, failures of the regular perturbation method occur when the *physical problem is characterized by multiple (time or length) scales*. It is just in these cases that the singular perturbation method has to be applied.

According to Prandtl's hypothesis, the solution of (9.75) changes rapidly in a narrow interval or *inner region*, corresponding to the boundary layer, and more gently in the *outer region*. It follows that in order to analyze the flow in the inner region, a proper *rescaling* is necessary.

The final step will be to match the inner and the outer solution, so that an approximate solution uniformly applicable over the complete region can be obtained.

For the sake of simplicity, the procedure will be applied below to a 2D flow, parallel to Ox. The axis Ox is supposed to be the trace of a boundary plate, and, due to symmetry arguments, the analysis will be restricted to the sector $x \geq 0$, $y \geq 0$.

Accordingly, the velocity field is expressed by

$$\mathbf{v} = U\mathbf{i}, \tag{9.91}$$

where U is a constant. Then, two reference lengths L and δ are introduced, together with two reference velocities U and ϵU, in order to take into account the different changes of velocity components u and v along the x and y axes. Furthermore, let

$$x = x^* L, \quad y = y^* \delta, \quad u = u^* U, \quad v = v^* \epsilon U. \tag{9.92}$$

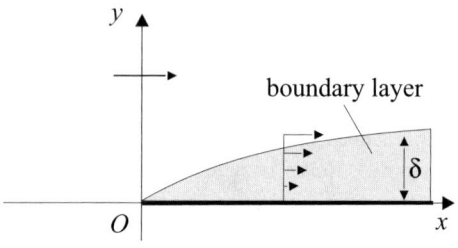

Figure 9.17

The complex velocity at $T'(2l, 0)$ in the z plane is written as

$$\left(\frac{dF}{dz}\right)_{z=(2l,0)} = \left(\frac{d\hat{F}}{d\zeta}\right)_{\zeta=(l,0)} \frac{d\zeta}{dz}.$$

Equations (9.88) become

$$u\frac{\partial u}{\partial x} + v\frac{\partial u}{\partial y} = -\frac{1}{\rho}\frac{\partial p}{\partial x} + \nu\left(\frac{\partial^2 u}{\partial x^2} + \frac{\partial^2 u}{\partial y^2}\right),$$

$$u\frac{\partial v}{\partial x} + v\frac{\partial v}{\partial y} = -\frac{1}{\rho}\frac{\partial p}{\partial y} + \nu\left(\frac{\partial^2 v}{\partial x^2} + \frac{\partial^2 v}{\partial y^2}\right),$$

$$\frac{\partial u}{\partial x} + \frac{\partial v}{\partial y} = 0, \tag{9.93}$$

which in dimensionless form are

$$\frac{U^2}{L}u^*\frac{\partial u^*}{\partial x^*} + \frac{U^2\epsilon}{\delta}v^*\frac{\partial u^*}{\partial y^*} = -\frac{P}{\rho L}\frac{\partial p^*}{\partial x^*} + \nu\left(\frac{U}{L^2}\frac{\partial^2 u^*}{\partial x^{*2}} + \frac{U}{\delta^2}\frac{\partial^2 u^*}{\partial y^{*2}}\right),$$

$$\frac{U^2\epsilon}{L}u^*\frac{\partial v^*}{\partial x^*} + \frac{U^2\epsilon^2}{\delta}v^*\frac{\partial v^*}{\partial y^*} = -\frac{P}{\rho\delta}\frac{\partial p^*}{\partial y^*} + \nu\left(\frac{U\epsilon}{L^2}\frac{\partial^2 v^*}{\partial x^{*2}} + \frac{U\epsilon}{\delta^2}\frac{\partial^2 v^*}{\partial y^{*2}}\right),$$

$$\frac{U}{L}\frac{\partial u^*}{\partial x^*} + \frac{U\epsilon}{\delta}\frac{\partial v^*}{\partial y^*} = 0. \tag{9.94}$$

By deleting starred symbols since there is no danger of confusion, the previous system can be written in the form

$$u\frac{\partial u}{\partial x} + \frac{L\epsilon}{\delta}v\frac{\partial u}{\partial y} = -\frac{P}{\rho U^2}\frac{\partial p}{\partial x} + \frac{\nu}{UL}\frac{\partial^2 u}{\partial x^2} + \frac{\nu L}{U\delta^2}\frac{\partial^2 u}{\partial y^2},$$

9.11. Boundary Layer

$$u\frac{\partial v}{\partial x} + \frac{L\epsilon}{\delta}v\frac{\partial v}{\partial y} = -\frac{P}{\rho U^2}\frac{L}{\epsilon\delta}\frac{\partial p}{\partial y} + \frac{\nu}{UL}\frac{\partial^2 v}{\partial x^2} + \frac{\nu L}{U\delta^2}\frac{\partial^2 v}{\partial y^2},$$

$$\frac{\partial u}{\partial x} + \frac{L\epsilon}{\delta}\frac{\partial v}{\partial y} = 0. \tag{9.95}$$

It can be noted that, if the term $\epsilon L/\delta$ is not on the order of unity, then either u is independent of x, or v is independent of y. Neither of these possibilities has any physical meaning, so it emerges that

$$\frac{L\epsilon}{\delta} \simeq 1. \tag{9.96}$$

This conclusion allows us to write the system (9.95), assuming that $P \simeq \rho U^2$, in the form

$$u\frac{\partial u}{\partial x} + v\frac{\partial u}{\partial y} = -\frac{\partial p}{\partial x} + \frac{1}{R}\left(\frac{\partial^2 u}{\partial x^2} + \frac{L^2}{\delta^2}\frac{\partial^2 u}{\partial y^2}\right),$$

$$u\frac{\partial v}{\partial x} + v\frac{\partial v}{\partial y} = -\frac{L^2}{\delta^2}\frac{\partial p}{\partial y} + \frac{1}{R}\left(\frac{\partial^2 v}{\partial x^2} + \frac{L^2}{\delta^2}\frac{\partial^2 v}{\partial y^2}\right),$$

$$\frac{\partial u}{\partial x} + \frac{\partial v}{\partial y} = 0. \tag{9.97}$$

The first two equations show that the condition of preserving the viscosity effect when $R \to \infty$ requires that

$$\frac{1}{R}\frac{L^2}{\delta^2} \simeq 1 \Rightarrow \delta \simeq \frac{L}{\sqrt{R}}, \tag{9.98}$$

and finally equations (9.97) become

$$u\frac{\partial u}{\partial x} + v\frac{\partial u}{\partial y} = -\frac{\partial p}{\partial x} + \frac{1}{R}\frac{\partial^2 u}{\partial x^2} + \frac{\partial^2 u}{\partial y^2},$$

$$\frac{1}{R}\left(u\frac{\partial v}{\partial x} + v\frac{\partial v}{\partial y}\right) = -\frac{\partial p}{\partial y} + \frac{1}{R^2}\frac{\partial^2 v}{\partial x^2} + \frac{1}{R}\frac{\partial^2 v}{\partial y^2},$$

$$\frac{\partial u}{\partial x} + \frac{\partial v}{\partial y} = 0. \tag{9.99}$$

We can now apply the perturbation theory, i.e., we can search for a solution in terms of power series of the parameter $1/R$. In particular, when $R \to \infty$ the first term of this series must satisfy **Prandtl's equations**

$$u\frac{\partial u}{\partial x} + v\frac{\partial u}{\partial y} = -\frac{\partial p}{\partial x} + \frac{\partial^2 u}{\partial y^2},$$

$$\frac{\partial p}{\partial y} = 0,$$

$$\frac{\partial u}{\partial x} + \frac{\partial v}{\partial y} = 0. \tag{9.100}$$

Equation (9.100)$_2$ implies that $p = p(x)$, so that the pressure assumes the same value along the normal at $(x, 0)$. In addition, since Bernoulli's theorem (9.22) states that in the unperturbed region the pressure is uniform, we conclude that the pressure is uniform in the entire domain.

Equation (9.100)$_3$ implies (see (9.40)) the existence of the Stokes potential $\psi(x, y)$, so that
$$u = \frac{\partial \psi}{\partial y}, \qquad v = -\frac{\partial \psi}{\partial x},$$
and (9.100)$_1$ becomes
$$\frac{\partial \psi}{\partial y}\frac{\partial^2 \psi}{\partial x \partial y} - \frac{\partial \psi}{\partial x}\frac{\partial^2 \psi}{\partial y^2} = \frac{\partial^3 \psi}{\partial y^3}. \tag{9.101}$$

This equation must be integrated in the region $x \geq 0$, $y \geq 0$ with the following boundary conditions:
$$u = v = 0, \qquad \text{if } y = 0;$$
$$\lim_{y \to \infty} u = 1, \qquad \lim_{y \to \infty} v = 0,$$
which, in terms of ψ, are
$$\psi(x, 0) = 0, \qquad \frac{\partial \psi}{\partial y}(x, 0) = 0,$$
$$\lim_{y \to \infty} \frac{\partial \psi}{\partial y}(x, y) = 1. \tag{9.102}$$

It can be noted by inspection that if $\psi(x, y)$ is a solution of (9.101), then so is the function
$$\psi(x, y) = c^{1-n}\psi\left(\frac{x}{c}, \frac{y}{c^n}\right). \tag{9.103}$$
Moreover, it can be proved that the functions satisfying (9.103) have the form
$$\psi(x, y) = x^{1-n}\psi\left(1, \frac{y}{x^n}\right) \equiv x^{1-n} f(\eta).$$
The derivative of this function with respect to y is
$$\frac{\partial \psi}{\partial y} = x^{1-2n} f'(\eta),$$
so that the condition (9.102)$_3$ only holds for $n = 1/2$ and
$$\psi(x, y) = \sqrt{x} f(\eta), \tag{9.104}$$
where $\eta = y/\sqrt{x}$. Substituting (9.104) into (9.101), and taking into account the boundary conditions (9.102), we obtain the **Blasius equation**:
$$f''' + f f'' = 0 \tag{9.105}$$

with the boundary conditions

$$f(0) = 0, \quad f'(0) = 0, \quad \lim_{\eta \to \infty} f'(\eta) = 1. \qquad (9.106)$$

The problem (9.105), (9.106), for which a theorem of existence and uniqueness holds, can only be solved numerically. The pattern of the solution is shown in Figure 9.18 (see Exercise 3).

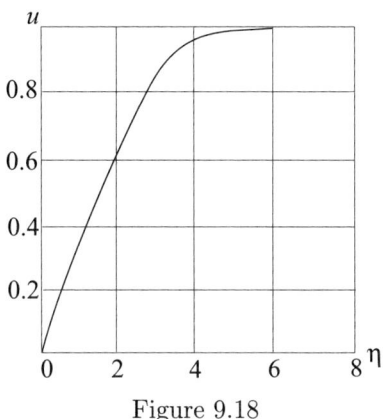

Figure 9.18

9.12 Motion of a Viscous Liquid around an Obstacle

The aim of this section is to apply the previous results to the *irrotational steady motion* of a viscous liquid around an obstacle of arbitrary shape.

According to the dimensional analysis of the Navier–Stokes equations, when the Reynolds number R attains high values the assumption of a perfect incompressible fluid is justified in all the domain except for the boundary layer, in which the viscosity is needed in order to satisfy the no-slip condition $\mathbf{v} = \mathbf{0}$ at the boundary. When integrating the motion equations in this layer, a problem of singular perturbations arises, so that a convenient change of variables is required. In particular, we introduce new variables ξ and η such that

1. the curve $\eta = 0$ describes the profile of the obstacle;
2. curves $\xi = const$ are normal to the obstacle;
3. the boundary layer is described by changing η from 0 to ∞.

For instance, in the previous example, the old dimensionless coordinates x and y were replaced with x and $y' = y/(1/R)$. With this change, the straight line $y = 0$ gives the obstacle profile, the straight lines $x = const$ are orthogonal to the plate and, given the high value of the Reynolds number, if $0 < y < \delta$, where $\delta = L/\sqrt{R}$ is the thickness of the boundary layer, the variable y' changes from 0 to $\sqrt{R} \to \infty$.

In the general case of a body S of arbitrary cross section, recalling that outside the obstacle the model of a perfect liquid is applicable, we must find a complex potential $F(z) = \varphi + i\psi$, where $\varphi(x,y)$ and $\psi(x,y)$ denote the kinetic and Stokes potential, respectively. Then the curvilinear coordinates (φ, ψ)

$$\begin{aligned} \varphi &= \varphi(x,y), \\ \psi &= \psi(x,y), \end{aligned} \qquad (9.107)$$

are introduced. Consequently, the family $\varphi = const$ contains the profile of the obstacle, whereas the curves $\psi = const$ are normal to them (see Figure 9.19). In the next step, we need to define new coordinates $\psi^* = \psi/\delta$, where δ is an estimate of the boundary layer thickness, and we need to write the motion equations in this new coordinate system. In the final step, the perturbation method is applied in order to obtain an approximate solution.

Let $F(z) = \varphi + i\psi$ be a complex potential for the field $\mathbf{v}_0$ of the inviscid flow. Then, the square of the infinitesimal distance between two points is written as[3]

$$ds^2 = |dz|^2 = \frac{|dF|^2}{|F'(z)|^2} = \frac{d\varphi^2 + d\psi^2}{\left|\frac{\partial \varphi}{\partial x} + i\frac{\partial \psi}{\partial x}\right|^2}. \qquad (9.108)$$

By taking into account of (9.43), the previous expression becomes

$$ds^2 = \frac{d\varphi^2 + d\psi^2}{|\bar{u}_0 - i\bar{v}_0|^2} = \frac{1}{v_0^2}(d\varphi^2 + d\psi^2), \qquad (9.109)$$

---

[3] We recall that the derivative at z_0 of a function $F(z)$ of a complex variable z is defined by

$$F'(z_0) = \lim_{z \to z_0} \frac{F(z) - F(z_0)}{z - z_0}.$$

This limit must be independent of the path $z \to z_0$ and, in particular, if $z = z_0 + \Delta x$, it holds that

$$F'(z_0) = \lim_{\Delta x \to 0} \frac{F(z) - F(z_0)}{\Delta x} = \frac{\partial \varphi}{\partial x} + i\frac{\partial \psi}{\partial x}.$$

9.12. Motion of a Viscous Liquid around an Obstacle

where $\bar{u}_0$ and $\bar{v}_0$ are the components of $\mathbf{v}_0$, and $v_0 = |\mathbf{v}_0|$.

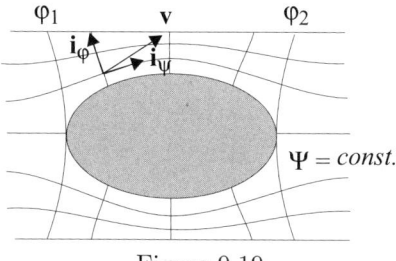

Figure 9.19

From (9.109), the following expressions can be derived (the reader should refer to (2.61)–(2.63)):

$$\nabla f = v_0 \left(\frac{\partial f}{\partial \varphi} \mathbf{i}_\varphi + \frac{\partial f}{\partial \psi} \mathbf{i}_\psi \right), \tag{9.110}$$

$$\omega = \frac{1}{2} \nabla \times \mathbf{v} = \frac{1}{2} v_0^2 \left[\frac{\partial}{\partial \varphi} \left(\frac{v_\psi}{v_0} \right) - \frac{\partial}{\partial \psi} \left(\frac{v_\varphi}{v_0} \right) \right] \mathbf{k} \equiv \omega \mathbf{k}, \tag{9.111}$$

$$\nabla \times \omega = v_0 \left(\frac{\partial \omega}{\partial \psi} \mathbf{i}_\varphi - \frac{\partial \omega}{\partial \varphi} \mathbf{i}_\psi \right), \tag{9.112}$$

$$\nabla \cdot \mathbf{v} = v_0^2 \left[\frac{\partial}{\partial \varphi} \left(\frac{v_\varphi}{v_0} \right) + \frac{\partial}{\partial \psi} \left(\frac{v_\psi}{v_0} \right) \right], \tag{9.113}$$

$$\mathbf{v} \times \omega = \omega (v_\psi \mathbf{i}_\varphi - v_\varphi \mathbf{i}_\psi), \tag{9.114}$$

where v_φ and v_ψ are the dimensionless velocity components for the viscous flow, $\mathbf{i}_\varphi$ and $\mathbf{i}_\psi$ are the unit vectors tangent to the curves φ and ψ are variable, while $\mathbf{k}$ is the unit vector normal to the plane of the motion.

Assuming that the viscous liquid is incompressible and the motion is steady, (9.88) becomes

$$\mathbf{v} \cdot \nabla \mathbf{v} = -\nabla p + \epsilon \Delta \mathbf{v},$$
$$\nabla \cdot \mathbf{v} = 0, \tag{9.115}$$

where, as usual, $\epsilon = 1/R$. In addition, from (4.18) it follows that

$$\mathbf{v} \cdot \nabla \mathbf{v} = \frac{1}{2} \nabla v^2 - 2\mathbf{v} \times \omega, \tag{9.116}$$

and, since $\nabla \cdot \mathbf{v} = 0$,

$$\Delta \mathbf{v} = \nabla (\nabla \cdot \mathbf{v}) - \nabla \times \nabla \times \mathbf{v} = -2\nabla \times \omega;$$

equations (9.115) become

$$\frac{1}{2}\nabla v^2 - 2\mathbf{v}\times\omega = -\nabla p - 2\epsilon\nabla\times\omega,$$
$$\nabla\cdot\mathbf{v} = 0. \tag{9.117}$$

By taking into account (9.110)–(9.114), it can now be proved that the previous equations assume the form

$$v_0\frac{\partial}{\partial\varphi}\left(\frac{v_\varphi^2+v_\psi^2}{2}\right) - 2v_\psi\omega = -v_0\frac{\partial p}{\partial\varphi} - 2\epsilon v_0\frac{\partial\omega}{\partial\psi},$$

$$v_0\frac{\partial}{\partial\psi}\left(\frac{v_\varphi^2+v_\psi^2}{2}\right) + 2v_\varphi\omega = -v_0\frac{\partial p}{\partial\psi} + 2\epsilon v_0\frac{\partial\omega}{\partial\varphi},$$

$$\frac{\partial}{\partial\varphi}\left(\frac{v_\varphi}{v_0}\right) + \frac{\partial}{\partial\psi}\left(\frac{v_\psi}{v_0}\right) = 0. \tag{9.118}$$

Finally, by introducing the quantities

$$w_\varphi = \frac{v_\varphi}{v_0}, \quad w_\psi = \frac{v_\psi}{v_0}, \tag{9.119}$$

we get

$$\frac{\partial w_\varphi}{\partial\varphi} + \frac{\partial w_\psi}{\partial\psi} = 0, \tag{9.120}$$

$$w_\varphi\frac{\partial w_\varphi}{\partial\varphi} + w_\psi\frac{\partial w_\varphi}{\partial\psi} + (w_\varphi^2+w_\psi^2)\frac{\partial}{\partial\varphi}\log v_0 \tag{9.121}$$
$$= -\frac{1}{v_0^2}\frac{\partial p}{\partial\varphi} + \epsilon\left[\frac{\partial^2 w_\varphi}{\partial\varphi^2} + \frac{\partial^2 w_\varphi}{\partial\psi^2} - 2\frac{\partial}{\partial\psi}\log v_0\left(\frac{\partial w_\psi}{\partial\varphi} - \frac{\partial w_\varphi}{\partial\psi}\right)\right],$$

$$w_\varphi\frac{\partial w_\psi}{\partial\varphi} + w_\psi\frac{\partial w_\psi}{\partial\psi} + (w_\varphi^2+w_\psi^2)\frac{\partial}{\partial\psi}\log v_0 \tag{9.122}$$
$$= -\frac{1}{v_0^2}\frac{\partial p}{\partial\psi} + \epsilon\left[\frac{\partial^2 w_\psi}{\partial\varphi^2} + \frac{\partial^2 w_\psi}{\partial\psi^2} + 2\frac{\partial}{\partial\varphi}\log v_0\left(\frac{\partial w_\psi}{\partial\varphi} - \frac{\partial w_\varphi}{\partial\psi}\right)\right].$$

Equations (9.120), (9.122), and (9.123) form an elliptic system for the unknowns $w_\varphi(\varphi,\psi)$, $w_\psi(\varphi,\psi)$, and $p(\varphi,\psi)$, since the solution v_0, which refers to the problem of an inviscid liquid, has been supposed to be known.

Boundary conditions related to this system are the no-slip condition on the boundary of the obstacle and the condition that at infinity the velocity is given by $v_0\mathbf{i}_\psi$. Noting that the profile of the obstacle is defined by $\varphi_1 \le \varphi \le \varphi_2$, $\psi = 0$ (see Figure 9.19), we obtain

$$w_\varphi(\varphi,0) = w_\psi(\varphi,0) = 0, \quad \varphi_1 \le \varphi \le \varphi_2, \tag{9.123}$$

9.12. Motion of a Viscous Liquid around an Obstacle

$$\lim_{\varphi,\psi \to \infty} w_\varphi = 0, \quad \lim_{\varphi,\psi \to \infty} w_\psi = 1. \tag{9.124}$$

The pressure p_0 relative to the inviscid motion is given by Bernoulli's theorem. In fact, at infinity the velocity value is unity and the pressure is zero, so that

$$p_0 = \frac{1}{2}(1 - v_0^2). \tag{9.125}$$

As seen in the previous section, in the equations (9.120), (9.122), and (9.123) a small parameter ϵ appears, where it multiplies the highest derivatives. Consequently, here again we introduce two expansions of ϵ: the first one valid far from the obstacle (outer solution), and the second one applicable near to the obstacle (inner solution). The second expansion must satisfy the system (9.120), (9.122), and (9.123), provided that it is written in convenient coordinates, which expand the boundary layer and allow the velocity to change from zero to the value relative to the inviscid motion.

More specifically, by taking into account (9.123), (9.124), and (9.125) the outer solution can be given the form

$$w_\varphi = \epsilon w_\varphi^{(1)} + \cdots,$$
$$w_\psi = 1 + \epsilon w_\psi^{(1)} + \cdots,$$
$$p = \frac{1}{2}(1 - v_0^2) + \epsilon p^{(1)} + \cdots. \tag{9.126}$$

Furthermore, the convenient variable for expanding the boundary layer is

$$\psi_* = \frac{\psi}{\Delta(\epsilon)}, \tag{9.127}$$

where $\Delta(\epsilon)$ is a function of ϵ, determined in the following discussion.

In order to write (9.120), (9.122), and (9.123) in terms of the new variables φ and ψ_*, we observe that

$$\frac{\partial}{\partial \psi} = \frac{1}{\Delta(\epsilon)} \frac{\partial}{\partial \psi_*},$$
$$\frac{\partial^2}{\partial \psi^2} = \frac{1}{\Delta^2(\epsilon)} \frac{\partial^2}{\partial \psi_*^2}.$$

Consequently, by using the notation

$$w_\varphi(\varphi, \psi) = w_\varphi^*(\varphi, \psi_*),$$
$$w_\psi(\varphi, \psi) = w_\psi^*(\varphi, \psi_*),$$
$$p(\varphi, \psi) = p^*(\varphi, \psi_*),$$

the system (9.120), (9.122), and (9.123) is written as

$$\frac{\partial w_\varphi^*}{\partial \varphi} + \frac{1}{\Delta(\epsilon)} \frac{\partial w_\psi^*}{\partial \psi_*} = 0, \tag{9.128}$$

$$w_\varphi^* \frac{\partial w_\varphi^*}{\partial \varphi} + \frac{w_\psi^*}{\Delta(\epsilon)} \frac{\partial w_\varphi^*}{\partial \psi_*} + (w_\varphi^{*2} + w_\psi^{*2}) \frac{\partial}{\partial \varphi} \log v_0 = -\frac{1}{v_0^2} \frac{\partial p^*}{\partial \varphi} \qquad (9.129)$$
$$+ \epsilon \left[\frac{\partial^2 w_\varphi^*}{\partial \varphi^2} + \frac{1}{\Delta^2(\epsilon)} \frac{\partial^2 w_\varphi^*}{\partial \psi_*^2} - \frac{2}{\Delta(\epsilon)} \frac{\partial}{\partial \varphi} \log v_0 \left(\frac{\partial w_\psi^*}{\partial \varphi} - \frac{1}{\Delta(\epsilon)} \frac{\partial w_\varphi^*}{\partial \psi_*} \right) \right],$$

$$w_\varphi^* \frac{\partial w_\psi^*}{\partial \varphi} + \frac{w_\psi^*}{\Delta(\epsilon)} \frac{\partial w_\psi^*}{\partial \psi_*} + (w_\varphi^{*2} + w_\psi^{*2}) \frac{1}{\Delta(\epsilon)} \frac{\partial}{\partial \psi_*} \log v_0 = -\frac{1}{v_0^2} \frac{1}{\Delta(\epsilon)} \frac{\partial p^*}{\partial \psi_*}$$
$$+ \epsilon \left[\frac{\partial^2 w_\psi^*}{\partial \varphi^2} + \frac{1}{\Delta^2(\epsilon)} \frac{\partial^2 w_\psi^*}{\partial \psi_*^2} + 2 \frac{\partial}{\partial \varphi} \log v_0 \left(\frac{\partial w_\psi^*}{\partial \varphi} - \frac{1}{\Delta(\epsilon)} \frac{\partial w_\varphi^*}{\partial \psi_*} \right) \right]. \qquad (9.130)$$

If $\partial w_\varphi^*/\partial \varphi$ and $\partial w_\psi^*/\partial \psi_*$ were of the same order, recalling that $1/\Delta(\epsilon)$ goes to infinity when $\epsilon \longrightarrow 0$, we see that the continuity equation would give $\partial w_\psi^*/\partial \psi_* = 0$, so that w_ψ^* would be independent of ψ_*. Moreover, we have that $w_\psi^* = 0$ when $\psi_* = 0$ for arbitrary φ, so that we would have everywhere $w_\psi^* = 0$, despite the requirement that the component w_ψ^* changes from zero at the boundary to a finite value outside the layer.

Therefore, we need to have

$$w_\varphi^* = \hat{w}_\varphi(\varphi, \psi_*) + \cdots,$$
$$w_\psi^* = \Delta(\epsilon) \hat{w}_\psi(\varphi, \psi_*) + \cdots,$$
$$p = \hat{p}(\varphi, \psi_*) + \cdots,$$

so that the system (9.128), (9.130) and (9.130) is written as

$$\frac{\partial \hat{w}_\varphi}{\partial \varphi} + \frac{\partial \hat{w}_\psi}{\partial \psi_*} = 0, \qquad (9.131)$$

$$\hat{w}_\varphi \frac{\partial \hat{w}_\varphi}{\partial \varphi} + \hat{w}_\psi \frac{\partial \hat{w}_\varphi}{\partial \psi_*}$$
$$+ (\hat{w}_\varphi^2 + \Delta^2(\epsilon) \hat{w}_\psi^2) \frac{\partial}{\partial \varphi} \log v_0 = -\frac{1}{v_0^2} \frac{\partial \hat{p}}{\partial \varphi} \qquad (9.132)$$
$$+ \epsilon \left[\frac{\partial^2 \hat{w}_\varphi}{\partial \varphi^2} + \frac{1}{\Delta^2(\epsilon)} \frac{\partial^2 \hat{w}_\varphi}{\partial \psi_*^2} - 2 \frac{\partial}{\partial \varphi} \log v_0 \left(\frac{\partial \hat{w}_\psi}{\partial \varphi} - \frac{1}{\Delta(\epsilon)} \frac{\partial \hat{w}_\varphi}{\partial \psi_*} \right) \right],$$

$$\Delta(\epsilon) \hat{w}_\varphi \frac{\partial \hat{w}_\psi}{\partial \varphi} + \Delta(\epsilon) \hat{w}_\psi \frac{\partial \hat{w}_\psi}{\partial \psi_*}$$
$$+ (\hat{w}_\varphi^2 + \Delta^2(\epsilon) \hat{w}_\psi^2) \frac{1}{\Delta(\epsilon)} \frac{\partial}{\partial \psi_*} \log v_0 = -\frac{1}{v_0^2} \frac{1}{\Delta(\epsilon)} \frac{\partial \hat{p}}{\partial \psi_*} \qquad (9.133)$$
$$+ \epsilon \left[\Delta(\epsilon) \frac{\partial^2 \hat{w}_\psi}{\partial \varphi^2} + \frac{1}{\Delta(\epsilon)} \frac{\partial^2 \hat{w}_\psi}{\partial \psi_*^2} + 2 \frac{\partial}{\partial \varphi} \log v_0 \left(\Delta(\epsilon) \frac{\partial \hat{w}_\psi}{\partial \varphi} - \frac{1}{\Delta(\epsilon)} \frac{\partial w_\varphi^*}{\partial \psi_*} \right) \right].$$

9.13. Ordinary Waves in Perfect Fluids

On the other hand, it also holds that

$$v_0(\varphi, \psi) = v_0(\varphi, 0) + \left(\frac{\partial v_0}{\partial \psi}\right)_{\psi=0} \psi + \cdots$$

$$= v_0(\varphi, 0) + \Delta(\epsilon) \left(\frac{\partial v_0}{\partial \psi}\right)_{\psi=0} \psi_* + \cdots, \qquad (9.134)$$

where $v_0(\varphi, 0)$ is the velocity on the obstacle boundary for the inviscid flow. Consequently, for $\epsilon \longrightarrow 0$, the following parabolic system is obtained:

$$\frac{\partial \hat{w}_\varphi}{\partial \varphi} + \frac{\partial \hat{w}_\psi}{\partial \psi_*} = 0, \qquad (9.135)$$

$$\hat{w}_\varphi \frac{\partial \hat{w}_\varphi}{\partial \varphi} + \hat{w}_\psi \frac{\partial \hat{w}_\varphi}{\partial \psi_*} + \frac{\hat{w}_\varphi^2}{v_0} \frac{dv_0}{d\varphi} = -\frac{1}{v_0^2} \frac{\partial \hat{p}}{\partial \varphi} + \frac{\partial^2 \hat{w}_\varphi}{\partial \psi_*^2}, \qquad (9.136)$$

$$\frac{\partial \hat{p}}{\partial \psi_*} = 0. \qquad (9.137)$$

The boundary conditions are

$$\hat{w}_\varphi(\varphi, 0) = \hat{w}_\psi(\varphi, 0) = 0, \; \varphi_1 \leq \varphi \leq \varphi_2,$$
$$\hat{w}_\varphi(\varphi_1, \psi_*) = 1, \qquad (9.138)$$

the velocity attaining the value corresponding to the undisturbed motion ahead the obstacle.

The inner solution of (9.120), (9.122), and (9.123) must be compatible with the outer solution, so that the additional requirements are

$$\lim_{\psi_* \to \infty} \hat{w}_\varphi = 1, \; \hat{p} = \frac{1}{2}(1 - v_0)^2. \qquad (9.139)$$

The boundary value problem (9.135)–(9.139) is very complex, so that it is not surprising that a general solution is not yet available. When considering a 2D plate, we find that the problem reduces to the system (9.100), which is the origin of the Blausius equation. We remark that the compatibility condition $(9.139)_2$ allows us to eliminate the pressure in (9.136), since

$$\frac{\partial \hat{p}}{\partial \varphi} = \frac{d\hat{p}}{d\varphi} = -v_0 \frac{dv_0}{d\varphi}.$$

9.13 Ordinary Waves in Perfect Fluids

This section is devoted to wave propagation in perfect fluids. Two topics will be addressed: first, waves are considered to be small perturbations of an undisturbed state, corresponding to a homogeneous fluid at rest. This assumption allows us to linearize the motion equations and to apply elementary methods (see remarks in Section 8.6). Next, ordinary waves of discontinuity are analyzed, showing that the propagation velocity has the same expression obtained by the linearized theory. On the other hand, nonlinearity of the system influences the propagation of waves whose pattern is different whether we consider a linear case or not. Finally, the last section deals with shock waves in perfect fluids.

Consider a *compressible* perfect fluid at rest, with uniform mass density ρ_0, in absence of body forces. Assume that the motion is produced by a small perturbation, so that it is characterized by a velocity $\mathbf{v}$ and a density ρ that are only lightly different from $\mathbf{0}$ and ρ_0, respectively. More precisely, such an assumption states that quantities $\mathbf{v}$ and $\sigma = \rho - \rho_0$ are first-order quantities together with their first-order derivatives.

In this case, the motion equations

$$\rho \dot{\mathbf{v}} = -\nabla p(\rho) = -p'(\rho)\nabla \rho,$$
$$\dot{\rho} + \rho \nabla \cdot \mathbf{v} = 0, \tag{9.140}$$

can be linearized. In fact, neglecting first-order terms of $|\mathbf{v}|$ and σ, we get

$$p'(\rho) = p'(\rho_0) + p''(\rho_0)\sigma + \cdots,$$
$$\dot{\rho} = \dot{\sigma} = \frac{\partial \sigma}{\partial t} + \mathbf{v} \cdot \nabla \sigma = \frac{\partial \sigma}{\partial t} + \cdots,$$
$$\dot{\mathbf{v}} = \frac{\partial \mathbf{v}}{\partial t} + \mathbf{v} \cdot \nabla \mathbf{v} = \frac{\partial \mathbf{v}}{\partial t} + \cdots,$$

so that equations (9.140) become

$$\rho_0 \frac{\partial \mathbf{v}}{\partial t} = -p'(\rho_0)\nabla \sigma,$$
$$\frac{\partial \sigma}{\partial t} + \rho_0 \nabla \cdot \mathbf{v} = 0. \tag{9.141}$$

Consider now a sinusoidal 2D wave propagating in the direction of the vector $\mathbf{n}$ with velocity U and wavelength λ:

$$\mathbf{v} = \mathbf{a} \sin \frac{2\pi}{\lambda}(\mathbf{n} \cdot \mathbf{x} - Ut), \quad \sigma = b \sin \frac{2\pi}{\lambda}(\mathbf{n} \cdot \mathbf{x} - Ut).$$

We want a solution of the previous system having this form. To find this solution, we note that

$$\frac{\partial \mathbf{v}}{\partial t} = -\frac{2\pi}{\lambda} U \mathbf{a} \cos \frac{2\pi}{\lambda}(\mathbf{n} \cdot \mathbf{x} - Ut), \quad \frac{\partial \sigma}{\partial t} = -\frac{2\pi}{\lambda} Ub \cos \frac{2\pi}{\lambda}(\mathbf{n} \cdot \mathbf{x} - Ut),$$
$$\nabla \cdot \mathbf{v} = \frac{2\pi}{\lambda} \mathbf{n} \cdot \mathbf{a} \cos \frac{2\pi}{\lambda}(\mathbf{n} \cdot \mathbf{x} - Ut), \quad \nabla \sigma = \frac{2\pi}{\lambda} \mathbf{n} b \cos \frac{2\pi}{\lambda}(\mathbf{n} \cdot \mathbf{x} - Ut),$$

9.13. Ordinary Waves in Perfect Fluids

so that the system (9.141) becomes

$$\rho_0 U \mathbf{a} = p'(\rho_0)b\mathbf{n},$$
$$-Ub + \rho_0 \mathbf{a} \cdot \mathbf{n} = 0,$$

i.e.,

$$(p'(\rho_0)\mathbf{n} \otimes \mathbf{n} - U^2)\mathbf{I}\mathbf{a} = 0,$$
$$(p'(\rho_0) - U^2)b = 0.$$

Assuming a reference frame whose axis Ox is oriented along $\mathbf{n}$, we get

$$U = 0, \quad \mathbf{a} \perp \mathbf{n},$$
$$U = \pm\sqrt{p'(\rho_0)}, \quad \mathbf{a} = a\mathbf{n}, \quad (9.142)$$

where the eigenvalue 0 has multiplicity 2. Thus we prove the existence of **dilational waves** propagating at a speed

$$U = \sqrt{p'(\rho_0)}.$$

The velocity U is called the **sound velocity**, and the ratio $m = v/U$ is known as the **Mach number**. In particular, the motion is called **subsonic** or **supersonic**, depending on whether $m < 1$ or $m > 1$.

Let $S(t)$ be the wavefront of a singular wave for the nonlinear equations (9.107). In order to find the propagation speed c_n of $S(t)$, the jump system associated to (9.107) (see (4.38) and Chapter 8) can be written as

$$\rho(c_n - v_n)\mathbf{a} = p'(\rho)b\mathbf{n},$$
$$(c_n - v_n)b = \rho \mathbf{n} \cdot \mathbf{a}, \quad (9.143)$$

where $\mathbf{a}$ and b refer to the discontinuities of the first derivatives of $\mathbf{v}$ and ρ, respectively. By getting b from the second equation and substituting into the first one, it follows that

$$(c_n - v_n)^2 \mathbf{a} = p'(\rho)\mathbf{n}(\mathbf{n} \cdot \mathbf{a}),$$

i.e.,

$$(p'(\rho)\mathbf{n} \otimes \mathbf{n} - \mathbf{I}(c_n - v_n)^2)\mathbf{a} = 0. \quad (9.144)$$

Equation (9.144) shows that the squares of relative velocity are equal to the eigenvalues of the **acoustic tensor**

$$p'(\rho)\mathbf{n} \otimes \mathbf{n}$$

and the discontinuity vectors $\mathbf{a}$ are the related eigenvectors. The relative advancing velocity is

$$c_n = v_n \pm \sqrt{p'(\rho)}, \quad \mathbf{a} = a\mathbf{n}, \quad (9.145)$$

so that, if the undisturbed state is at rest, the same result $(9.142)_2$ is obtained.

Exercise 9.1
Starting from equation (9.72), we leave the reader to prove that in an incompressible viscous fluid no ordinary discontinuity wave exists.
Hint: Note that the equation we are considering contains second-order derivatives of velocity and first-order derivatives of density. This implies that the singular surface is of second-order with respect to $\mathbf{v}$, and of first-order with respect to ρ. In addition, jumps of second-order derivatives with respect to time, strictly related to c_n, do not appear in (9.72).

9.14 Shock Waves in Fluids

In this section we deal with some fundamental properties of compressible inviscid fluids, under the assumption that the shock is *adiabatic* i.e., it happens without absorbing or losing heat:

$$[[\mathbf{h}]] \cdot \mathbf{n} = 0. \tag{9.146}$$

It is well known that if v and η are chosen as thermodynamic variables, then the reduced dissipation inequality leads to the following relations:

$$v = \frac{\partial h}{\partial p}(p, \eta), \quad \theta = -\frac{\partial h}{\partial \eta}(p, \eta),$$

where the enthalpy $h = \epsilon + pv$. Therefore, if

$$\frac{\partial^2 h}{\partial p \eta} = -\frac{\partial \theta}{\partial p} \neq 0,$$

it is possible to find the relation

$$\eta = \eta(p, v). \tag{9.147}$$

In the following discussion, all the constitutive equations will be supposed to depend on v, η, and the entropy being expressed by (9.147).
Jump conditions $(5.22)_2$, (5.31), and (5.46), written for a perfect fluid,

$$[[\rho(c_n - v_n)]] = 0,$$
$$[[\rho\mathbf{v}(c_n - v_n) - p\mathbf{n}]] = 0,$$
$$\left[\left[\rho\left(\frac{1}{2}v^2 + \epsilon(\rho, p)\right)(c_n - v_n) - pv_n\right]\right] = 0, \tag{9.148}$$

9.14. Shock Waves in Fluids

give a system of one vector equation and two scalar equations with the unknowns $\mathbf{v}^-$, ρ^-, p^-, referring to the perturbed region, and the normal speed of propagation c_n of Σ since the values of $\mathbf{v}^+, \rho^+$, and p^+ are assigned.

If we decompose the velocity $\mathbf{v}$ into its normal v_n and tangential component $\mathbf{v}_\tau$,

$$\mathbf{v} = v_n \mathbf{n} + \mathbf{v}_\tau,$$

where

$$v^2 = v_n^2 + v_\tau^2,$$

then equations (9.148) become

$$[[\rho(c_n - v_n)]] = 0,$$
$$[[\rho\mathbf{v}_\tau(c_n - v_n)]] = \mathbf{0},$$
$$[[\rho v_n(c_n - v_n) - p]] = 0$$
$$\left[\left[\rho\left(\frac{1}{2}(v_n^2 + v_\tau^2) + \epsilon(\rho, p)\right)(c_n - v_n) - pv_n\right]\right] = 0, \quad (9.149)$$

From $(9.149)_{1,2}$ it follows that

$$[[\rho\mathbf{v}_\tau(c_n - v_n)]] = \rho^\pm(c_n - v_n^\pm)[[\mathbf{v}_\tau]] = \mathbf{0}, \quad (9.150)$$

so that, *if the shock wave is material*

$$c_n = v_n^\pm,$$

then from $(9.149)_{3,4}$ we get the conditions

$$[[p]] = 0, \ [[v_n]] = 0.$$

We conclude that the discontinuities only refer to density, tangential velocity, and internal energy.

On the other hand, if $c_n \neq v_n^\pm$, (9.150) states that the *continuity of the tangential components of velocity on the nonmaterial wavefront*

$$[[\mathbf{v}_\tau]] = \mathbf{0}, \quad (9.151)$$

and the system (9.149) reduces to the system

$$[[\rho(c_n - v_n)]] = 0,$$
$$[[\rho v_n(c_n - v_n) - p]] = 0,$$
$$\left[\left[\rho\left(\frac{1}{2}v_n^2 + \epsilon(\rho, p)\right)(c_n - v_n) - pv_n\right]\right] = 0. \quad (9.152)$$

of three equations for the four unknowns v_n^-, ρ^-, p^-, and c_n. This system can be solved provided that the constitutive equation $\epsilon(\rho, p)$ has been assigned together with the value of one of the above quantities.

There are some relevant consequences of the system (9.152). First, if we introduce at $\mathbf{x} \in \Sigma$ a reference system attached to the wavefront, the previous system is written as

$$\rho^- w^- = \rho^+ w^+ \equiv j,$$
$$\rho^-(w^-)^2 + p^- = \rho^+(w^+)^2 + p^+,$$
$$\rho^- \left(\frac{1}{2}(w^-)^2 + \epsilon^-\right) + p^- = \rho^+ \left(\frac{1}{2}(w^+)^2 + \epsilon^+\right) + p^+, \quad (9.153)$$

where w is the velocity of particles on the wavefront with respect to the new reference system. Taking into account the first of these equations and introducing the specific volume $v = 1/\rho$, we find that the system (9.153) can be written as

$$j^2 v^- + p^- = j^2 v^+ + p^+,$$
$$\frac{1}{2} j^2 (v^-)^2 + \epsilon^- = \frac{1}{2} j^2 (v^+)^2 + \epsilon^+. \quad (9.154)$$

The first equation allows us to derive the condition

$$j^2 = \frac{p^- - p^+}{v^+ - v^-}, \quad (9.155)$$

corresponding to the two possibilities

$$p^- > p^+ \quad \text{and} \quad v^- < v^+ \quad \text{or} \quad p^- < p^+ \quad \text{and} \quad v^- > v^+.$$

Finally, after taking into account (9.155), the equation (9.154) becomes

$$f(v^-, p^-) \equiv \epsilon^-(v^-, p^-) - \epsilon^+ + \frac{1}{2}(v^- - v^+)(p^- + p^+) = 0. \quad (9.156)$$

This equation is called **Hugoniot's equation** or the **adiabatic shock equation** and, given a pair p^+, v^+ for the undisturbed region, it allows us to find all values p^-, v^- which are compatible with jump conditions, but not necessarily possible. In fact we observe that, assuming that the shock wave exists, there must be a unique combination of v^+, v^-, p^+, p^-, once j or, equivalently, c_n has been assigned. This remark highlights that the jump system does not provide a complete description of shock phenomenon, since it admits more than one solution. Some of these solutions must be rejected on a physical basis.

If $\epsilon(v, p)$ is assumed to be a one-to-one function of each of its variables, then the equation (9.156) defines a curve γ in the plane (p, v) having the following properties:

- the straight line $v = v^+$ intersects the curve only at (v^+, p^+);

9.15. Shock Waves in a Perfect Gas

- the straight line $p = p^+$ intersects the curve only at (v^+, p^+);
- it is a convex curve.

This last property can be deduced by resorting to the implicit function theory and assuming some suitable properties of the constitutive equation.

A geometric interpretation of the curve defined by (9.156) highlights some aspects governing shock waves. The quantity j^2 is the opposite of the slope of the straight line connecting the point (v^+, p^+) with any other point (v^-, p^-) of the curve itself (see Figure 9.20).

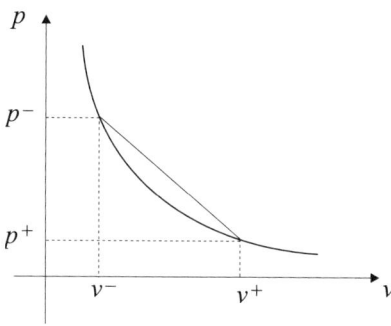

Figure 9.20

All the straight lines originating at the point (v^+, p^+) and intersecting the curve represent transitions which are allowed by the jump conditions. This conclusion indicates that we need to introduce a physical criterion to select the values (v^-, p^-) that are physically acceptable.

In the next section the case of a perfect gas will be considered in great detail.

9.15 Shock Waves in a Perfect Gas

In this section the shock wave equations (9.153) will be examined in the case of a perfect gas. For given quantities p^+ and v^+, the unknowns p^-, w^-, w^+ will be expressed in terms of the fourth unknown v^-.

The Clausius–Duhem inequality, suitably written for the case of a shock wave, will be introduced in order to select the physically acceptable values of v^+.

The constitutive equation of a perfect gas is

$$\epsilon(p, v^+) = \frac{1}{\gamma - 1} p v^+, \qquad (9.157)$$

where γ is expressed in terms of the universal gas constant R and the specific heat at constant volume c_v by the relation

$$\gamma = 1 + \frac{R}{c_v}.$$

In a monatomic gas $\gamma = 5/3$, and for a diatomic gas $\gamma = 7/5$.

Substituting (9.157) into (9.156), we obtain the explicit form of Hugoniot's curve:

$$\frac{p^-}{p^+} = \frac{v^+(\gamma+1) - v^-(\gamma-1)}{v^-(\gamma+1) - v^+(\gamma-1)}. \tag{9.158}$$

For a monatomic gas whose thermodynamic properties in the undisturbed region have been assigned, Hugoniot's curve has the behavior shown in Figure 9.21.

Substituting (9.158) into (9.155) leads to

$$j^2 = \frac{2\gamma p^+}{v^-(\gamma+1) - v^+(\gamma+1)}. \tag{9.159}$$

Therefore, from $(9.153)_1$, the velocities w^- and w^+ are derived:

$$\begin{aligned} w^- &= jv^- = v^- \sqrt{\frac{2\gamma p^+}{v^-(\gamma+1) - v^+(\gamma+1)}}, \\ w^+ &= jv^+ = v^+ \sqrt{\frac{2\gamma p^+}{v^-(\gamma+1) - v^+(\gamma+1)}}. \end{aligned} \tag{9.160}$$

On the other hand, the Clausius–Duhem inequality leads to the jump condition

$$\left[\!\left[\rho\eta(c_n - v_n) - \frac{\mathbf{h}}{\theta}\cdot\mathbf{n} \right]\!\right] \geq 0, \tag{9.161}$$

which, as the shock is adiabatic ($\mathbf{h} = \mathbf{0}$), reduces to

$$[\![\rho\eta(c_n - v_n)]\!] \geq 0;$$

in the proper frame of the wave front, this assumes the form

$$j\,[\![\eta]\!] \leq 0. \tag{9.162}$$

9.15. Shock Waves in a Perfect Gas

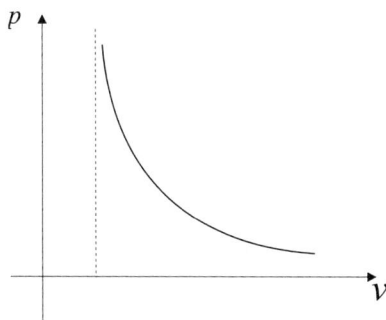

Figure 9.21

To within an arbitrary constant, the specific entropy of a perfect gas is

$$\eta = c_v \ln pv^\gamma. \tag{9.163}$$

Using (9.159) and (9.163), it becomes

$$\eta = \sqrt{\frac{2\gamma p^+}{v^-(\gamma+1) - v^+(\gamma+1)}} \; c_v \ln \frac{p^-(v^-)^\gamma}{p^+(v^+)^\gamma} \geq 0. \tag{9.164}$$

The first factor exists if

$$v^- > \frac{(\gamma-1)}{(\gamma+1)} v^+, \tag{9.165}$$

and the left-hand side of (9.164) has the same sign as the second factor. This is positive if

$$v^- \leq v^+. \tag{9.166}$$

Owing to (9.165) and (9.166), it is clear that the Clausius–Duhem inequality imposes the following restriction on the values of v^-:

$$\frac{(\gamma-1)}{(\gamma+1)} v^+ < v^- \leq v^+. \tag{9.167}$$

Many relevant conclusions can be deduced from (9.167). First of all, the specific volume v across a shock wave decreases. Therefore, from (9.166) it follows that

$$p^- > p^+,$$

which means that only compressive shock waves are physically acceptable for a perfect gas. It is worthwhile to remark that, under suitable assumptions, this result represents a general feature of shock waves.

Furthermore, the specific volume does not decrease indefinitely, but rather approaches a finite value. This limit volume is $v^+/4$ in the case of a monatomic gas and $v^+/6$ for a diatomic gas.

Finally, let us compare the gas velocities w^- and w^+ with the corresponding speeds of sound U. For a perfect gas, we have

$$U^2 = \left(\frac{\partial p}{\partial \rho}\right)_\eta = \gamma p \upsilon.$$

From (9.160) and (9.158) we have

$$\left(\frac{w^-}{U^-}\right)^2 = \frac{2}{\frac{\upsilon^-}{\upsilon^+}(\gamma-1) - (\gamma-1)},$$

$$\left(\frac{w^-}{U^-}\right)^2 = \frac{2}{\frac{\upsilon^-}{\upsilon^+}(\gamma-1) - (\gamma-1)}.$$

(9.168)

It can be easily proved that when the values of υ^- satisfy (9.166), we obtain

$$\left(\frac{w^-}{U^-}\right)^2 < 1, \qquad \left(\frac{w^+}{U^+}\right)^2 > 1.$$

Therefore, it can be concluded that

$$w^- < U^-, \qquad w^+ > U^+,$$

i.e., the shock waves travel with a supersonic velocity with respect to the gas in the undisturbed gas, and they propagate with a subsonic velocity with respect to the compressed gas.

9.16 Exercises

1. Find Archimedes' force on the cylinder of radii $R_1 < R_2$, height h and density ρ, with vertical axis, immersed in water and filled with water.

2. Find the density of a sphere of radius R in order for it to be floating with half of it immersed in water.

3. Find the total force acting on a body S immersed in a perfect gas G. Assuming that the axis Ox_3 is oriented upward, the pressure p on S due to the presence of G is (see (9.11)

$$p(x_3) = p_0 \exp\left(-\frac{gx_3}{R\theta}\right) = \rho_0 R\theta \exp\left(-\frac{gx_3}{R\theta}\right).$$

9.16. Exercises

From this we derive the expansion

$$p(x_3) = p_0 - \rho_0 g x_3 + \frac{1}{2}\rho_0 \frac{g^2}{R\theta}x_3^2 + O(3), \qquad (9.169)$$

in which the constitutive equation $p = \rho R\theta$ has been taken into account. Substituting (9.169) into the expression for the total force acting on S

$$\mathbf{F} = -p_0 \int_\sigma \exp\left(-\frac{g x_3}{R\theta}\right) \mathbf{N}\, d\sigma, \qquad (9.170)$$

where σ is the surface of S and $\mathbf{N}$ is the exterior unit normal to σ, and repeating the considerations leading to Archimedes' principle for liquid, we derive

$$\mathbf{F} = \rho_0 g k V - \frac{1}{2}\rho_0 \frac{g^2}{R\theta}\int_C x^2\, dc + O(4), \qquad (9.171)$$

where V is the volume of the region occupied by S and $\mathbf{k}$ is the unit upward-oriented vector. Formula (9.171) shows that we have again a total force which is equivalent to an upward-driven force with a value, in general, that differs from (9.14).

4. Find a solution of Blausius's equation (9.105) in terms of power series in the neighborhood of the origin, with the boundary conditions (9.106)$_1$. Use an analytic expansion (by introducing the variable $\xi = 1/x$) in order to satisfy the second condition.

9.17 The Program Potential

Aim of the Program Potential

The program Potential, evaluates the complex velocity, the kinetic potential, and Stokes potential when the complex potential $F = F(z)$ is given. Moreover, it plots the level curves of the above potentials.

Description of the Problem and Relative Algorithm

The algorithm calculates the real and imaginary parts of the complex potential $F = F(z)$, with $z = x + iy$. We recall that these parts coincide with the kinetic potential $\varphi = \varphi(x, y)$ and the Stokes potential $\psi = \psi(x, y)$, respectively. Then the program evaluates the derivatives of φ and ψ with respect to x in order to find the complex velocity $V = (\partial\varphi/\partial x) + i(\partial\psi/\partial x)$. Finally, the program plots the level curves $\varphi = const$ and $\psi = const$ contained in a given rectangle with vertices $A_1 \equiv (a, 0)$, $A_2 \equiv (b, 0)$, $A_3 \equiv (0, c)$, and $A_4 \equiv (0, d)$.

Command Line of the Program Potential

Potential[F, {a, b}, {c, d}, {val0K, val1K, stepK}, {val0S, val1S, stepS}, points, option]

Parameter List

Input Data

F = complex potential $F = F(z)$;

{a, b}, {c,d} = determine the vertices $A_1 \equiv (a, 0)$, $A_2 \equiv (b, 0)$, $A_3 \equiv (0, c)$, $A_4 \equiv (0, d)$ of the rectangle in which the potential level curves have to be drawn;

{val0K, val1K, stepK} = lowest and highest values of the kinetic level curves; stepK is the step to draw them;

{val0S, val1S, stepS} = lowest and highest values of the Stokes level curves; stepS is the step to draw them;

points = number of plot points;

option = plot option. If option = Kinetic, the program plots only the kinetic level curves; if option = Stokes, it plots only the Stokes level curves. When option = All, both the curves are represented.

9.17. The Program Potential

Output Data

The complex potential $F = \varphi(x,y) + i\psi(x,y)$;

the complex velocity $V = \dfrac{\partial \varphi}{\partial x} + i\dfrac{\partial \psi}{\partial x}$;

the plot of the level curves $\varphi = const$ and $\psi = const$.

Worked Examples

1. **Uniform motion.** Consider the complex potential

$$F(z) = 2z.$$

To apply the program Potential it is sufficient to input the data

F = 2z;

$\{a, b\} = \{-1, 1\}$;

$\{c, d\} = \{-1, 1\}$;

$\{val0K, val1K, stepK\} = \{-10, 10, 0.5\}$;

$\{val0S, val1S, stepS\} = \{-10, 10, 0.5\}$;

points = 100;

option = All;

Potential[F, $\{a, b\}$, $\{c, d\}$, $\{val0K, val1K, stepK\}$,

$\{val0S, val1S, stepS\}$, points, option]

The corresponding output is

Complex potential

F = 2x + 2iy

Complex velocity

V = 2

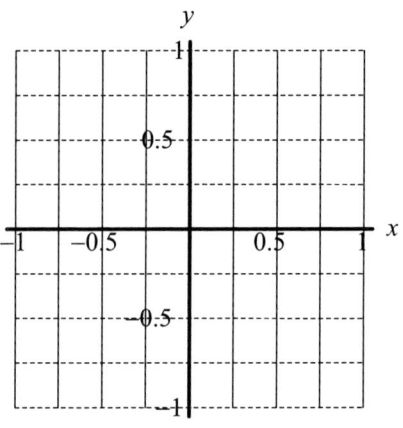

Figure 9.22

– – – Level line of the Stokes potential ψ
——— Level line of the kinetic potential φ

2. **Vortex potential.** Consider the complex potential

$$F(z) = -i \log z,$$

describing a planar flow in which the fluid particles uniformly rotate around an axis containing the origin and orthogonal to the plane Oxy. Applying the program `Potential` with the input data[4]

F = -ILog[z];
{a, b} = {-1, 1};
{c, d} = {-1, 1};
{valOK, val1K, stepK} = {-1, 1, 0.5};
{valOS, val1S, stepS} = {-2, 2, 0.4};
points = 100;
option = All;
Potential[F, {a, b}, {c, d},
{valOK, val1K, stepK}, {valOS, val1S, stepS}, points, option],

---

[4] We note that according to the usual syntax of Mathematica, all the mathematical constants and the elementary functions are written with a capital letter. Consequently, I denotes the imaginary unity.

9.17. The Program Potential

we obtain

Complex potential

$$F = \text{ArcTan}\left[\frac{y}{x}\right] - \frac{1}{2}i\text{Log}[x^2 + y^2]$$

Complex velocity

$$V = \frac{ix}{x^2 + y^2} - \frac{y}{x^2\left(1 + \frac{y^2}{x^2}\right)}$$

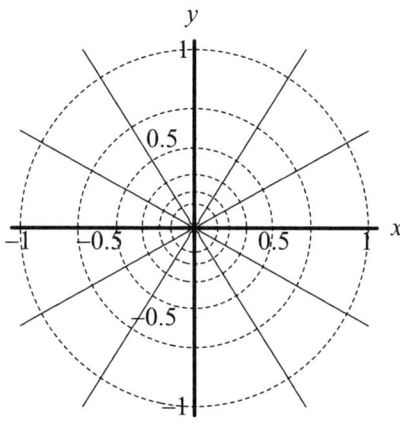

Figure 9.23

- - - Level line of the Stokes potential ψ
—— Level line of the kinetic potential φ

3. **Sources and Sinks.** The flow produced by a source or a sink that has an intensity Q is described by the complex potential

$$F(z) = \frac{Q}{2\pi} \log z,$$

where Q is positive for a source and negative for a sink. When $Q = 2\pi$, the program Potential with the input

$F = \text{Log}[z];$
$\{a, b\} = \{-2.8, 2.8\};$
$\{c, d\} = \{-2.8, 2.8\};$
$\{\text{val0K}, \text{val1K}, \text{stepK}\} = \{-1, 1, 0.25\};$
$\{\text{val0S}, \text{val1S}, \text{stepS}\} = \{-1, 1, 0.25\};$

```
points = 100;
option = All;
Potential[F, {a, b}, {c, d}, {val0K, val1K, stepK},
{val0S, val1S, stepS}, points, option]
```

gives the output

Complex potential

$$F = I \mathrm{ArcTan}\left[\frac{y}{x}\right] + \frac{1}{2}\mathrm{Log}[x^2 + y^2]$$

Complex velocity

$$V = \frac{x}{x^2 + y^2} - \frac{Iy}{x^2\left(1 + \frac{y^2}{x^2}\right)}$$

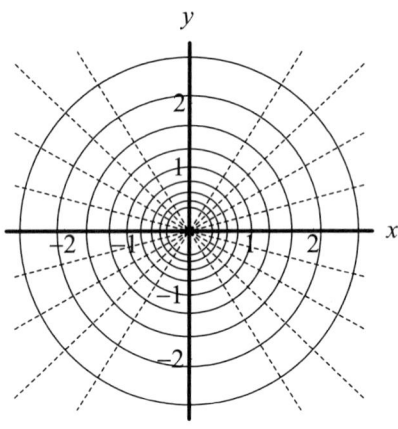

Figure 9.24

--- Level line of the Stokes potential ψ
— Level line of the kinetic potential φ

4. **Doublet.** The flow produced by a source–sink dipole with intensity $Q = 2\pi$ is described by the complex potential

$$F(z) = \frac{1}{z}.$$

The program Potential, with the input

```
F = 1/z;
```

9.17. The Program Potential

```
{a, b} = {-2, 2};
{c, d} = {-2, 2};
{valOK, val1K, stepK} = {-0.5, 0.5, 0.2};
{valOS, val1S, stepS} = {-0.5, 0.5, 0.2};
points = 100;
option = All;
Potential[F, {a, b}, {c, d}, {valOK, val1K, stepK},
   {valOS, val1S, stepS}, points, option],
```

gives the output

Complex potential

$$F = \frac{x}{x^2+y^2} - I\frac{y}{x^2+y^2}$$

Complex velocity

$$V = -\frac{2x^2}{(x^2+y^2)^2} + \frac{2Ixy}{(x^2+y^2)^2} + \frac{1}{x^2+y^2}$$

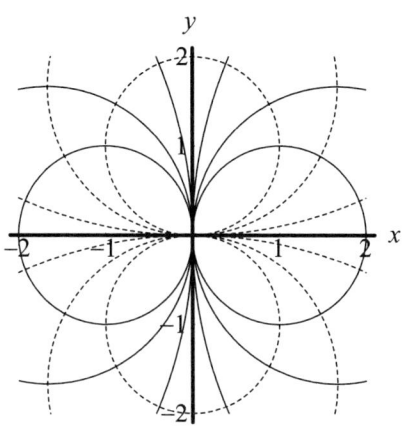

Figure 9.25

– – – Level line of the Stokes potential ψ
——— Level line of the kinetic potential φ

5. Consider the complex potential

$$F(z) = z + \frac{1}{z}$$

describing the planar flow produced by the superposition of a uniform motion along the x-axis and a flow generated by a source–sink dipole with intensity 1. In order to represent the level curves of $\psi = \psi(x, y)$, it is sufficient to input

```
F = z + 1/z;
{a, b} = {-2, 2};
{c, d} = {-2, 2};
{val0K, val1K, stepK} = {-0.5, 0.5, 0.1};
{val0S, val1S, stepS} = {-0.5, 0.5, 0.1};
points = 100;
option = Stokes;
Potential[F, {a, b}, {c, d}, {val0K, val1K, stepK},
  {val0S, val1S, stepS}, points, option].
```

In output we have

Complex potential

$$F = x\left(1 + \frac{1}{x^2 + y^2}\right) - I\left(y - \frac{y}{x^2 + y^2}\right)$$

Complex velocity

$$V = 1 - \frac{2x^2}{(x^2 + y^2)^2} + \frac{2Ixy}{(x^2 + y^2)^2} + \frac{1}{x^2 + y^2}$$

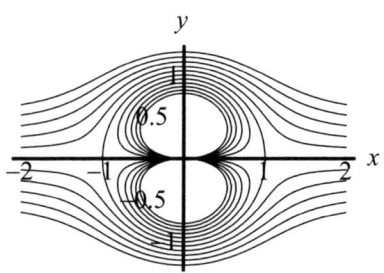

Figure 9.26

Level line of the Stokes potential ψ

Exercises

Apply the program Potential to the following complex potentials.

9.17. The Program Potential

1. $F(z) = z + \dfrac{1}{z} + i\alpha \log z$, with $\alpha = 1, 2, 2.2$.

2. $F(z) = \dfrac{\exp(i\pi/4)}{z}$.

3. $F(z) = \sqrt{z}$.

4. $F(z) = \sqrt{1 - z^2}$.

5. $F(z) = z + \sqrt{1 - z^2}$.

9.18 The Program Wing

Aim of the Program Wing

The program Wing draws the curve Γ which represents the image under the Joukowsky map of the unit circle with a given center as well as the straight line r containing the point $T' \equiv (2l, 0)$ and forming an angle β with respect to Ox. Moreover, it gives the parametric equations of both Γ and r and the values of l and β.

Description of the Problem and Relative Algorithm

Given the center $C \equiv (x_C, y_C)$ of the unit circle, the program applies the Joukowsky transformation

$$z = \zeta + \frac{l^2}{\zeta},$$

where

$$l = x_C + \sqrt{1 - y_C}.$$

Then it determines the corresponding curve Γ and draws the straight line r containing T', which coincides with the exit point of the wing, and forms an angle $\beta = \arcsin y_C$ with the Ox-axis. Moreover, it gives the parametric equations of Γ and r.

Command Line of the Program Wing

Wing[xc_,yc_]

Parameter List

Input Data

xc = abscissa of the center C of the unit circle;

yc = ordinate of the center C of the unit circle.

Output Data

parametric equation of Γ;

parametric equation of the straight line r containing T' and forming an angle β with Ox;

plot of Γ, r, and the circumference C;

values of l and β.

Worked Examples

1. Apply Joukowsky's transformation to the unit circle with the center $C \equiv (0.2, 0.1)$; i.e., enter as input the command

   `Wing[0.2, 0.1].`

   The corresponding output is[5]

   `Parametric equations of the wing`

   $$x(t) = 0.2 + \text{Cos}[t] + \frac{1.42799(0.2 + \text{Cos}[t])}{(0.2 + \text{Cos}[t])^2 + (0.1 + \text{Sin}[t])^2}$$

   $$y(t) = 0.1 + \text{Sin}[t] - \frac{1.42799(0.1 + \text{Sin}[t])}{(0.2 + \text{Cos}[t])^2 + (0.1 + \text{Sin}[t])^2}$$

   Parametric equations of the straight line r containing the point $T' \equiv (21, 0)$ and forming an angle β with Ox-axis

   $x(t) = 2.38997 - 0.994987t$

   $y(t) = 0.1t$

   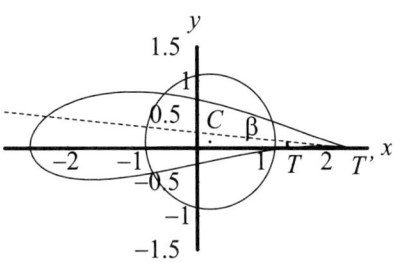

   Figure 9.27

   ——— Wing profile
   — — — Straight line r

   $\beta = 5.73917°$

   $l = 1.19499$

2. Consider Joukowsky's transformation relative to the unit circle with its center at $C \equiv (0.1, 0.15)$ and use the program `Wing`, entering

---

[5] We note that, for typographical reasons, the plots of this section are presented in a different form with respect to the plots which are obtained by launching the package **Mechanics.m**.

Wing[0.1, 0.15].

In output we obtain

Parametric equations of the wing
$$x(t) = 0.1 + \text{Cos}[t] + \frac{1.18524(0.1 + \text{Cos}[t])}{(0.1 + \text{Cos}[t])^2 + (0.15 + \text{Sin}[t])^2}$$
$$y(t) = 0.15 + \text{Sin}[t] - \frac{1.18524(0.15 + \text{Sin}[t])}{(0.1 + \text{Cos}[t])^2 + (0.15 + \text{Sin}[t])^2}$$

Parametric equations of the straight line r containing the point $T' \equiv (21, 0)$ and forming an angle β with Ox-axis
$$x(t) = 2.17737 - 0.988686t$$
$$y(t) = 0.15t$$

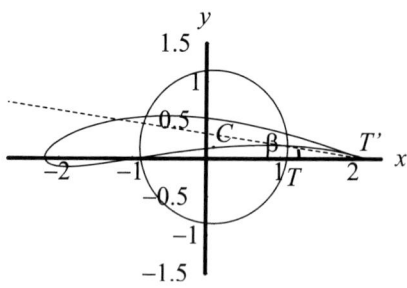

Figure 9.28

——— Wing profile
- - - Straight line r

$\beta = 8.62693°$

$1 = 1.08869$

Exercises

Apply the program Wing to the unit circle with the center at $C \equiv (x_C, y_C)$ for the following values of x_C and y_C

1. $x_C = 0.1$, $y_C = 0.2$.
2. $x_C = -0.18$, $y_C = 0.2$.
3. $x_C = 0.19$, $y_C = 0.21$.
4. $x_C = 0.18$, $y_C = 0.18$.

9.19 The Program Joukowsky

Aim of the Program Joukowsky

When the complex potential $F = F(z)$ is given by the conformal Joukowsky transformation, the program Joukowsky draws the streamlines both around the unit circle and the wing profile, which is the image of the circle under the above transformation.

Description of the Problem and Relative Algorithm

Given the center $C \equiv (x_C, y_C)$ of the unit circle and Joukowsky's transformation, the program determines the complex potential $F = F(z)$ of the flow around the cylinder (see Section 9.5). Moreover, it draws the streamlines around both the cylinder profile and the wing. The above streamlines are the integral curves of the field

$$\mathbf{X} = \left(\frac{\partial \mathrm{Re}(F)}{\partial x}, \frac{\partial \mathrm{Im}(F)}{\partial y} \right)$$

corresponding to a given set of initial data; that is, they are solutions of the Cauchy problems

$$\begin{cases} \dot{x} = \dfrac{\partial \mathrm{Re}(F)}{\partial x}, \\ \dot{y} = \dfrac{\partial \mathrm{Im}(F)}{\partial y}, \quad i = 1, \cdots, n. \\ x(t_0) = x_{0i}, \\ y(t_0) = y_{0i}, \end{cases} \quad (9.172)$$

Command Line of the Program Joukowsky

Joukowsky[φ, xc, yc, $\{$a, b$\}$, $\{$c, d$\}$, indata, steps, T1, T2]

Parameter List

Input Data

φ = angle of attack;

xc = abscissa of the center C of the unit circle;

yc = ordinate of the center C of the unit circle;

{a, b}, {c, d} = definition of the graphic window in which to represent the streamlines;

indata = set of initial data for numerically integrating the differential system (9.172);

steps = steps of the numerical integration;

T1, T2 = lowest and highest extrema of the integration interval.

Output Data

streamlines around the cylinder;

streamlines around the wing profile.

Worked Examples

1. Consider the unit circle with its center at $C \equiv (-0.2, 0.1)$ and let $\varphi = 10°$ be the angle of attack. In order to show the streamlines around the cylinder and the wing, it is sufficient to enter [6]

    $\varphi = 10 \text{Degree}$;

    $\{\text{xc}, \text{yc}\} = \{-0.2, 0.1\}$;

    $\{\text{a}, \text{b}\} = \{-3, 3\}$;

    $\{\text{c}, \text{d}\} = \{-1.5, 1.6\}$;

    $\text{indata} = \text{Join}[\text{Table}[\{-3, -1.4 + 1.4 \text{i}/15\}, \{\text{i}, 0, 15\}],$

    $\text{Table}[\{-3 + 2 * \text{i}/5, -1.5\}, \{\text{i}, 0, 5\}]]$;

    $\text{steps} = 10000$;

    $\{\text{T1}, \text{T2}\} = \{0, 10\}$;

    $\text{Joukowsky}[\varphi, \text{xc}, \text{yc}, \{\text{a}, \text{b}\}, \{\text{c}, \text{d}\}, \text{indata}, \text{steps}, \text{T1}, \text{T2}]$.

---

[6] We note that Mathematica uses 10 Degree instead of 10°. Moreover, the initial data are given by resorting to the built-in functions Join and Table, which respectively denote the union operation and the indexed list (see on-line help of Mathematica).

9.19. The Program Joukowsky

The corresponding output is

```
Streamlines around the cylinder
```

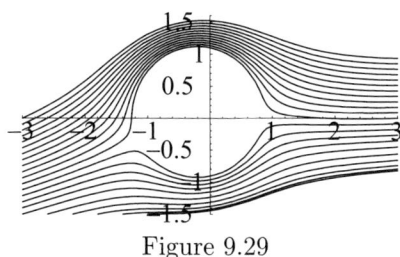

Figure 9.29

```
Streamlines around the wing profile
```

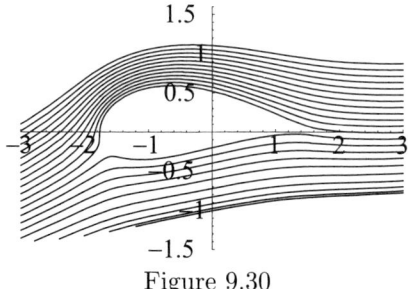

Figure 9.30

Exercises

Determine the set of initial conditions `indata` and the other input parameters that allow us to apply the program `Joukowsky` in the following cases:

1. $C \equiv (-0.2, 0.1), \quad \varphi = 29°$.
2. $C \equiv (-0.18, 0.2), \quad \varphi = 8°$.
3. $C \equiv (-0.22, 0.18), \quad \varphi = 20°$.

9.20 The Program JoukowskyMap

Aim of the Program JoukowskyMap

When a (closed, open, or piecewise defined) curve or a point set are assigned, the program JoukowskyMap determines the corresponding image under Joukowsky's transformation of a unit circle with a given center.

Description of the Problem and Relative Algorithm

Given the center $C \equiv (x_C, y_C)$ of the unit circle and a (closed, open, or piecewise defined) curve or a point set Γ, the program applies Joukowsky's transformation and determines the image Γ' of Γ. In particular, if Γ is a curve, it determines the parametric equations of Γ' as well as its plot; instead, if Γ is a finite set of points, it determines the image of each of them.

Command Line of the Program JoukowskyMap

JoukowskyMap[xc, yc, curve, data, range, option]

Parameter List

Input Data

{xc, yc} = coordinates of the center C of the unit circle;

curve = option relative to the curve Γ for which the choices closed, open, piecewise, and points are possible;

data = parametric equations of Γ or point list;

range = variability range of the parameter t in data or null if Γ is a point set;

option = parametric; for this choice the program gives the parametric equations or the coordinates of Γ'; for a different choice of option it shows only the plot.

Output Data

Plots of Γ, Γ', and the unit circle in the same graphic window;

parametric equations of Γ';

coordinates of the points of Γ'.

9.20. The Program JoukowskyMap

Use Instructions

For the input datum `curve` the following choices are possible:

`curve = closed` if Γ is a closed curve defined by two parametric equations $x = x(t)$, $y = y(t)$, where the parameter t varies in the interval $[\tau_1, \tau_2]$, which in input is given by `range = {`τ_1, τ_2`}`;

`curve = open` if Γ is an open curve defined by two parametric equations $x = x(t)$, $y = y(t)$, where the parameter t varies in the interval $[\tau_1, \tau_2]$, which in input is given by `range = {`τ_1, τ_2`}`;

`curve = piecewise` if Γ is a curve defined by more parametric equations $x_i = x_i(t)$, $y_i = y_i(t)$, $i = 1, \ldots, n$, where the parameter t varies in one or more intervals which are given by `range = {`τ_1, τ_2`}`; or `range = {{`$\tau_{1,1}, \tau_{1,2}$`}, ..., {`$\tau_{n,1}, \tau_{n,2}$`}}`;

`curve = points` if Γ is the finite set of points; in this case `range = null`.

Moreover, if the input datum `option = parametric`, the program also gives the parametric equations of Γ' or the coordinates of the points of Γ'.

Worked Examples

1. Let Γ be the square internal to the unit circle with its center at $C \equiv (0.1, 0.2)$ and vertices $A \equiv (0.5, -0.5)$, $B \equiv (0.5, 0.5)$, $D \equiv (-0.5, 0.5)$, $E \equiv (-0.5, -0.5)$. The program JoukowskyMap shows the image of Γ under Joukowsky's transformation given the following data:

{xc, yc} = {0.1, 0.2};
curve = piecewise;
{xa, ya} = {xe1, ye1} = {0.5, −0.5};
{xa1, ya1} = {xb, yb} = {0.5, 0.5};
{xb1, yb1} = {xd, yd} = {−0.5, 0.5};
{xd1, yd1} = {xe, ye} = {−0.5, −0.5};
data = {{xa + t(xa1 − xa), ya + t(ya1 − ya)},
{xb + t(xb1 − xb), yb + t(yb1 − yb)},
{xd + t(xd1 − xd), yd + t(yd1 − yd)},
{xe + t(xe1 − xe), ye + t(ye1 − ye)}};
range = {0, 1};
option = null;

JoukowskyMap[xc, yc, curve, data, range, option].

The corresponding output is[7]

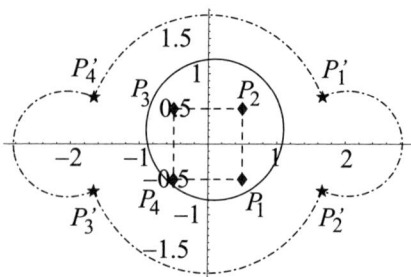

Figure 9.31

———— Unit circle
– – – – Input curve or list of points
– · – · – Output curve or list of points

2. Let Γ be an ellipse which is external to the unit circle with the center at $C \equiv (0.1, 0.2)$ and parametric equations

$$\begin{cases} x(t) = x_C + 2\cos t, \\ y(t) = y_C + \sin t, \end{cases} \quad t \in [0, 2\pi].$$

To see the image curve Γ' of Γ under Joukowsky's transformation, the program JoukowskyMap can be used by entering

{xc, yc} = {0.1, 0.2};

curve = closed;

data = {xc + 2Cos[t], yc + Sin[t]};

range = {0, 2π};

option = null;

JoukowskyMap[xc, yc, curve, data, range, option].

---

[7]The plots of this section, for typographic reasons, differ from the plots obtained by the program JoukowskyMap of the package **Mechanics.m**.

9.20. The Program JoukowskyMap

The corresponding output is

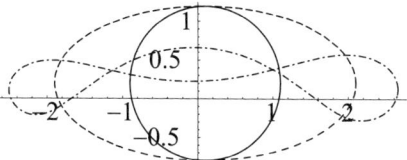

Figure 9.32

——— Unit circle
– – – – Input curve or list of points
– · – · – Output curve or list of points

3. Consider the set of points $\Gamma = \{P_1 \equiv (0, 1.5),\ P_2 \equiv (0.5, 0),\ P_3 \equiv (0.75, 0.75),\ P_4 \equiv (0.75, 0),\ P_5 \equiv (1, 0),\ P_6 \equiv (1, 1)\}$. To determine the image of Γ under Joukowsky's transformation relative to the unit circle with the center at $C \equiv (0.2, 0.1)$, we have to enter

$\{xc, yc\} = \{0.2, 0.1\}$;

curve = points;

data = $\{\{0, 1.5\}, \{0.5, 0\}, \{0.75, 0.75\}, \{0.75, 0\}, \{1, 0\}, \{1, 1\}\}$;

option = null;

JoukowskyMap[xc, yc, curve, data, range, option],

to obtain the output

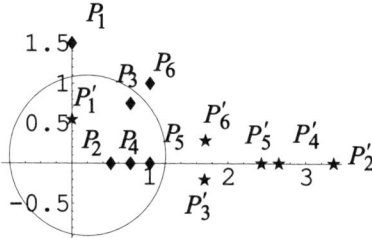

Figure 9.33

——— Unit circle
◇ ◇ ◇ ◇ Input list of points
★ ★ ★ ★ Output list of points

Exercises

Apply the program JoukowskyMap to a unit circle with the center at $C \equiv (0.2, 0.1)$ relative to the following curves or sets of points Γ.

1. Γ: radius of the unit circle along the Ox-axis.

2. Γ: triangle internal to the circle with vertices $A \equiv (0.1, 0), B \equiv (1.1, 0), D \equiv (0.2, 0.1)$.

3. Γ: segment parallel to the Oy-axis with end points $A \equiv (1, 0), B \equiv (1, 1.1)$.

4. Γ: segment with end points $A \equiv (1, 0), C \equiv (0.2, 0.1)$.

5. Γ: finite set of points on the Ox-axis. Verify that Joukowsky's transformation leaves the above points on the Ox-axis.

6. Γ: finite set of points on the Oy-axis. Verify that Joukowsky's transformation leaves the above points on the Oy-axis.

7. Γ: circle with the center at the origin and radius $r = 0.6$.

8. Γ: circle with the center at the origin and radius $r = 2$.

9. Γ: circle with the center C and radius $r = 2$.

10. Γ: circle with the center C and radius $r = 0.8$.

11. Γ: square with vertices $A \equiv (1.5, -1.5), B \equiv (1.5, 1.5), D \equiv (-1.5, 1.5), E \equiv (-1.5, -1.5)$.

12. Γ: polygonal with vertices $A \equiv (0, 0.5), B \equiv (1, 0.2), D \equiv (0, 1), E \equiv (0.2, 1)$.

13. Γ: polygonal with vertices $A \equiv (1.1, 1.2), B \equiv (1.2, 1.2), D \equiv (1.2, 0)$.

14. Γ: semicircle with the center $C \equiv (0.2, 0.1)$, radius $r = 0.5$, and chord parallel to the Ox-axis.

15. Γ: semicircle with the center $C \equiv (0.2, 0.1)$, radius $r = 2$, and chord parallel to the Ox-axis.

Chapter 10

Linear Elasticity

10.1 Basic Equations of Linear Elasticity

This chapter deals with some fundamental problems of **linear elasticity** and the material it covers is organized as follows. Since the existence and uniqueness theorems play a fundamental role, the first sections will be devoted to these theorems. The existence and the uniqueness of a solution can be proved in a *classical* way, by requiring that the solution exists and is unique within a class of regular functions; but it can also be proved in a *weak* sense (see Appendix) if it is the unique solution of a suitable integral problem.

After discussing some elementary solutions, we present a method for analyzing beam systems, according to the Saint-Venant assumption, whose validity is not yet completely proved.

Finally, we will discuss the Fourier method as a tool for analyzing problems characterized by simple geometry.

More general cases require the use of numerical methods whose reliability is based on the following requirements:

- for the problem we are going to analyze numerically, a *theorem of existence and uniqueness* exists;

- the involved procedures allow us to control the errors.

Additional remarks concern the fact that, when dealing with a linearized theory obtained by replacing the nonlinear equations with the corresponding linearized ones, the implicit assumption is that small perturbations produce small effects. This is not always the case, and it is relevant

- to investigate the condition of stability of an equilibrium configuration of the elastic system, in whose proximity it is of interest to linearize the equations;

- to express nonlinear equations in a dimensionless form;
- to define the circumstances, related to initial and loading conditions, under which the linearized equations give an appropriate description of the motion or the equilibrium.

In this respect, in the following discussion it is assumed, without any further insight, that the linearization provides correct results if

1. loads, displacement **u**, displacement gradient **H**, velocity **v**, and velocity gradient $\nabla \mathbf{v}$ are all first-order quantities, so that their products or powers can be neglected;

2. the configuration C_* around which the linearization is applied is an unstressed equilibrium configuration;

3. all the processes take place under constant and uniform temperature.

With the above remarks in mind, we recall that in linear elasticity there is no further need to distinguish between the Piola–Kirchhoff tensor and the Cauchy stress tensor (see (6.28)); in addition, if the initial state is unstressed and the evolution takes place at constant and uniform temperature, then the stress tensor is written as

$$T_{ij} = \mathbb{C}_{ijhk} E_{hk}, \qquad (10.1)$$

where the ***linear elasticity tensor*** $\mathbb{C}_{ijhk}$ is characterized by the following properties of symmetry:

$$\mathbb{C}_{ijhk} = \mathbb{C}_{jihk} = \mathbb{C}_{ijkh} = \mathbb{C}_{hkij}. \qquad (10.2)$$

In particular, for a linear elastic and isotropic medium, by taking into account (10.1) and (7.18), the elasticity tensor becomes

$$\mathbb{C}_{ijhk} = \lambda \delta_{ij}\delta_{hk} + \mu(\delta_{ih}\delta_{jk} + \delta_{jh}\delta_{ik}). \qquad (10.3)$$

Then the equations of linear elasticity are obtained by substituting the stress tensor (10.1) into the motion equation (5.69) and observing that, if $\mathbf{v} = \partial \mathbf{u}(\mathbf{X}, t)/\partial t$ is the velocity field, then in the linearization we have

$$\mathbf{a} = \frac{\partial^2 \mathbf{u}}{\partial t^2} + \dot{\mathbf{u}} \cdot \nabla \dot{\mathbf{u}} \cong \frac{\partial^2 \mathbf{u}}{\partial t^2}. \qquad (10.4)$$

Moreover, the tractions $\mathbf{t}$ and $\mathbf{t}_*$ as well as the unit vectors $\mathbf{N}$ and $\mathbf{N}_*$ that are normal to the boundaries ∂C and ∂C_* only differ from each other by terms higher than the first-order terms of the displacement and displacement gradients (see (5.72), (5.73)).

10.1. Basic Equations of Linear Elasticity

In general, a dynamical problem of linear elasticity can be formulated as a ***mixed boundary value problem***: find a displacement field $\mathbf{u}(\mathbf{X},t)$ such that

$$\begin{cases} \rho_* \dfrac{\partial^2 u_i}{\partial t^2} = \dfrac{\partial}{\partial X_j}\left(\mathbb{C}_{ijhk}\dfrac{\partial u_h}{\partial X_k}\right) + \rho_* b_i & \text{on } C_*, \\[2mm] \mathbb{C}_{ijhk}\dfrac{\partial u_h}{\partial X_k} N_j = t_i(\mathbf{X},t) & \text{on } \Sigma_1, \\[2mm] \mathbf{u}(\mathbf{X},t) = 0 & \text{on } \Sigma_2, \\[2mm] \mathbf{u}(\mathbf{X},0) = \mathbf{u}_0(\mathbf{X}), \quad \dot{\mathbf{u}}(\mathbf{X},0) = \dot{\mathbf{u}}_0(\mathbf{X}) & \text{on } C_*, \end{cases} \quad (10.5)$$

where Σ_1 is the portion of the boundary ∂C_* on which the traction $\mathbf{t}(\mathbf{X},t)$ is prescribed, Σ_2 is the fixed portion of ∂C_*, $\Sigma_1 \cup \Sigma_2 = \partial C_*$, and $\mathbf{N}$ is the outward unit vector normal to ∂C_*.

In particular, if $\Sigma_2 = \emptyset$, then the boundary problem is a ***stress boundary value problem***; on the other hand, if $\Sigma_1 = \emptyset$, then the problem reduces to a ***displacement boundary value problem***.

A static mixed boundary value problem is formulated in this way:

$$\begin{cases} \dfrac{\partial}{\partial X_j}\left(\mathbb{C}_{ijhk}\dfrac{\partial u_h}{\partial X_k}\right) + \rho_* b_i = 0 & \text{on } C_*, \\[2mm] \mathbb{C}_{ijhk}\dfrac{\partial u_h}{\partial X_k} N_j = t_i(\mathbf{X}) & \text{on } \Sigma_1, \\[2mm] u_i(\mathbf{X}) = 0 & \text{on } \Sigma_2. \end{cases} \quad (10.6)$$

If the solid is homogeneous and isotropic, then the elasticity tensor $\mathbb{C}$ assumes the form (10.3), where λ and μ are constant, so that the system (10.5) is written as

$$\begin{cases} \rho_* \dfrac{\partial^2 \mathbf{u}}{\partial t^2} = \mu \Delta \mathbf{u} + (\lambda+\mu)\nabla\nabla\cdot\mathbf{u} + \rho_* b_i & \text{on } C_*, \\[2mm] (2\mu\mathbf{E} + \lambda(\operatorname{tr}\mathbf{E})\mathbf{I})\,\mathbf{N} = \mathbf{t}(\mathbf{X}) & \text{on } \Sigma_1, \\[2mm] \mathbf{u}(\mathbf{X}) = 0 & \text{on } \Sigma_2, \\[2mm] \mathbf{u}(\mathbf{X},0) = \mathbf{u}_0(\mathbf{X}), \quad \dot{\mathbf{u}}(\mathbf{X},0) = \dot{\mathbf{u}}_0(\mathbf{X}), & \text{on } C_*. \end{cases} \quad (10.7)$$

When dealing with a stress boundary value problem, the traction $\mathbf{t}$ and the body force $\mathbf{b}$ have to satisfy the overall equilibrium equations of the system S:[1]

$$\begin{cases} \int_{\partial C_*} \mathbf{t}\,d\sigma + \int_{C_*} \rho_* \mathbf{b}\,dc = \mathbf{0}, \\ \int_{\partial C_*} \mathbf{r}\times\mathbf{t}\,d\sigma + \int_{C_*} \rho_* \mathbf{r}\times\mathbf{b}\,dc = \mathbf{0}. \end{cases} \quad (10.8)$$

The first condition is a restriction on the prescribed values of $\mathbf{t}$ and $\mathbf{b}$. Moreover, introducing the decomposition

$$\mathbf{r} = \mathbf{r}_* + \mathbf{u}, \quad (10.9)$$

where $\mathbf{r}_*$ is the position vector in the reference configuration, and recalling that $\mathbf{t}, \mathbf{b}$, and $\mathbf{u}$ are first-order quantities, we find that $(10.8)_2$ can be written in the form

$$\int_{\partial C_*} \mathbf{r}_* \times \mathbf{t}\,d\sigma + \int_{C_*} \rho_* \mathbf{r}_* \times \mathbf{b}\,dc = \mathbf{0}. \quad (10.10)$$

This condition must be satisfied both from data and solution, i.e., it is a *compatibility condition*.

Sometimes it is convenient to assume as unknowns the stresses rather then the displacements. In a traction problem, it holds that

$$\begin{aligned} \nabla_{\mathbf{X}} \cdot \mathbf{T} + \rho_* \mathbf{b} &= \mathbf{0} \quad \text{on } C_*, \\ \mathbf{TN} &= \mathbf{t} \quad \text{on } \partial C_*. \end{aligned} \quad (10.11)$$

The solutions $\mathbf{T}(\mathbf{X})$ of this system have to satisfy the integrability condition of the deformation tensor $\mathbf{E} = \mathbb{C}^{-1}\mathbf{T}$. Such a condition (see Section 3.8) is written as

$$\nabla \times \nabla \times \mathbf{E} = \mathbf{0}$$

and, due to the constitutive law (10.1), the equilibrium equations become

$$\begin{aligned} \nabla_{\mathbf{X}} \cdot \mathbf{T} + \rho_* \mathbf{b} &= \mathbf{0} \quad \text{on } C_*, \\ \nabla \times \nabla \times (\mathbb{C}^{-1}\mathbf{T}) &= \mathbf{0} \quad \text{on } C_*, \\ \mathbf{TN} &= \mathbf{t} \quad \text{on } \partial C_*. \end{aligned} \quad (10.12)$$

---

[1] If the portion Σ_2 is nonempty, then conditions (10.8) are replaced by

$$\begin{cases} \int_{\Sigma_1} \mathbf{t}\,d\sigma + \int_{C_*} \rho_* \mathbf{b}\,dc + \int_{\Sigma_2} \boldsymbol{\phi}\,d\sigma = \mathbf{0}, \\ \int_{\Sigma_1} \mathbf{r}\times\mathbf{t}\,d\sigma + \int_{C_*} \rho_* \mathbf{r}\times\mathbf{b}\,dc + \int_{\Sigma_1} \mathbf{r}\times\boldsymbol{\phi}\,d\sigma = \mathbf{0}, \end{cases}$$

where $\boldsymbol{\phi}$ is the reaction provided by the constraints fixing the portion Σ_2 of the boundary. These conditions of global equilibrium are certainly satisfied due to the presence of the above reactions, provided that they are assured by the constraints.

10.2 Uniqueness Theorems

This section is aimed at proving the uniqueness of a classical solution of mixed boundary problems; in the next section we will discuss the existence and uniqueness of weak solutions of mixed or pure traction problems.

In order to reach this goal, it is relevant first to prove that, in any arbitrary motion, the following integral relation holds:

$$\frac{d}{dt}\int_{C_*} \rho_* \left(\frac{1}{2}\dot{u}^2 + \Psi\right) dc_* = \int_{\partial C_*} \mathbf{t}\cdot\dot{\mathbf{u}}\,d\sigma_* + \int_{C_*} \rho_* \mathbf{b}\cdot\dot{\mathbf{u}}\,dc_*, \qquad (10.13)$$

where

$$\rho_* \Psi = \frac{1}{2}\mathbb{C}_{ijhk} E_{ij} E_{hk} \qquad (10.14)$$

is the **energy density of elastic deformation.**

The motion equation $(10.5)_1$, when multiplied by $\dot{\mathbf{u}}$, gives

$$\frac{d}{dt}\int_{C_*} \rho_* \frac{1}{2}\dot{u}^2 dc_* = \int_{C_*} \left(\dot{u}_i \frac{\partial}{\partial X_j}(\mathbb{C}_{ijhk} E_{hk}) + \rho_* b_i \dot{u}_i\right) dc_*$$

$$= -\int_{C_*} \mathbb{C}_{ijhk} \dot{E}_{ij} E_{hk} dc_*$$

$$+ \int_{C_*} \left(\frac{\partial}{\partial X_j}(\mathbb{C}_{ijhk} E_{hk} \dot{u}_i) + \rho_* b_i \dot{u}_i\right) dc_*,$$

so that, by using (10.14) as well as by applying Gauss's theorem, we get

$$\frac{d}{dt}\int_{C_*} \rho_* \left(\frac{1}{2}\dot{u}^2 + \Psi\right) dc_* = \int_{\partial C_*} \mathbb{C}_{ijhk} E_{hk} \dot{u}_i N_j\,d\sigma_* + \int_{C_*} \rho_* b_i \dot{u}_i\,dc_*.$$

This relation, with the boundary conditions $(10.5)_{2,3}$, allows us to obtain (10.13).

The following uniqueness theorem can now be proved:

Theorem 10.1
If the elasticity tensor is positive definite, so that

$$\mathbb{C}_{ijhk} E_{ij} E_{hk} > 0, \qquad (10.15)$$

and a solution $\mathbf{u}(\mathbf{X}, t)$ of the mixed problem (10.5) exists, then this solution is unique.

PROOF If $\mathbf{u}_1$ and $\mathbf{u}_2$ are two solutions of the mixed problem (10.5), then the difference $\mathbf{u} = \mathbf{u}_1 - \mathbf{u}_2$ will be a solution of (10.5) corresponding to the following homogeneous data:

$$\mathbf{u} = \mathbf{0} \quad \text{on } \Sigma_1 \quad \mathbf{t} = \mathbf{0} \quad \text{on } \Sigma_2,$$
$$\mathbf{u}(\mathbf{X}, 0) = \dot{\mathbf{u}}(\mathbf{X}, 0) = \mathbf{0} \quad \text{on } C_*.$$

It follows that for the motion **u** the integral condition (10.13) gives

$$\frac{d}{dt} \int_{C_*} \rho_* \left(\frac{1}{2} \dot{u}^2 + \Psi \right) dc_* = 0. \tag{10.16}$$

Moreover, when the initial conditions for **u** are taken into account, the integral appearing in (10.16) must initially vanish so that during the subsequent evolution we have

$$\int_{C_*} \rho_* \left(\frac{1}{2} \dot{u}^2 + \Psi \right) dc_* = 0.$$

Since all the quantities under the integral are nonnegative and the functions are of class C^2, at any instant and at any point it must hold that

$$\dot{\mathbf{u}}(\mathbf{X}, t) = \mathbf{0}, \ \Psi(\mathbf{X}, t) = 0;$$

since initially $\mathbf{u}(\mathbf{X}, 0) = \mathbf{0}$, it follows that **u** vanishes at any instant and at any point, so that $\mathbf{u}_1 = \mathbf{u}_2$. ∎

We use the same procedure to prove the uniqueness of an equilibrium problem. In fact, if the equilibrium equation $(10.6)_1$, previously multiplied by **u**, is integrated over C_*, then when (10.13) is again taken into account, we obtain the following relation:

$$\int_{C_*} \rho_* \Psi \, dc_* = \int_{\partial C_*} \mathbf{t} \cdot \mathbf{u} \, d\sigma_* + \int_{C_*} \rho_* \mathbf{b} \cdot \mathbf{u} \, dc_*. \tag{10.17}$$

Now the following theorem can be proved.

Theorem 10.2
Let the elasticity tensor be positive definite (see (10.15)). Then, if a solution $\mathbf{u}(\mathbf{X})$ of a mixed problem or a displacement problem exists, it is unique. For a problem of pure traction, the solution is unique, apart from an arbitrary infinitesimal rigid displacement.

PROOF If $\mathbf{u}_1$ and $\mathbf{u}_2$ are two solutions, then their difference must also be a solution satisfying the homogeneous data

$$\mathbf{u} = \mathbf{0} \quad \text{on } \Sigma_1, \quad \mathbf{t} = \mathbf{0} \quad \text{on } \Sigma_2.$$

It follows from (10.17) that

$$\int_{C_*} \rho_* \Psi \, dc_* = 0,$$

and since C is positive definite, it also follows that $\Psi = 0$, i.e., $\mathbf{E} = \mathbf{0}$. This proves that $\mathbf{u}$ is a rigid displacement. If Σ_1 does not reduce to a point or a straight line, then the condition $\mathbf{u} = \mathbf{0}$ on Σ_1 implies that $\mathbf{u} = \mathbf{0}$ in all the region C_*. ∎

Remark When considering an isotropic material, the condition (10.15) is fulfilled if and only if

$$\lambda + 2\mu > 0, \qquad \mu > 0. \tag{10.18}$$

This remark is proved by the following chain of equalities:

$$2\Psi = \mathbb{C}_{ijhk} E_{ij} E_{hk} = (\lambda \delta_{ij} \delta_{hk} + \mu(\delta_{ih}\delta_{jk} + \delta_{jh}\delta_{ik})) E_{ij} E_{hk}$$
$$= \lambda (\operatorname{tr} \mathbf{E})^2 + 2\mu \operatorname{tr}(\mathbf{E}^2) = (\lambda + 2\mu)(\operatorname{tr} \mathbf{E})^2 + 2\mu(E_{12}^2 + E_{13}^2 + E_{23}^2),$$

so that Ψ is positive definite if and only if the conditions (10.18) are satisfied.

10.3 Existence and Uniqueness of Equilibrium Solutions

In this section, in order to prove an existence and uniqueness theorem for a mixed boundary problem (see (10.5)), a weak formulation is used (see Appendix).

Let

$$\mathcal{U}_0 = (H(C_*))^3 \tag{10.19}$$

be the vector space of all the vector functions $\mathbf{v}(\mathbf{X})$ whose components

- are square integrable functions vanishing on Σ_1;
- have weak first derivatives in C_*.

Then $\mathcal{U}_0$ is a complete vector space with respect to the Sobolev norm

$$\|\mathbf{v}\|^2_{\mathcal{U}_0} = \sum_{i=1}^{3} \int_{C_*} v_i^2 \, dc_* + \sum_{i,j=1}^{3} \int_{C_*} v_{i,j}^2 \, dc_*, \tag{10.20}$$

and it is a Hilbert space with respect to the scalar product

$$(\mathbf{v}, \mathbf{u})_{\mathcal{U}_I} = \sum_{i=1}^{3} \int_{C_*} v_i u_i \, dc_* + \sum_{i,j=1}^{3} \int_{C_*} v_{i,j} u_{i,j} \, dc_*, \qquad (10.21)$$

where $a_{i,j} = \partial a_i / \partial X_j$.

Introducing the notations

$$B(\mathbf{u}, \mathbf{v}) = \int_{C_*} \mathbb{C}_{ijhk} u_{h,k} v_{i,j} \, dc_*,$$

$$F(\mathbf{v}) = \int_{C_*} \rho_* b_i v_i \, dc_* + \int_{\Sigma_2} t_i v_i \, d\sigma_*, \qquad (10.22)$$

the weak formulation of the problem (10.6) gives:

$$B(\mathbf{u}, \mathbf{v}) = F(\mathbf{v}) \qquad \forall \mathbf{v} \in \mathcal{U}_0. \qquad (10.23)$$

Now the following theorem can be proved:

Theorem 10.3
Let C_* be a compact region of $\Re^3$ bounded by a regular surface and assume that $b_i(\mathbf{X}) \in L_2(C_*)$, $t_i(\mathbf{X}) \in L_2(\Sigma_2)$. If the elasticity tensor satisfies the conditions

1. $\mathbb{C}_{ijhk}(\mathbf{X}) \in L_1(C_*)$, $|\mathbb{C}_{ijhk}| < K$, $K > 0$;

2. *it is elliptic; i.e.,* $\mathbb{C}_{ijhk} \xi_{ij} \xi_{hk} > \mu \sum_{i,j} \xi_{ij}^2$, $\mu > 0$, *for any symmetric tensor* ξ_{ij};

then the weak solution $\mathbf{u} \in \mathcal{U}_0$ of the problem (10.6) is unique.

PROOF The definition of $\mathcal{U}_0$ and the assumptions about b_i, t_i, and $\mathbb{C}_{ijhk}$ allow us to state that B is a bilinear continuous form on $\mathcal{U}_0 \times \mathcal{U}_0$ and F is a linear continuous form on $\mathcal{U}_0$. In fact, the first condition gives

$$\left(\int_{C_*} \mathbb{C}_{ijhk} u_{i,j} v_{h,k} \, dc_* \right)^2 \leq K^2 \left(\sum_{i,j} \int_{C_*} u_{i,j} v_{i,j} \, dc_* \right)^2.$$

Since $\mathbf{v}, \mathbf{u} \in \mathcal{U}_0$, the squared components v_i and u_i can be summed up as well as their first derivatives, and Schwarz's inequality (see (1.10)) can be applied to the integral on the left-hand side in $L_2(C_*)$, so that we obtain

10.3. Existence and Uniqueness of Equilibrium Solutions

the relation

$$\left(\int_{C_*} \mathbb{C}_{ijhk} u_{i,j} v_{h,k} \, dc_*\right)^2 \leq K^2 \sum_{i,j} \int_{C_*} u_{i,j}^2 \, dc_* \int_{C_*} v_{i,j}^2 \, dc_*$$

$$\leq K^2 \left(\sum_i \int_{C_*} u_i^2 \, dc_* + \sum_{i,j} \int_{C_*} u_{i,j}^2 \, dc_*\right) \left(\sum_i \int_{C_*} v_i^2 \, dc_* + \sum_{i,j} \int_{C_*} v_{i,j}^2 \, dc_*\right),$$

which in a compact form is written as

$$|B(\mathbf{u}, \mathbf{v})| \leq K \, \|\mathbf{u}\|_{\mathcal{U}_0} \, \|\mathbf{v}\|_{\mathcal{U}_0} \, .$$

This proves the continuity of B; in a similar way, the continuity of F can be verified.

At this point, the theorem is a consequence of the Lax–Milgram theorem (see Appendix) if it is proved that B is strongly coercive or strongly elliptic. Then the symmetry properties of $\mathbb{C}$ allow us to write

$$\mathbb{C}_{ijhk} u_{i,j} u_{h,k} = \mathbb{C}_{ijhk} E_{ij} E_{hk} > \mu \sum_{i,j} E_{ij}^2,$$

where $\mathbf{E}$ is the tensor of infinitesimal deformations. From the previous inequality it also follows that

$$\int_{C_*} \mathbb{C}_{ijhk} u_{i,j} u_{h,k} \, dc_* > \mu \sum_{i,j} \int_{C_*} E_{ij}^2 \, dc_*.$$

By recalling Korn's inequality (see Appendix), we see that

$$\sum_{i,j} \int_{C_*} E_{ij}^2 \, dc_* \geq \sigma \, \|\mathbf{u}\|_{\mathcal{U}_0}^2 \, ,$$

where σ is a positive constant, and the theorem is proved. ∎

If $\mathbb{C}, t_i$, and b_i are conveniently regular functions, then the weak solution is also a regular solution for the mixed boundary problem (10.6).

For the problem of pure traction

$$\begin{aligned} (\mathbb{C}_{ijhk} u_{h,k})_{,j} &= -\rho_* b_i \quad \text{on } C_*, \\ \mathbb{C}_{ijhk} u_{h,k} N_{*j} &= t_i \quad \text{on } \partial C_*, \end{aligned} \quad (10.24)$$

the following theorem holds:

Theorem 10.4
Let the hypotheses about $\mathbb{C}, \mathbf{b},$ and $\mathbf{t}$ of the previous theorem be satisfied, together with the global equilibrium conditions

$$\int_{\partial C_*} \mathbf{t}\, d\sigma + \int_{C_*} \rho_* \mathbf{b}\, dc_* = \mathbf{0},$$

$$\int_{\partial C_*} \mathbf{r}_* \times \mathbf{t}\, d\sigma + \int_{C_*} \rho_* \mathbf{r}_* \times \mathbf{b}\, dc_* = \mathbf{0}. \qquad (10.25)$$

Then the problem (10.24) has a unique solution, apart from an arbitrary infinitesimal rigid displacement.

PROOF Only the main steps of the proof are given. Let

$$\mathcal{U} = (H^1(C_*))^3$$

be the vector space of the functions $\mathbf{v}(\mathbf{x})$ whose components belong to $L_2(C_*)$ together with their derivatives. It is a Banach space with respect to the Sobolev norm (10.20), and a Hilbert space with respect to the scalar product (10.21). Once again, (10.23) represents the weak formulation of the equilibrium problem (10.24) provided that Σ_2 in (10.21) is replaced with ∂C_*. Moreover, it is possible to prove, by the same procedure followed in the previous theorem, that $B(\mathbf{u}, \mathbf{v})$ and $F(\mathbf{v})$ are continuous in $\mathcal{U} \times \mathcal{U}$ and $\mathcal{U}$, respectively. However, $B(\mathbf{u}, \mathbf{v})$ is not strongly coercive. In fact, from condition (2) we derive

$$B(\mathbf{u}, \mathbf{u}) = \int_{C_*} \mathbb{C}_{ijhk} E_{ij} E_{hk}\, dc_* \geq \mu \sum_{i,j} \int_{C_*} E_{ij}^2\, dc_*,$$

so that

$$B(\mathbf{u}, \mathbf{u}) = 0 \iff \sum_{i,j} \int_{C_*} E_{ij}^2\, dc_* = 0. \qquad (10.26)$$

Consequently, if the subspace R of $\mathcal{U}$ given by

$$R = \{\hat{\mathbf{u}} \in \mathcal{U},\quad \hat{\mathbf{u}} = \mathbf{a} + \mathbf{b} \times \mathbf{r}_*,\quad \mathbf{a}, \mathbf{b} \in \Re^3\} \qquad (10.27)$$

of rigid displacements is introduced, it follows that the condition (10.26) is satisfied for any element of R and therefore it does not imply that $\mathbf{u} = \mathbf{0}$.

It is possible to verify that any Cauchy sequence $\{\hat{\mathbf{u}}_n\}$ of elements belonging to R converges to an element of R, so that this subspace is closed. Owing to known results of analysis, the quotient space

$$\mathcal{H} = \mathcal{U}/R$$

is a Banach space with respect to the norm

$$\|[\mathbf{u}]\|_{\mathcal{H}} = \inf_{\hat{\mathbf{u}} \in R} \|\mathbf{u} + \hat{\mathbf{u}}\|_{\mathcal{U}}, \qquad (10.28)$$

where $[\mathbf{u}]$ denotes the equivalence class of $\mathcal{U}$. It is also a Hilbert space with respect to the scalar product

$$([\mathbf{u}], [\mathbf{v}])_{\mathcal{H}} = \inf_{\hat{\mathbf{u}}, \hat{\mathbf{v}} \in R} (\mathbf{u} + \hat{\mathbf{u}}, \mathbf{v} + \hat{\mathbf{v}})_{\mathcal{U}}. \qquad (10.29)$$

On the other hand, the positions

$$\dot{B}([\mathbf{u}], [\mathbf{v}]) \equiv B(\mathbf{u}, \mathbf{v}) = \int_{C_*} \mathbb{C}_{ijhk} E_{ij}(\mathbf{u}) E_{hk}(\mathbf{v}) \, dc_*,$$

$$\dot{F}([\mathbf{v}]) = F(\mathbf{v}), \qquad (10.30)$$

define a bilinear form on $\mathcal{H} \times \mathcal{H}$ and a linear form on $\mathcal{H}$, respectively. In fact, $E_{ij}(\mathbf{u}) = E_{ij}(\mathbf{u}')$ if and only if $\mathbf{u}$ and $\mathbf{u}'$ differ for a rigid displacement, that is, belong to the same equivalence class. Moreover,

$$F(\mathbf{v}) - F(\mathbf{v}') = \mathbf{a} \cdot \left(\int_{\partial C_*} \mathbf{t} \, d\sigma + \int_{C_*} \rho_* \mathbf{b} \, dc_* \right)$$
$$+ \mathbf{b} \cdot \left(\int_{\partial C_*} \mathbf{r}_* \times \mathbf{t} \, d\sigma + \int_{C_*} \rho_* \mathbf{r}_* \times \mathbf{b} \, dc_* \right) = 0,$$

due to the conditions (10.25). It is also easy to verify that $\dot{B}(\mathbf{u}, \mathbf{v})$ and $F(\mathbf{v})$ are continuous and that the weak formulation of the equilibrium boundary problem is written as

$$\dot{B}([\mathbf{u}], [\mathbf{v}]) = \dot{F}([\mathbf{v}]) \qquad \forall [\mathbf{v}] \in \mathcal{H}.$$

Finally, owing to the inequality (see Appendix)

$$\sum_{i,j} \int_{C_*} E_{ij}^2 \, dc_* \geq c \, \|[\mathbf{u}]\|_{\mathcal{H}}^2,$$

the bilinear form $\dot{B}([\mathbf{u}], [\mathbf{v}])$ is strongly coercive and the theorem is proved.
∎

10.4 Examples of Deformations

A deformation of the system S is defined to be **homogeneous** if its Jacobian matrix is independent of the coordinates. In this case, the deformation tensor is constant and, presuming the material is homogeneous, the stress state is constant and the equilibrium equations are fulfilled.

We now discuss some examples of homogeneous deformations.

1. **Uniform expansion**

    As a consequence of the displacement field

    $$u_i = \alpha X_i, \; i = 1, 2, 3,$$

    any point **X** moves along the straight line passing through the origin and the point **X**, independently of the direction. This deformation is infinitesimal if α is a first-order quantity. If $\alpha > 0$, the deformation is an uniform expansion; if $\alpha < 0$, it is a volume contraction. The displacement and deformation gradient tensors are

    $$\mathbf{H} = \mathbf{E} = \begin{pmatrix} \alpha & 0 & 0 \\ 0 & \alpha & 0 \\ 0 & 0 & \alpha \end{pmatrix},$$

    and the stress tensor is given by

    $$\mathbf{T} = (3\lambda + 2\mu)\alpha \mathbf{I},$$

    and represents a uniform pressure if $\alpha < 0$, and a tension if $\alpha > 0$.

2. **Homotetic deformation**

    Consider the displacement field

    $$u_1 = \alpha_1 X_1, \quad u_2 = \alpha_2 X_2, \quad u_3 = \alpha_3 X_3,$$

    and the corresponding tensors

    $$\mathbf{H} = \mathbf{E} = \begin{pmatrix} \alpha_1 & 0 & 0 \\ 0 & \alpha_2 & 0 \\ 0 & 0 & \alpha_3 \end{pmatrix},$$

    as well as the stress tensor

    $$\mathbf{T} = \lambda(\alpha_1 + \alpha_2 + \alpha_3)\mathbf{I} + 2\mu\mathbf{E}.$$

    In particular, if $\alpha_2 = \alpha_3 = 0$ and $\alpha_1 > 0$, then the stress field reduces to an uniaxial traction.

3. **Pure distortion**

    The displacement field

    $$u_1 = kX_2, \; u_2 = u_3 = 0,$$

implies that
$$\mathbf{E} = \begin{pmatrix} 0 & k/2 & 0 \\ 0 & 0 & 0 \\ 0 & 0 & 0 \end{pmatrix}$$
and
$$\mathbf{T} = \begin{pmatrix} 0 & \mu k & 0 \\ 0 & 0 & 0 \\ 0 & 0 & 0 \end{pmatrix}.$$

10.5 The Boussinesq–Papkovich–Neuber Solution

When the problem does not suggest any symmetry, the approach usually followed is the Boussinesq–Papkovich–Neuber method. In the absence of body forces and at equilibrium, $(10.7)_1$ assumes the form

$$\nabla \nabla \cdot \mathbf{u} + (1 - 2\nu)\Delta \mathbf{u} = 0, \tag{10.31}$$

where

$$\nu = \frac{\lambda}{2(\lambda + \mu)}. \tag{10.32}$$

Then the following theorem holds:

Theorem 10.5
The displacement fields

$$\mathbf{u}_1 = \nabla \varphi, \quad \text{with } \Delta \varphi = 0, \tag{10.33}$$

and

$$\mathbf{u}_2 = 4(1 - \nu)\boldsymbol{\Phi} - \nabla(\boldsymbol{\Phi} \cdot \mathbf{X}), \quad \text{with } \Delta \boldsymbol{\Phi} = 0,$$

are two solutions of the equilibrium equation (10.31), *for any choice of the harmonic functions φ and $\boldsymbol{\Phi}$. Since* (10.31) *is linear, any linear combination of these solutions still represents a solution.*

PROOF To prove that $\mathbf{u}_1$ is a solution of (10.31), it is sufficient to note that

$$\nabla \cdot \mathbf{u}_1 = \Delta \varphi = 0, \quad \Delta(\mathbf{u}_1)_i = \sum_{j=1}^{3} \frac{\partial^2}{\partial X_j^2}\left(\frac{\partial \varphi}{\partial X_i}\right) = 0,$$

since the differentiation order with respect to the variables X_i and X_j can be reversed, and φ is a harmonic function.

When considering the second displacement field, we get

$$u_{2j} = 4(1-\nu)\Phi_j - \Phi_j - X_h\frac{\partial \Phi_h}{\partial X_j} = (3-4\nu)\Phi_j - X_h\frac{\partial \Phi_h}{\partial X_j},$$

so that

$$\nabla \cdot \mathbf{u}_2 = (3-4\nu)\frac{\partial \Phi_j}{\partial X_j} - \frac{\partial \Phi_j}{\partial X_j} - X_h\frac{\partial^2 \Phi_h}{\partial X_j^2} = 2(1-2\nu)\nabla \cdot \mathbf{\Phi},$$

since $\mathbf{\Phi}$ is harmonic. Moreover,

$$\Delta u_{2i} = \sum_{j=1}^{3}\frac{\partial^2 u_{2i}}{\partial X_j^2} = (3-4\nu)\sum_{j=1}^{3}\frac{\partial^2 \Phi_i}{\partial X_j^2} - \frac{\partial}{\partial X_j}\frac{\partial}{\partial X_j}\left(X_h\frac{\partial \Phi_h}{\partial X_i}\right)$$

and, since

$$\frac{\partial}{\partial X_j}\frac{\partial}{\partial X_j}\left(X_h\frac{\partial \Phi_h}{\partial X_i}\right) = \frac{\partial}{\partial X_j}\left(\frac{\partial \Phi_j}{\partial X_i} + X_h\frac{\partial^2 \Phi_h}{\partial X_j \partial X_i}\right),$$

it holds that

$$\Delta \mathbf{u}_2 = -\nabla\nabla \cdot \mathbf{\Phi}.$$

It is now easy to verify that $\mathbf{u}_2$ is a solution of (10.31). ∎

By introducing convenient hypotheses concerning the elastic coefficients as well as the differentiability of the functions, we can prove that any solution of (10.31) can be expressed as a combination of $\mathbf{u}_1$ and $\mathbf{u}_2$.

In addition, we can also verify that, if the boundary surface has a symmetry axis X_3 and the displacement components in cylindrical coordinates (r, ψ, X_3) are

$$u_r = u_r(r, X_3), \quad u_\psi = 0, \quad u_3 = u_3(r, X_3),$$

then there are two *scalar* functions φ and Φ such that the displacement can be written as $\mathbf{u} = \mathbf{u}_1 + \mathbf{u}_2$, with $\mathbf{\Phi} = \Phi \mathbf{a}_3$, where $\mathbf{a}_3$ is the unit vector of the axis X_3.[2]

10.6 Saint-Venant's Conjecture

In the previous section we proved two fundamental theorems which assure the existence and uniqueness of the solution of two boundary value

---
[2] For further insight into these topics the reader is referred to [27], [28].

10.6. Saint-Venant's Conjecture

problems of linear elasticity. However, we said nothing about the form of the solution itself. In the introductory notes, we stressed the need to resort to numerical methods to find an approximate form of the solution or to Fourier's approach, when the region C_* has a simple form. These considerations show the importance of a method which allows us to find analytical solutions of equilibrium problems, which we often find in applications (see [32]).

In the engineering sciences, it is very important to find the equilibrium configuration of elastic isotropic systems S satisfying the two following conditions:

- the unstressed reference configuration C_* of S has a dimension which is much greater than the others; that is, C_* is a cylinder whose length L satisfies the condition

$$L \gg a, \tag{10.34}$$

where a is a characteristic dimension of a section orthogonal to the direction along which the length L is evaluated;

- the body forces are absent and the surface forces act only on the bases of the above cylinder.

A system satisfying these conditions is called a *beam*. Let C_* be the cylindrical reference configuration of the elastic isotropic system S, σ_1 and σ_2 its simply connected bases, σ_l the lateral surface of C_*, and $OX_1X_2X_3$ an orthogonal coordinate system having the origin in the center of mass of σ_2 and the OX_3-axis along the length L of the cylinder (see Figure 10.1).

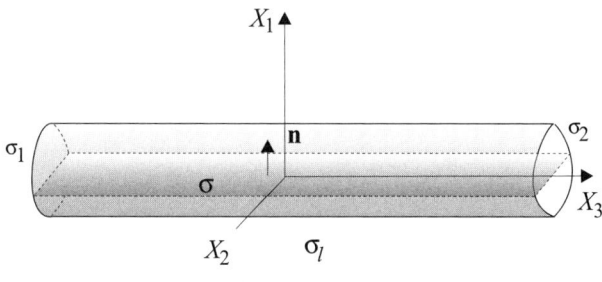

Figure 10.1

The method, due to Saint-Venant, which allows us to find explicit solutions of the equilibrium problems relative to a beam can be described in the following way:

1. First, it is assumed that any longitudinal planar section σ of C_* containing the axis X_3 is subjected to a stress $\mathbf{t}_\sigma$, whose component along

the X_3-axis is the only component different from zero; that is,
$$\mathbf{t}_\sigma = \mathbf{Tn} = \mathbf{t}_\sigma \mathbf{e}_3 \qquad \forall \mathbf{n} \perp \mathbf{e}_3, \tag{10.35}$$
or equivalently
$$T_{11} = \mathbf{e}_1 \cdot \mathbf{Te}_1 = 0, \quad T_{12} = \mathbf{e}_1 \cdot \mathbf{Te}_2 = 0, \quad T_{22} = \mathbf{e}_2 \cdot \mathbf{Te}_2 = 0. \tag{10.36}$$
Then the most general solution satisfying these conditions is determined by the equilibrium equations. This procedure is called a ***semi-inverse method***.

2. The solutions found in step (1) are used to determine the *total* force and torque acting on the bases. By using the **Saint-Venant conjecture**, which is discussed below, they are supposed to represent a good approximation of the equilibrium solutions corresponding to force distributions on the bases with the *same* total force and torque.

Let us now examine these steps in detail. In our hypotheses, the equilibrium equations can be written as
$$\mu \Delta \mathbf{u} + (\lambda + \mu) \nabla \nabla \cdot \mathbf{u} = 0 \quad \text{on } C_*, \tag{10.37}$$
and it is possible to prove (see [29], [30], and [32]) that the most general solution of (10.37) satisfying the conditions (10.36) is given by the following functions:
$$\begin{aligned}
u_1 = {} & c_1 + \gamma_3 X_2 - \gamma_2 X_3 \\
& - k \left\{ a_0 X_1 + \frac{1}{2} a_1 (X_1^2 - X_2^2) + a_2 X_1 X_2 \right. \\
& \left. + X_3 [b_0 X_1 + \frac{1}{2} b_1 (X_1^2 - X_2^2) + b_2 X_1 X_2] \right\} \\
& + \alpha\, X_2 X_3 - \frac{1}{2} a_1 X_3^2 - \frac{1}{6} b_1 X_3^3,
\end{aligned} \tag{10.38}$$

$$\begin{aligned}
u_2 = {} & c_2 + \gamma_1 X_3 - \gamma_3 X_1 \\
& - k \left\{ a_0 X_2 + \frac{1}{2} a_2 (X_2^2 - X_1^2) + a_1 X_1 X_2 \right. \\
& \left. + X_3 [b_0 X_2 + \frac{1}{2} b_2 (X_2^2 - X_1^2) + b_1 X_1 X_2] \right\} \\
& - \alpha\, X_1 X_3 - \frac{1}{2} a_2 X_3^2 - \frac{1}{6} b_2 X_3^3,
\end{aligned} \tag{10.39}$$

$$\begin{aligned}
u_3 = {} & c_3 + \gamma_2 X_1 - \gamma_1 X_2 + X_3(a_0 + a_1 X_1 + a_2 X_2) \\
& + \frac{1}{2} X_3^2 (b_0 + b_1 X_1 + b_2 X_2) - \frac{1}{2} b_0 (X_1^2 + X_2^2) \\
& - b_1 X_1 X_2^2 - b_2 X_1^2 X_2 + \alpha \psi(X_1, X_2) \\
& + b_1 \eta_1(X_1, X_2) + b_2 \eta_2(X_1, X_2),
\end{aligned} \tag{10.40}$$

10.6. Saint-Venant's Conjecture

where $c_1, c_2, c_3, \gamma_1, \gamma_2, \gamma_3, a_0, a_1, a_2, b_0, b_1, b_2,$ and α are arbitrary constants and $\psi(X_1, X_2)$, $\eta_1(X_1, X_2)$, and $\eta_2(X_1, X_2)$ are arbitrary harmonic functions; i.e.,
$$\Delta \psi = \Delta \eta_1 = \Delta \eta_2 = 0.$$

If the axes OX_1 and OX_2 are chosen to coincide with the inertial axes of the bases, we have
$$\int_{\sigma_2} X_1 \, d\sigma = \int_{\sigma_2} X_2 \, d\sigma = \int_{\sigma_2} X_1 X_2 \, d\sigma = 0,$$

and it is possible to verify that *all* the solutions (10.38)–(10.40) satisfy the boundary condition
$$\mathbf{TN} = \mathbf{0} \quad \text{on } \sigma_1, \tag{10.41}$$

for *any* choice of the constants $c_1, c_2, c_3, \gamma_1, \gamma_2, \gamma_3, a_0, a_1, a_2, b_0, b_1, b_2,$ and α, if $b_0 = 0$ and the functions $\psi(X_1, X_2), \eta_1(X_1, X_2),$ and $\eta_2(X_1, X_2)$ satisfy the following Neumann boundary value problems:

$$\Delta \psi = 0 \quad \text{on } \sigma_2,$$
$$\frac{d\psi}{dN} = -X_2 N_1 + X_1 N_2 \quad \text{on } \partial \sigma_2,$$
$$\Delta \eta_1 = 0 \quad \text{on } \sigma_2,$$
$$\frac{d\eta_1}{dN} = \frac{kX_1^2 + (2-k)X_2^2}{2} N_1 + (2+k) X_1 X_2 N_2 \quad \text{on } \partial \sigma_2,$$
$$\Delta \eta_2 = 0 \quad \text{on } \sigma_2,$$
$$\frac{d\eta_2}{dN} = \frac{kX_2^2 + (2-k)X_1^2}{2} N_2 + (2+k) X_1 X_2 N_1 \quad \text{on } \partial \sigma_2.$$

Suppose that the functions $\psi(X_1, X_2), \eta_1(X_1, X_2), \eta_2(X_1, X_2)$ have been determined (up to a constant) by solving the previous boundary problems. Then the family of solutions (10.38)–(10.40) satisfies the equilibrium condition (10.37) and the lateral boundary condition (10.41) for any choice of the constants $c_1, c_2, c_3, \gamma_1, \gamma_2, \gamma_3, a_0, a_1, a_2, b_1, b_2,$ and α, since $b_0 = 0$, in order to satisfy (10.41).

It remains to impose the remaining boundary conditions
$$\mathbf{TN}_1 = \mathbf{t}_1 \quad \text{on } \sigma_1, \quad \mathbf{TN}_2 = \mathbf{t}_2 \quad \text{on } \sigma_2,$$

on the bases of the beam. It is quite evident that it is impossible to satisfy these *functional* conditions with the remaining *constants* $c_3, \gamma_1, \gamma_2, \gamma_3, a_0, a_1, a_2, b_1, b_2,$ and α. Therefore these constants are determined by requiring that *the total force* $\mathbf{R}$ *and torque* $\mathbf{M}$ *evaluated starting from the solution* (10.38)–(10.40) *coincide with the total force* $\mathbf{G}$ *and torque* $\mathbf{L}$ *of the force distributions* $\mathbf{t}(\mathbf{X})$ (see (7.20)) *on the bases of the beam*. In this regard, we remark that, owing to the conditions (10.8) and (10.10), the forces acting on the bases are equilibrated so that

$$\mathbf{R} = \mathbf{R}_1 + \mathbf{R}_2 = \mathbf{0}, \ \mathbf{M} = \mathbf{M}_1 + \mathbf{M}_2 = \mathbf{0},$$

where $\mathbf{R}_i$ and $\mathbf{M}_i$ denote the total force and torque on the basis σ_i, respectively. Consequently, the total force and torque of the surface forces, evaluated starting from the solution (10.38)–(10.40), will balance the applied forces on the bases if and only if

$$\mathbf{R}_2 = \mathbf{G}_2, \ \mathbf{M}_2 = \mathbf{L}_2. \tag{10.42}$$

If we introduce the notation

$$\rho_1^2 = \frac{1}{A} \int X_1^2 \, d\sigma, \ \rho_2^2 = \frac{1}{A} \int X_2^2 \, d\sigma,$$

where A is the area of σ_2, it is possible to verify (see [32]) that conditions (10.42) lead to the following expressions for the undetermined constants:

$$b_1 = \frac{R_1}{AE_Y \rho_1^2}, \ b_2 = \frac{R_2}{AE_Y \rho_2^2}, \tag{10.43}$$

$$a_0 = \frac{R_3}{AE_Y}, \ a_1 = -\frac{M_2}{AE_Y \rho_1^2}, \ a_2 = \frac{M_2}{AE_Y \rho_2^2}, \tag{10.44}$$

$$M_3 = -\frac{E}{4(1+k)} \left\{ 2A(\rho_1^2 + \rho_2^2)\alpha - b_1 \int_{\sigma_2} [(2-k)X_2^3 - (4+k)X_1^2 X_2] \, d\sigma \right.$$

$$- b_2 \int_{\sigma_2} [(2-k)X_1^3 - (4+k)X_1 X_2^2] \, d\sigma$$

$$- 2 \int_{\sigma_2} \left[X_1 \left(\alpha \frac{\partial \psi}{\partial X_2} + b_1 \frac{\partial \eta_1}{\partial X_2} + b_2 \frac{\partial \eta_2}{\partial X_2} \right) \right.$$

$$\left. \left. - X_2 \left(\alpha \frac{\partial \psi}{\partial X_1} + b_1 \frac{\partial \eta_1}{\partial X_1} + b_2 \frac{\partial \eta_2}{\partial X_1} \right) \right] d\sigma \right\}, \tag{10.45}$$

where, R_1, R_2, R_3, M_1, M_2, and M_3 are the components of $\mathbf{R}_2$ and $\mathbf{M}_2$, respectively; moreover, E_Y is Young's modulus (see (7.20)) and A is the area of σ_2. The coefficient q of α appearing in (10.45), which has the form

$$q = \frac{E}{2(1+k)} \left\{ 2A(\rho_1^2 + \rho_2^2) - \int_{\sigma_2} \left[X_1 \frac{\partial \psi}{\partial X_2} - X_2 \frac{\partial \psi}{\partial X_1} \right] d\sigma \right\},$$

is called the *torsional stiffness* and can be proved to be positive.

Finally, we note that the constants c_1, c_2, c_3, γ_1, γ_2, and γ_3 are not relevant, since they define an infinitesimal rigid displacement.

In conclusion, the previous procedure has led us to a set of three functions satisfying the equilibrium equations, the boundary condition of vanishing forces on the lateral surface of the beam, and the further condition that the total forces acting on the bases are globally equilibrated and have the same resultant and torque of the surface loads, applied on the bases. It is evident that these functions *do not represent a solution of the posed boundary value problem* since they do not satisfy the boundary conditions on the bases *at any point of σ_1 and σ_2*.

The **Saint-Venant conjecture** consists of supposing that these functions represent a good approximation of the effective solution of the above boundary value problem. More precisely, Saint-Venant hypothesized that *forces having the same resultant and torque produce the same effect on the beam regardless of their distribution on the bases, except for a small region near the bases themselves*.

This conjecture allows us to consider the three functions we found as the solution of our equilibrium problem. The conjecture has not yet been proved in the above form (see the end of this section); however, because of its fundamental role in linear elasticity, it is also called the **Saint-Venant principle**.

10.7 The Fundamental Saint-Venant Solutions

The solution proposed by Saint-Venant depends on six constants a_0, a_1, a_2, b_1, b_2, and α. In other words, by setting all but one of these constants equal to zero, six families of solutions are determined. Due to the linearity of the equilibrium boundary value problem, any linear combination of these six solutions is also a solution.

Let us now discuss the meaning of some deformation patterns.

1. **Pure (compression) extension**

   The constant $a_0 \neq 0$ and all the other constants vanish, so that we have (see (10.38)–(10.40) and (10.43)–(10.45))

   $$u_1 = -ka_0 X_1, \quad u_2 = -ka_0 X_2, \quad u_3 = -ka_0 X_3,$$
   $$R_1 = R_2 = 0, \quad R_3 = AE_Y a_0, \quad M_1 = M_2 = M_3 = 0. \tag{10.46}$$

   In other words, the beam is subjected to forces orthogonal to the bases, stretching or compressing the beam (see Figure 10.2). Each planar section remains planar and the deformation is independent of the form of the basis σ_2.

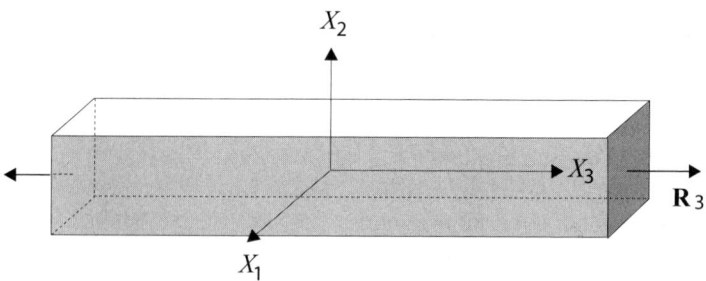

Figure 10.2

2. Uniform bending

The constant $a_1 \neq 0$ (or $a_2 \neq 0$) and all the other ones vanish so that we have (see (10.38)–(10.40))

$$u_1 = -\frac{1}{2}a_1[k(X_1^2 - X_2^2) + X_3^2], \quad u_2 = -a_1 X_1 X_2, \quad u_3 = a_1 X_1 X_2,$$
$$R_1 = R_2 = 0, \quad R_3 = 0, \quad M_1 = 0, \quad M_2 = -A\rho_1^2 a_1, \quad M_3 = 0. \tag{10.47}$$

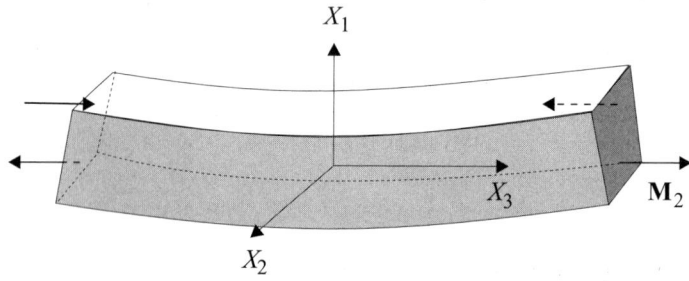

Figure 10.3

The forces acting on each base are equivalent to a couple lying in the plane OX_1X_3 (see Figure 10.3). The rulings of the cylinder become parabolas in this plane. The sections of the beam which are orthogonal to OX_3 remain planar and all their planes contain the point $(-a_1^{-1}, 0, 0)$ on the X_1-axis. Finally, the deformation is again independent of the form of σ_2.

10.7. The Fundamental Saint-Venant Solutions

3. **Nonuniform bending**

   The constant $b_1 \neq 0$ (or $b_2 \neq 0$) and all other ones vanish so that we have (see (10.38)–(10.40))

   $$u_1 = -\frac{1}{2}b_1\left[kX_3(X_1^2 - X_2^2) - \frac{1}{3}X_3^3\right], \quad u_2 = -kb_1 X_1 X_2 X_3,$$

   $$u_3 = b_1\left[\frac{1}{2}X_1 X_3^2 - X_1 X_2^2 + \eta_1(X_1, X_2)\right],$$

   $$R_1 = AE_Y \rho_1^2 b_1, \quad R_2 = R_3 = 0, \quad M_1 = M_2 = 0,$$

   $$\frac{4(1+k)}{E_Y} M_3 = b_1 \int_{\sigma_2} [(2-k)X_2^3 - (4+k)X_1^2 X_2] \, d\sigma$$

   $$+ 2b_1 \int_{\sigma_2} \left(X_1 \frac{\partial \eta_1}{\partial X_2} - X_2 \frac{\partial \eta_1}{\partial X_1}\right) d\sigma. \quad (10.48)$$

   The forces, which are equivalent to a unique force parallel to the OX_1-axis plus a couple contained in σ_2 (see Figure 10.4), act on each base. The sections of the beam undergo a bending depending on η_1; that is, of the form of σ_2.

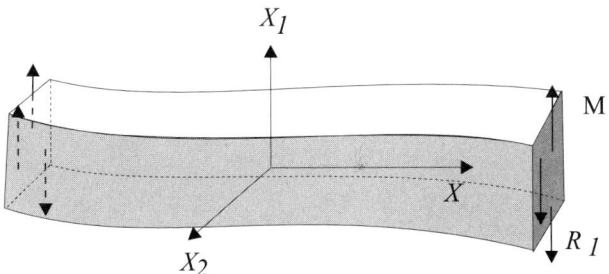

Figure 10.4

4. **Torsion**

   The constant $\alpha \neq 0$ and the others vanish so that we have (see (10.38)–(10.40))

   $$u_1 = \alpha X_1 X_3, \quad u_2 = -\alpha X_1 X_3, \quad u_3 = \alpha \psi(X_1, X_2),$$
   $$R_1 = R_2 = R_3 = 0, \quad M_1 = M_2 = 0,$$

   $$M_3 = \alpha \frac{E_Y}{2(1+k)} \left\{ A(\rho_1^2 + \rho_2^2) - \int_{\sigma_2} \left(X_1 \frac{\partial \psi}{\partial X_2} - X_2 \frac{\partial \psi}{\partial X_1}\right) d\sigma \right\}.$$
   $$(10.49)$$

   On each base acts a force which is equivalent to a couple lying on the base itself. Every ruling becomes a cylindrical helix and αl is a measure of the angle between two sections which are l apart (see Figure 10.5).

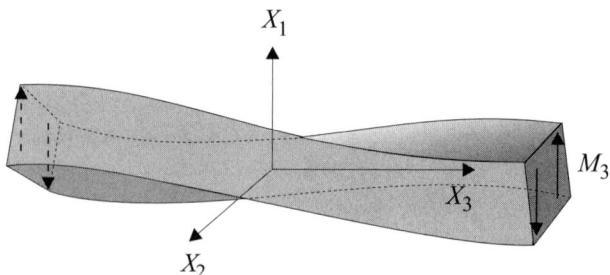

Figure 10.5

As already mentioned, Saint-Venant's conjecture is so useful when dealing with elastic systems whose geometry is such that one dimension dominates the others that it is often referred as a *principle*. In this context, we ask ourselves if Saint-Venant's solution does represent a sufficient approximation of the true solution and if a rigorous mathematical proof of this principle can be given. For instance, for a beam with sharp edges, the first derivatives of the effective solution are singular at such points; therefore, Saint-Venant's solution, which is supposed to be regular, cannot be a convenient approximation.

Finally, we close this section with a result due to Toupin (see [31]), which represents a partial proof of Saint-Venant's conjecture.

Theorem 10.6
Let S be a cylinder of length $2l$ and denote as $C(Z_1, Z_2)$ that part of it which is obtained by varying the coordinate X_3 from Z_1 to Z_2. If

$$U(X_3) = \int_{C(-l, X_3)} (2\mu E_{ij} E_{ij} + \lambda E_{ii} E_{jj}) \, dc$$

is the elastic energy of the portion from the section σ_1 (of coordinate $-l$) to the section of coordinate X_3 and, moreover, the section σ_1 is free, then

$$\frac{U(X_3)}{U(l)} \leq \gamma_0 \exp\left(-\frac{l - X_3}{\gamma}\right),$$

where γ_0 and γ depend on the geometry and elastic properties of S.

This theorem states that the elastic energy relative to the portion between the coordinates $-l, z$, compared to the energy relative to the whole cylinder, decays exponentially as the section of coordinate X_3 moves away from σ_2, towards the free section σ_1.

10.8 Ordinary Waves in Elastic Systems

In order to describe the wave propagation in linear elastic and *anisotropic systems*, it is enough to apply the general theory presented in Chapter 8 $(10.6)_1$, which allows us to state the following:

Theorem 10.7
The amplitude of an ordinary wave propagating in the direction $\mathbf{n}$ *is an eigenvector of the acoustic tensor* $\mathbf{Q}(\mathbf{n})$, *and the normal velocities* c_n *are linked to the positive eigenvalues* Λ *of* $\mathbf{Q}(\mathbf{n})$ *through the formula*

$$c_n = \pm\sqrt{\frac{\Lambda}{\rho_*}}. \tag{10.50}$$

PROOF When (8.44) and (8.45) are referred to the system $(10.6)_1$, we obtain the jump system

$$\rho_*\left[\left[\frac{\partial^2 u_i}{\partial t^2}\right]\right] = \mathbb{C}_{ijhk}\left[\left[\frac{\partial^2 u_j}{\partial x_h \partial x_k}\right]\right]. \tag{10.51}$$

By applying the kinetic conditions (4.40)–(4.42), we get the following eigenvalue problem:

$$(\mathbf{Q}(\mathbf{n}) - \rho_* c_n^2 \mathbf{I})\mathbf{a} \equiv (\mathbf{Q}(\mathbf{n}) - \Lambda \mathbf{I})\mathbf{a} = \mathbf{0}, \tag{10.52}$$

where $\mathbf{a}$ is the amplitude of the discontinuity, $\mathbf{n}$ is the unit vector normal to the surface $\sigma(t)$, and $\mathbf{Q}(\mathbf{n})$ is the ***acoustic tensor***

$$Q_{ij}(\mathbf{n}) = \mathbb{C}_{ijhk} n_h n_k. \tag{10.53}$$

∎

In Section 10.2 we proved some theorems of existence and uniqueness that are based on the assumption that the elasticity tensor $\mathbb{C}$ is positive definite. For other theorems of existence and uniqueness, relative to the solution of the equilibrium boundary problems, it was necessary to require that $\mathbb{C}$ be strongly elliptic. The following theorem proves that the assumption that $\mathbb{C}$ is positive definite is also fundamental when dealing with wave propagation.

Theorem 10.8
If the elasticity tensor is positive definite, then the acoustic tensor $\mathbf{Q}(\mathbf{n})$ *is symmetric and positive definite.*

PROOF   First, we note that the acoustic tensor is *symmetric*, due to the symmetry properties (10.2) of the elasticity tensor. Furthermore, these symmetry properties of C allow us to write, for any **v**,

$$Q_{ij}(\mathbf{n})v_iv_j = \mathbb{C}_{ijhk}v_iv_jn_hn_k = \frac{1}{4}\mathbb{C}_{ijhk}(v_in_h + v_hn_i)(v_jn_k + v_kn_j).$$

Since the tensor $v_in_h + v_hn_i$ is symmetric and C is positive definite, then

$$Q_{ij}(\mathbf{n})v_iv_j > 0.$$

∎

Given a propagation direction **n**, a wave is said to be a ***longitudinal*** or ***dilational wave*** if

$$\mathbf{a}\|\mathbf{n}; \qquad (10.54)$$

it is called a ***transverse*** or ***shear wave*** if

$$\mathbf{a}\cdot\mathbf{n} = 0. \qquad (10.55)$$

With this in mind, we prove the following:

Theorem 10.9
Let $\mathbb{C}$ be symmetric and positive definite. Then, given a propagation direction **n**, the following cases are possible:

1. There are ordinary waves propagating in the direction **n** with three different velocities; moreover, their amplitudes of discontinuity are mutually orthogonal;

2. There are waves propagating in the direction **n**, with only two different velocities; one of these is related to waves whose amplitude has a fixed direction, whereas the other velocity is related to waves having amplitudes in any direction, but orthogonal to the first one;

3. There are waves propagating in the direction **n** with only one velocity and amplitude vector in any direction;

4. Finally, there is at least one direction $\mathbf{n}_1$ characterized by longitudinal and transverse waves.

PROOF   The points (1), (2), and (3) derive from Theorem 10.7 and from the fact that the acoustic tensor is symmetric and positive definite. To prove the property (4), note that, given (10.54) and (10.52), the ordinary wave

10.8. Ordinary Waves in Elastic Systems

propagating along $\mathbf{n}$ is longitudinal if $\mathbf{n}$ is an eigenvector of the acoustic tensor
$$(\mathbf{Q}(\mathbf{n}) - \rho c_n^2 \mathbf{I})\mathbf{n} = \mathbf{0},$$
so that we have to prove that the mapping $\mathbf{n} \longrightarrow \mathbf{Q}(\mathbf{n})\mathbf{n}$ has at least one eigenvector.

Since $\mathbf{Q}(\mathbf{n})$ is positive definite, it holds that
$$\mathbf{n} \cdot \mathbf{Q}(\mathbf{n})\mathbf{n} > 0, \qquad \mathbf{Q}(\mathbf{n})\mathbf{n} \neq \mathbf{0}. \qquad (10.56)$$

As a consequence, the function
$$\mathbf{l}(\mathbf{n}) = \frac{\mathbf{Q}(\mathbf{n})\mathbf{n}}{|\mathbf{Q}(\mathbf{n})\mathbf{n}|} \qquad (10.57)$$

maps unit vectors $\mathbf{n}$ into unit vectors. However, it cannot be that
$$\mathbf{l}(\mathbf{n}) = -\mathbf{n},$$
corresponding to
$$\frac{\mathbf{n} \cdot \mathbf{Q}(\mathbf{n})\mathbf{n}}{|\mathbf{Q}(\mathbf{n})\mathbf{n}|} = -1,$$
since this contradicts (10.56). By invoking the so-called theorem of fixed points, the function (10.57) has at least one fixed point; i.e., there exists a unit vector $\mathbf{n}_1$ such that
$$\mathbf{l}(\mathbf{n}_1) = \mathbf{n}_1.$$
By considering (10.57), we see that
$$\mathbf{Q}(\mathbf{n}_1)\mathbf{n}_1 = |\mathbf{Q}(\mathbf{n}_1)\mathbf{n}_1|\,\mathbf{n}_1 \equiv \lambda \mathbf{n}_1,$$
and the theorem is proved. ∎

Only if the material exhibits additional symmetries, i.e., only if the material has an inherent reference frame are there preferential directions of polarization.

With this in mind, we now consider the case of transverse anisotropy, also called orthotropy of revolution. This is of particular relevance for geological media.

This assumption means that there is a plane $(0x_1, 0x_2)$ in which all the directions are equivalent and there is an axis $0x_3$ of orthotropy. The number of independent elastic constant reduces to 5, namely:
$$\mathbb{C}_{1111} = \mathbb{C}_{2222}, \quad \mathbb{C}_{2233} = \mathbb{C}_{1133}, \quad \mathbb{C}_{1313} = \mathbb{C}_{2323},$$
$$\mathbb{C}_{3333} \quad \text{and} \quad \mathbb{C}_{1122}, \qquad (10.58)$$
$$\mathbb{C}_{1212} = \tfrac{1}{2}(\mathbb{C}_{1111} - \mathbb{C}_{1122}).$$

All the other constants are equal to zero.

If the direction of propagation is oriented along the orthotropic axis, i.e. $\mathbf{n} = \mathbf{n}_3$, then we can infer the existence of a longitudinal wave moving with speed of propagation
$$c_P = \sqrt{\mathbb{C}_{3333}/\rho}$$
and two transverse waves moving with speed of propagation
$$c_S = \sqrt{\mathbb{C}_{1313}/\rho} = \sqrt{\mathbb{C}_{2323}/\rho}.$$

If $\mathbf{n} \cdot \mathbf{n}_3 = 0$, i.e., the propagation vector belongs to the plane of isotropy, then there is a longitudinal wave propagating with speed
$$c_P = \sqrt{\mathbb{C}_{1111}/\rho} = \sqrt{\mathbb{C}_{2222}/\rho}$$
and two transverse waves. One of the latter waves is polarized in the direction of $\mathbf{n}_3$, i.e., it has its amplitude in this direction, and propagates with speed
$$c_S = \sqrt{\mathbb{C}_{1313}/\rho} = \sqrt{\mathbb{C}_{2323}/\rho},$$
whereas the other is polarized in the isotropic plane and moves with normal velocity
$$c_S = \sqrt{\mathbb{C}_{1212}/\rho}.$$

These aspects are particularly useful when using wave propagation as a tool for identifying the elastic properties of geological media.

If the elastic system is isotropic and homogeneous, then the elasticity tensor assumes the form (10.3) and the acoustic tensor is written as
$$\mathbf{Q}(\mathbf{n}) = (\lambda + \mu)\mathbf{n} \otimes \mathbf{n} + \mu\mathbf{I}. \tag{10.59}$$

It is important to determine the conditions which the Lame coefficients λ and μ must satisfy in order that the elasticity tensor be strongly elliptic and the acoustic tensor be positive definite.

The answer relies on the following theorem:

Theorem 10.10
Given an isotropic, linear elastic material, if
$$\lambda + 2\mu > 0, \qquad \mu > 0, \tag{10.60}$$
then the elasticity tensor is positive definite and the acoustic tensor is symmetric and positive definite. Moreover, in any direction $\mathbf{n}$ there are longitudinal waves whose velocity is
$$c_P = \pm\sqrt{\frac{\lambda + 2\mu}{\rho_*}} \tag{10.61}$$

10.8. Ordinary Waves in Elastic Systems

and transverse waves whose velocity is

$$c_S = \pm\sqrt{\frac{\mu}{\rho_*}}. \tag{10.62}$$

Moreover, the amplitudes of these last waves lie in a plane normal to **n**.

PROOF By considering (10.3), we see that the elasticity tensor becomes

$$\mathbb{C}_{ijhk}\xi_{ij}\xi_{hk} = (\lambda + 2\mu)\sum_{i=1}^{3}\xi_{ii}^2 + 2\mu(\xi_{12}^2 + \xi_{13}^2 + \xi_{13}^2)$$

so that, if (10.60) holds then $\mathbb{C}$ is positive definite. It follows that the acoustic tensor is symmetric and positive definite. In addition, the eigenvalues problem for the acoustic tensor $\mathbf{Q}(\mathbf{n})$ is

$$((\lambda+\mu)\mathbf{n}\otimes\mathbf{n} + \mu\mathbf{I})\mathbf{a} = \rho_* c_n^2 \mathbf{a}.$$

Assuming an orthogonal base system with origin at the point $\mathbf{x}$ of the wavefront and the OX_1-axis along $\mathbf{n}$, we can write the previous equation as

$$\begin{pmatrix} \lambda + 2\mu & 0 & 0 \\ 0 & \mu & 0 \\ 0 & 0 & \mu \end{pmatrix} \begin{pmatrix} a_1 \\ a_2 \\ a_3 \end{pmatrix} = \rho_* c_n^2 \begin{pmatrix} a_1 \\ a_2 \\ a_3 \end{pmatrix}.$$

The characteristic equation becomes

$$(\lambda + 2\mu - \rho_* c_n^2)(\mu - \rho_* c_n^2)^2 = 0$$

and (10.61), (10.62) are proved. In addition, it can be shown that the eigenvectors corresponding to the first value of the velocity have the form $(a_1, 0, 0)$, so that the wave is longitudinal; the eigenvectors associated with the second value of the velocity have the form $(0, a_1, a_2)$, so that the wave is transverse and orthogonal to $\mathbf{n}$. ∎

When c_n is known, the evolution of the wavefront $f = const$ can be derived (see Chapter 4) from the equation

$$c_n = -\frac{1}{|\nabla f|}\frac{\partial f}{\partial t}.$$

When dealing with isotropic, linear elastic materials, the results of the previous theorem can also be obtained by using the decomposition suggested by Helmholtz's theorem:

Theorem 10.11
A vector field **u**, *which is finite, uniform and regular and vanishes at infinity, can be expressed as a sum of a gradient of a scalar potential* ϕ *and the curl of a vector* $\mathbf{\Psi}$ *whose divergence is zero:*

$$\mathbf{u} = \mathbf{u}_1 + \mathbf{u}_2,$$

where
$$\mathbf{u}_1 = \nabla \phi, \quad \mathbf{u}_2 = \nabla \times \mathbf{\Psi}, \quad \nabla \cdot \mathbf{\Psi} = 0. \tag{10.63}$$

We use this theorem without proving it. Substitute (10.63) into the (10.7) to obtain

$$\rho_* \ddot{\mathbf{u}} = (\lambda + 2\mu) \nabla (\nabla \cdot \mathbf{u}) - \mu \nabla \times (\nabla \times \mathbf{u}), \tag{10.64}$$

and rearrange the operators to obtain the equations

$$\nabla \left[\rho_* \frac{\partial^2 \phi}{\partial t^2} - (\lambda + 2\mu) \Delta \phi \right] + \nabla \times \left[\rho_* \frac{\partial^2 \mathbf{\Psi}}{\partial t^2} - \mu \Delta \mathbf{\Psi} \right] = 0. \tag{10.65}$$

Equation (10.65) implies that the potentials ϕ and $\mathbf{\Psi}$ satisfy D'Alembert's equation, i.e.,

$$\rho_* \frac{\partial^2 \phi}{\partial t^2} = (\lambda + 2\mu) \Delta \phi, \tag{10.66}$$

$$\rho_* \frac{\partial^2 \mathbf{\Psi}}{\partial t^2} = \mu \Delta \mathbf{\Psi}. \tag{10.67}$$

The first displacement field $\mathbf{u}_1$, obtained from the potential ϕ as a solution of (10.66), represents a wave propagating at a normal speed

$$c_P = \sqrt{\frac{\lambda + 2\mu}{\rho_*}},$$

and, since
$$\nabla \cdot \mathbf{u}_1 \neq 0, \quad \nabla \times \mathbf{u}_1 = 0,$$

we conclude that an irrotational displacement produced by the wave generates volume changes. Furthermore, if we consider a plane wave represented by the solution

$$\phi = \phi(\mathbf{n} \cdot \mathbf{X} - c_P t),$$

then the displacement $\mathbf{u}_1 = \nabla \phi = \mathbf{n} \phi$ is parallel to the propagation direction, so that the wave is longitudinal.

The second field $\mathbf{u}_2 = \nabla \times \mathbf{\Psi}$ describes waves propagating at a normal speed

$$c_S = \sqrt{\frac{\mu}{\rho_*}}$$

and satisfying the conditions

$$\nabla \cdot \mathbf{u_2} = 0, \qquad \nabla \times \mathbf{u_2} \neq \mathbf{0}.$$

Again, by considering a plane wave $\boldsymbol{\Psi} = \boldsymbol{\Psi}(\mathbf{n} \cdot \mathbf{X} - c_S t)$ and the corresponding displacement

$$\mathbf{u_2} = \nabla \times \boldsymbol{\Psi} = \mathbf{n} \times \boldsymbol{\Psi},$$

we have that $\mathbf{n} \cdot \mathbf{u_2} = 0$ and the wave is transverse.

10.9 Plane Waves

In this section we consider plane harmonic waves whose displacement field is represented by

$$\mathbf{u}(\mathbf{X}, t) = \mathbf{A} \cos\left(\frac{2\pi}{\lambda} \mathbf{n} \cdot \mathbf{X} - \omega t + \varphi\right) \equiv \mathbf{A} \cos(\mathbf{k} \cdot \mathbf{X} - \omega t + \varphi). \quad (10.68)$$

The absolute value of the constant $\mathbf{A}$ is called the *amplitude*, $\mathbf{n}$ is the propagation unit vector, ω is the *circular frequency*,

$$\frac{|\mathbf{k}|}{2\pi}$$

is the *wave number*, λ is the *wavelength*, and φ is the phase. In the following discussion,

$$f = \frac{\omega}{2\pi}, \quad T = \frac{1}{f},$$

denote the frequency and the period, respectively, and

$$\alpha = f\lambda = \frac{\omega}{k} \quad (10.69)$$

is the *phase velocity*.

The propagation of these waves will be analyzed in spatial regions with the following geometries:

1. a homogeneous and isotropic half-space $y \geq 0$, whose boundary $y = 0$ is a free surface, i.e.,

$$\mathbf{TN} = \mathbf{0}, \quad (10.70)$$

where $\mathbf{N}$ is the unit vector normal to the surface $y = 0$. This case is relevant when we investigate **reflection phenomena** and prove the existence of Rayleigh surface waves;

2. a domain formed by two homogeneous and isotropic media, whose interface $y = 0$ is characterized by the jump conditions

$$[[\mathbf{u}]] = \mathbf{0}, [[\mathbf{TN}]] = \mathbf{0}. \tag{10.71}$$

This case is relevant when we investigate **refraction phenomena**;

3. a layer of limited depth. This case is relevant when we investigate the Rayleigh–Lamb **dispersion relation**.

In all the above cases, a boundary problem is formulated and we search for a solution by using the *method of separation of variables*. Without loss of generality, the direction of propagation defined by the unit vector $\mathbf{n}$ is supposed to lie on the plane xy. This is a *vertical plane*, whereas the xz-plane, which represents the boundary surface of the half-space, is the horizontal plane. The displacement u_n produced by longitudinal waves is in the direction $\mathbf{n}$, so that it lies in the vertical plane, whereas the displacement produced by transverse waves is normal to $\mathbf{n}$. The components u_v and u_h lie on the vertical plane and in the normal plane to $\mathbf{n}$, respectively (see Figure 10.6).

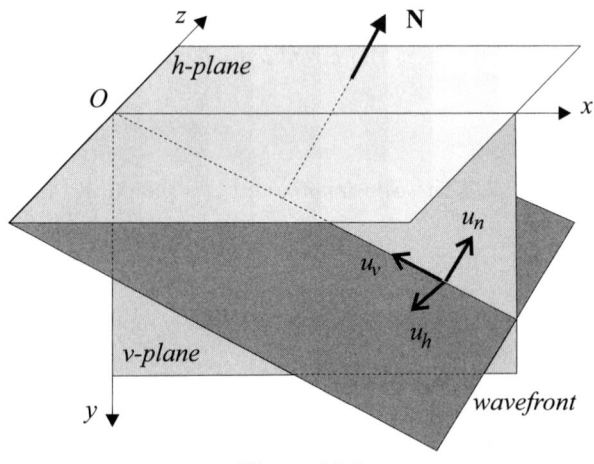

Figure 10.6

The selected reference system is such that the vector $\mathbf{k}$ has only two components (k, p), so that

$$\mathbf{k} \cdot \mathbf{X} = kx + py.$$

Furthermore, by using Helmholtz's theorem, the displacement $\mathbf{u}$ is expressed by (10.63). In particular, since the motion is independent

10.9. Plane Waves

of z,[3] the displacement components are

$$u_x = \frac{\partial \phi}{\partial x} + \frac{\partial \Psi_3}{\partial y}, \qquad (10.72)$$

$$u_y = \frac{\partial \phi}{\partial y} - \frac{\partial \Psi_3}{\partial x},$$

$$u_z = -\frac{\partial \Psi_1}{\partial y} + \frac{\partial \Psi_2}{\partial x},$$

and the condition $\nabla \cdot \boldsymbol{\Psi} = 0$ becomes

$$\frac{\partial \Psi_1}{\partial x} + \frac{\partial \Psi_2}{\partial y} = 0. \qquad (10.73)$$

The mathematical problem of determining solutions in the half-space $y \geq 0$ reduces to finding the solution of the equations (10.66) and (10.67):

$$\frac{\partial^2 \phi}{\partial x^2} + \frac{\partial^2 \phi}{\partial y^2} = \frac{1}{c_p^2} \frac{\partial^2 \phi}{\partial t^2}, \qquad (10.74)$$

$$\frac{\partial^2 \Psi_\alpha}{\partial x^2} + \frac{\partial^2 \Psi_\alpha}{\partial y^2} = \frac{1}{c_s^2} \frac{\partial^2 \Psi_\alpha}{\partial t^2}, \quad \alpha = 1, 2, 3, \qquad (10.75)$$

when the initial and boundary conditions are given.

To do this, consider the equation (10.74),[4] and search for a solution in the form

$$\phi(x, y, t) = X(x) Y(y) T(t). \qquad (10.76)$$

By substituting (10.76) into (10.74) we get

$$\frac{X''}{X} + \frac{Y''}{Y} = \frac{1}{c_p^2} \frac{T''}{T} \equiv \frac{a}{c_p^2}, \qquad (10.77)$$

where a is a constant. Then from

$$\frac{T''}{T} = a,$$

by assuming $a = -\omega^2$, we derive the solution

$$T(t) = A_t \sin \omega t + B_t \cos \omega t, \qquad (10.78)$$

---

[3] This is related to the assumption of plane waves, so that all the physical quantities are independent of z.
[4] The same argument applies to (10.75).

which expresses the harmonic character of the wave under consideration. Furthermore, from (10.77) it follows that

$$\frac{X''}{X} = -\frac{\omega^2}{c_p^2} - \frac{Y''}{Y} \equiv b,$$

and, since we are searching for harmonic solutions in at least one of the directions x and y, by assuming $b = -k^2$, the following system holds:

$$\frac{X''}{X} = -k^2,$$
$$\frac{Y''}{Y} = k^2 - \frac{\omega^2}{c_p^2}. \tag{10.79}$$

The first equation represents the propagation in the x-axis direction

$$X(x) = A_x \sin kx + B_x \cos kx, \tag{10.80}$$

and the solution (10.76) becomes

$$\phi(x, y, t) = Y(y)(A_x \sin kx + B_x \cos kx)(A_t \sin \omega t + B_t \cos \omega t). \tag{10.81}$$

In order to find the function $Y(y)$, two cases have to be distinguished:

$$k^2 - \frac{\omega^2}{c_p^2} \geq 0,$$
$$k^2 - \frac{\omega^2}{c_p^2} < 0. \tag{10.82}$$

Similarly, for the equation (10.75) we get

$$\Psi_\alpha(x, y, t) \equiv Y_\alpha(y) X_\alpha(x) T_\alpha(t)$$
$$= Y_\alpha(y)(A_{x\alpha} \sin k_\alpha x + B_{x\alpha} \cos k_\alpha x)(A_{t\alpha} \sin \omega_\alpha t + B_{t\alpha} \cos \omega_\alpha t), \tag{10.83}$$

where $\alpha = 1, 2, 3$, and any function $Y_\alpha(y)$ depends on the choice of one of the two cases

$$k_\alpha^2 - \frac{\omega_\alpha^2}{c_s^2} \geq 0,$$
$$k_\alpha^2 - \frac{\omega_\alpha^2}{c_s^2} < 0. \tag{10.84}$$

In particular, by assuming $(10.82)_2$ and $(10.84\ )_2$, we have

$$Y(y) = A_y \sin py + B_y \cos py,$$
$$Y_\alpha(y) = A_{y\alpha} \sin q_\alpha y + B_{y\alpha} \cos q_\alpha y, \quad \alpha = 1, 2, 3, \tag{10.85}$$

10.9. Plane Waves

where

$$p = \sqrt{\frac{\omega^2}{c_p^2} - k^2},$$

$$p_\alpha = \sqrt{\frac{\omega_\alpha^2}{c_s^2} - k_\alpha^2}, \ \alpha = 1, 2, 3. \tag{10.86}$$

In the following discussion, it is convenient to use Euler's formulae

$$\cos\beta + i\sin\beta = e^{i\beta}, \ \cos\beta - i\sin\beta = e^{-i\beta}, \tag{10.87}$$

which allow us to write (10.85) in the exponential form

$$Y(y) = A_y e^{-ipy} + B_y e^{ipy},$$
$$Y_\alpha(y) = A_{y\alpha} e^{-ip_\alpha y} + B_{y\alpha} e^{ip_\alpha y}, \ \alpha = 1, 2, 3, \tag{10.88}$$

where the constants $A_y, B_y, A_{y\alpha}$, and $B_{y\alpha}$ are properly defined.[5] By applying the same exponential form to $X(x), T(t), X_\alpha(x)$, and $T_\alpha(t)$, (10.81) and (10.83) can be written as

$$\phi = (A_y e^{-ipy} + B_y e^{ipy})(A_x e^{-ikx} + B_x e^{ikx})(A_t e^{-i\omega t} + B_t e^{i\omega t}),$$
$$\Psi_\alpha = (A_{y\alpha} e^{-ip_\alpha y} + B_{y\alpha} e^{ip_\alpha y})(A_{x\alpha} e^{-ik_\alpha x} + B_{x\alpha} e^{ik_\alpha x}) \tag{10.89}$$
$$\times (A_{t\alpha} e^{-i\omega_\alpha t} + B_{t\alpha} e^{i\omega_\alpha t}).$$

where again all the constants have been defined.

In order to appreciate the mathematical aspects of the problem, we observe that the uniqueness of the solution in the half-space $y \geq 0$ needs to define the following:

1. the initial conditions, i.e., the values of the functions ϕ and Ψ_α and their time derivatives at the initial instant;

2. the asymptotic conditions, i.e., the behavior of ϕ and Ψ_α if $x \longrightarrow \pm\infty$ and $y \longrightarrow \infty$;

---

[5]It can be proved, starting from (10.87), that the expression

$$C\sin\beta + D\cos\beta$$

can be written as

$$C' e^{i\beta} + D' e^{-i\beta},$$

with

$$C' + D' = C, \ i(C' - D') = D.$$

3. the boundary condition (10.70), which is equivalent to three scalar conditions on the surface $y = 0$.

In addition, (10.73) must be taken into account.

In this way we obtain $8 + 12 + 3 + 1 = 24$ conditions for the 24 constants in (10.90). However, the periodic dependence on the variables x, y, and t rules out both the initial and asymptotic conditions. As far as the initial data are concerned, it is always possible to eliminate the constants B_t and $B_{t\alpha}$ and to put A_t and $A_{t\alpha}$ into other constants, in order to reduce their number to 16. Moreover, in the absence of asymptotic conditions, the remaining constants are still undetermined. In conclusion, we are faced with a problem in which 16 constants are limited by 4 conditions so that it admits ∞^{12} solutions. We will need to give to some of these constants a convenient value, depending on the relevant aspects of the problem under consideration.

In order to evaluate (10.70) and (10.71), it is useful to write down the components of the stress tensor. From (7.18), (10.72), since

$$\Delta \phi = \left(\frac{\partial^2 \phi}{\partial x^2} + \frac{\partial^2 \phi}{\partial y^2} \right),$$

it follows that

$$\begin{aligned}
T_{11} &= \lambda \Delta \phi + 2\mu \left(\frac{\partial^2 \phi}{\partial x^2} + \frac{\partial^2 \Psi_3}{\partial x \partial y} \right), \\
T_{22} &= \lambda \Delta \phi + 2\mu \left(\frac{\partial^2 \phi}{\partial y^2} - \frac{\partial^2 \Psi_3}{\partial x \partial y} \right), \\
T_{33} &= \lambda \Delta \phi, \\
T_{12} &= 2\mu \left[\frac{\partial^2 \phi}{\partial x \partial y} + \frac{1}{2} \left(\frac{\partial^2 \Psi_3}{\partial y^2} - \frac{\partial^2 \Psi_3}{\partial x^2} \right) \right], \\
T_{23} &= \mu \left(\frac{\partial^2 \Psi_2}{\partial y \partial x} - \frac{\partial^2 \Psi_1}{\partial y^2} \right), \\
T_{13} &= \mu \left(-\frac{\partial^2 \Psi_1}{\partial y \partial x} + \frac{\partial^2 \Psi_2}{\partial x^2} \right).
\end{aligned} \quad (10.90)$$

10.10 Reflection of Plane Waves in a Half-Space

In the discussion below, longitudinal waves will be denoted by P, whereas transverse waves will be denoted by SH if they are polarized in the horizontal plane (Figure 10.7) or by SV if they are polarized in the vertical plane (Figure 10.8).

10.10. Reflection of Plane Waves in a Half-Space

When the interface between two media is free, then there arises a simple reflection characterized by the *conversion of modes*, i.e., an incident longitudinal or transverse plane wave gives rise to two reflected waves, one longitudinal and the other transverse.

On the basis of the previous remarks, it is always possible to set $B_t = 0$. Then, by developing the products in $(10.90)_1$ and by conveniently denoting the multiplicative constants, we obtain 4 terms whose wave vectors are

$$(k, -p), \quad (k, p), \quad (-k, p), \quad (-k, -p).$$

Apart a symmetry with respect to the Oy-axis, the last two wave numbers refer to waves propagating in the same manner as the first two. Since the same arguments apply to Ψ_α, we lose no generality by writing

$$\phi(x, y, t) = Ae^{i(kx - py - \omega t)} + Be^{i(kx + py - \omega t)},$$
$$\Psi_\alpha(x, y, t) = A_\alpha e^{i(k_\alpha x - p_\alpha y - \omega_\alpha t)} + B_\alpha e^{i(k_\alpha x + p_\alpha y - \omega_\alpha t)}, \quad (10.91)$$

where $\alpha = 1, 2, 3$. These functions must be determined through the conditions

$$T_{21} = T_{22} = T_{23} = 0 \quad \text{for } y = 0, \quad (10.92)$$

which are deduced from $\mathbf{TN} = \mathbf{0}$ for $y = 0$, where $\mathbf{N}$ has components $(0, 1, 0)$.

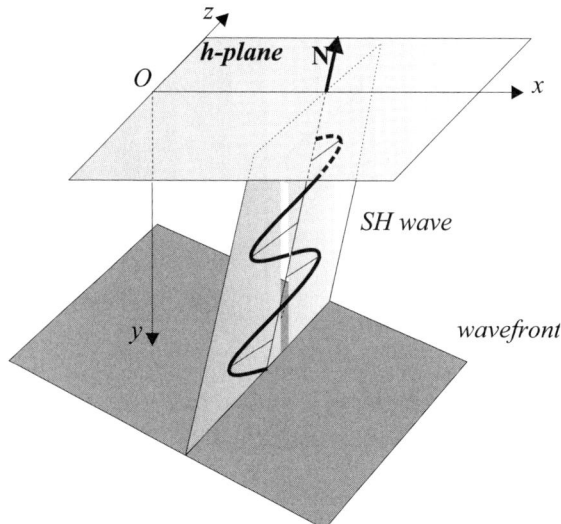

Figure 10.7

From (10.72) we deduce that u_x and u_y depend on ϕ and Ψ_3, which are the potentials describing P and SV waves; similarly, u_z only depends on

Ψ_1 and Ψ_2, which describe SH waves. Accordingly, it is convenient to split the problem into two other problems, the first related to the displacement field represented by $(u_x, u_y, 0)$ and the second described by $(0, 0, u_z)$.

(a) P and SV waves

$$u_x = \frac{\partial \phi}{\partial x} + \frac{\partial \Psi_3}{\partial y}, \quad u_y = \frac{\partial \phi}{\partial y} - \frac{\partial \Psi_3}{\partial x}, \quad u_z = 0, \tag{10.93}$$

$$T_{21} = 0, \quad T_{22} = 0 \quad \text{for y} = 0.$$

(b) SH waves

$$u_x = u_y = 0, \quad u_z = -\frac{\partial \Psi_1}{\partial y} + \frac{\partial \Psi_2}{\partial x},$$

$$\frac{\partial \Psi_1}{\partial x} + \frac{\partial \Psi_2}{\partial y} = 0, \tag{10.94}$$

$$T_{23} = 0 \quad \text{for y} = 0.$$

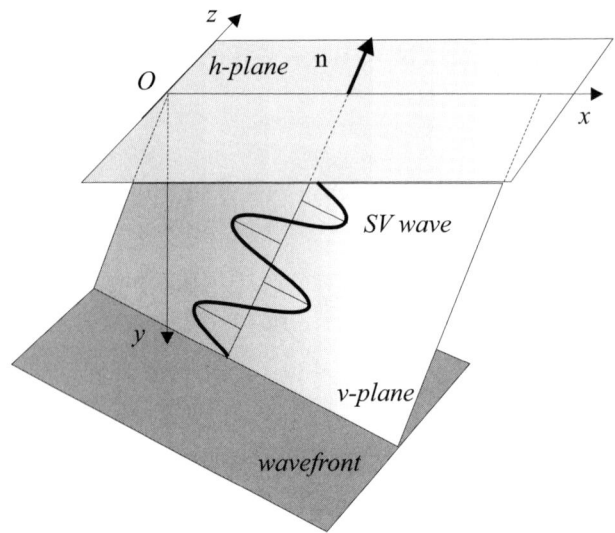

Figure 10.8

P and SV Waves

From (10.91) and (10.93) it follows that

$$\phi(x, y, t) = A(I_P)e^{i(kx - py - \omega t)} + A(R_P)e^{i(kx + py - \omega t)},$$
$$\Psi_3(x, y, t) = A(I_{SV})e^{i(k_3 x - p_3 y - \omega_3 t)} + A(R_{SV})e^{i(k_3 x + p_3 y - \omega_3 t)}, \tag{10.95}$$

where the constants A and B have been changed in order to highlight their physical meaning: the first term in $(10.95)_1$ represents the amplitude of the P wave incident in the direction $\mathbf{k}_P = (k, -p, 0)$, whereas $A(R_P)$ is the amplitude of the P wave reflected in the direction $\mathbf{k}'_P = (k, p, 0)$. A similar meaning is attributed to $A(I_S)$, which denotes the amplitude of a SV wave incident in the direction $\mathbf{k}_{SV} = (k_3, -p_3, 0)$, whereas $A(R_{SV})$ is the amplitude of the SV wave reflected in the direction $\mathbf{k}'_{SV} = (k_3, p_3, 0)$. In this way, $(10.95)_1$ describes two plane waves; the first one is incident with respect to the surface $y = 0$ at an angle θ_p, called the *angle of incidence*, and the second one is the reflected wave, with the corresponding angle of the same amplitude, called the *angle of reflection*. If θ_{SV} denotes the angle between the SV ray and the normal $\mathbf{N}$ to the plane of incidence, then the same considerations apply to $(10.95)_2$. The above functions represent a solution of the problem posed if the constants in (10.95) can be found such that they satisfy the boundary conditions $(10.93)_2$. From the above remarks, it follows that

$$k = k_P \sin\theta_p,\ p = k_P \cos\theta_p,$$
$$k_3 = k_{SV} \sin\theta_{SV},\ p_3 = k_{SV} \cos\theta_{SV}, \qquad (10.96)$$

where

$$k_P = \sqrt{k^2 + p^2},\ k_{SV} = \sqrt{k_3^2 + p_3^2}.$$

By substituting (10.95) into the first part of $(10.93)_2$ and by also taking into account (10.90), we can prove that the following equation holds:

$$k_P^2 \sin(2\theta_P) e^{i\mu}(A(I_P) - A(R_P))$$
$$- k_{SV}^2 \cos(2\theta_{SV}) e^{i\mu_3}(A(I_{SV}) + A(R_{SV})) = 0, \qquad (10.97)$$

where

$$\mu = k_P x \sin\theta_P - \omega t,\ \mu_3 = k_{SV} x \sin\theta_{SV} - \omega_3 t. \qquad (10.98)$$

Since x and t are arbitrary, (10.97) is satisfied only if

$$k_P x \sin\theta_P - \omega t = k_{SV} x \sin\theta_{SV} - \omega_3 t,$$
$$k_P^2 \sin(2\theta_P)(A(I_P) - A(R_P)) \qquad (10.99)$$
$$- k_{SV}^2 \cos(2\theta_{SV})(A(I_{SV}) + A(R_{SV})) = 0.$$

Similarly, from $(10.99)_1$ it follows that

$$\omega = \omega_3,$$
$$k_P \sin\theta_P = k_{SV} \sin\theta_{SV}. \qquad (10.100)$$

Substituting (10.95) into the second part of $(10.93)_2$ and considering (10.90) and (10.100), we see that

$$\cos(2\theta_{SV})(A(I_P) + A(R_P))$$
$$+ \sin(2\theta_{SV})(A(I_{SV}) - A(R_{SV})) = 0. \qquad (10.101)$$

Furthermore, if we introduce the ratio

$$\delta \equiv \frac{k_{SV}}{k_P} = \frac{c_P}{c_S} > 1,$$

where c_P and c_S are the normal speed of the longitudinal and transverse wave, then $(10.100)_1$ allows us to give the boundary conditions the form

$$\omega = \omega_3, \quad \frac{\sin\theta_P}{\sin\theta_{SV}} = \delta > 1,$$

$$\sin(2\theta_P)(A(I_P) - A(R_P)) - \delta^2 \cos(2\theta_{SV})(A(I_{SV}) + A(R_{SV})) = 0,$$
$$\cos(2\theta_{SV})(A(I_P) + A(R_P)) + \sin(2\theta_{SV})(A(I_{SV}) - A(R_{SV})) = 0. \tag{10.102}$$

The conditions (10.102) give rise to the following remarks.

Both waves have the same frequency. The condition $(10.102)_2$ gives the ratio of the angle of incidence of P and SV waves. The last two conditions allow us to link two of the four constants A to the other two, arbitrarily selected. Let us now consider some relevant cases.

- First, assume that $A(I_{SV}) = 0$. Then from $(10.102)_{2,3}$ it follows that

$$\frac{A(R_P)}{A(I_P)} = \frac{\sin(2\theta_P)\sin(2\theta_{SV}) - \delta^2 \cos^2(2\theta_{SV})}{\sin(2\theta_P)\sin(2\theta_{SV}) + \delta^2 \cos^2(2\theta_{SV})},$$
$$\frac{A(R_{SV})}{A(I_P)} = \frac{2\sin(2\theta_P)\cos(2\theta_{SV})}{\sin(2\theta_P)\sin(2\theta_{SV}) + \delta^2 \cos^2(2\theta_{SV})}, \tag{10.103}$$

where θ_{SV} is expressed in terms of θ_P and δ throughout $(10.102)_2$. This result shows that *an incident P wave whose angle of incidence is θ_P gives rise to two reflected waves; the first is a P wave having the reflected angle equal to the incident angle, and the second is an SV wave having a reflected angle $\theta_{SV} < \theta_P$*, according to $(10.102)_2$ (see Figure (10.9)).

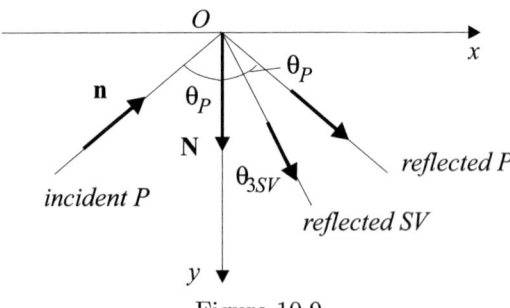

Figure 10.9

10.10. Reflection of Plane Waves in a Half-Space

In other words, the single P wave generates two reflected waves, P and SV; this phenomenon is referred to as **mode conversion**. Furthermore, from (10.103) it also follows that

- if the P wave is normal to the plane of incidence ($\theta_P = 0$), $A(R_{SV}) = 0$, and there is only one reflected P wave; and
- if the angle of incidence satisfies the condition

$$\sin(2\theta_P)\sin(2\theta_{SV}) - \delta^2 \cos^2(2\theta_{SV}) = 0,$$

then there is only the reflected SV wave.

- Let us now consider the case of an incident SV wave. Since $A(I_P) = 0$, $(10.102)_{3,4}$ give

$$\frac{A(R_{SV})}{A(I_{SV})} = \frac{\sin(2\theta_P)\sin(2\theta_{SV}) - \delta^2 \cos^2(2\theta_{SV})}{\sin(2\theta_P)\sin(2\theta_{SV}) + \delta^2 \cos^2(2\theta_{SV})},$$

$$\frac{A(R_P)}{A(I_{SV})} = -\frac{\delta^2 \sin(4\theta_{SV})}{\sin(2\theta_P)\sin(2\theta_{SV}) + \delta^2 \cos^2(2\theta_{3SV})}. \quad (10.104)$$

This result shows that *an incident SV wave whose angle of incidence is θ_{SV} gives rise to two reflected waves; the first one is an SV wave whose angle of reflection is equal to the angle of incidence, and the second is a P wave whose angle of reflection is $\theta_P > \theta_{SV}$, according to $(10.102)_2$*.

If $\theta_{SV} = 0, \pi/4$, then the reflected waves are SV waves of amplitude $-A(I_{SV}), A(I_{SV})$. Furthermore, from $(10.102)_2$ follows the existence of a limit value for the angle of reflection of an SV wave:

$$(\theta_{SV})_{\lim} = \arcsin \frac{1}{\delta}, \quad (10.105)$$

for which the reflected P wave moves in the x-direction. If

$$\sin(2\theta_P)\sin(2\theta_{SV}) - \delta^2 \cos^2(2\theta_{SV}) = 0,$$

then the SV wave is reflected as a P wave.

SH Waves

From $(10.91)_2$ it follows that

$$\Psi_1(x, y, t) = A_1 e^{i(k_1 x - p_1 y - \omega_1 t)} + B_1 e^{i(k_1 x + p_1 y - \omega_1 t)}, \quad (10.106)$$

$$\Psi_2(x, y, t) = A_2 e^{i(k_2 x - p_2 y - \omega_2 t)} + B_2 e^{i(k_2 x + p_2 y - \omega_2 t)}.$$

Since these functions must satisfy the boundary condition $(10.94)_3$, we get

$$p_1^2(A_1 + B_1)e^{i(k_1 x - \omega_1 t)} + k_2 p_2(A_2 - B_2)e^{i(k_2 x - \omega_2 t)} = 0,$$

and, by noting that x and t can be arbitrarily chosen, when $(10.86)_2$ is taken into account, it follows that

$$k_1 = k_2, \quad \omega_1 = \omega_2,$$
$$p_1(A_1 + B_1) = k_1(A_2 - B_2). \tag{10.107}$$

By introducing Helmholtz's decomposition $(10.94)_2$, we get

$$k_1 A_1 = p_1 A_2, \quad k_1 B_1 = -p_1 B_2, \tag{10.108}$$

so that $(10.107)_2$ becomes

$$(A_1 + B_1)(p_1^2 - k_1^2) = 0,$$

and finally

$$A_1 = -B_2, \quad \text{or} \quad p_1^2 = k_1^2. \tag{10.109}$$

By paying attention to the first relation of (10.109), the only one of practical interest, and also taking into account (10.108) and (10.109), we see that (10.106) can be written as

$$\Psi_1(x, y, t) = A_1 e^{i(k_1 x - p_1 y - \omega_1 t)} - A_1 e^{i(k_1 x + p_1 y - \omega_1 t)}, \tag{10.110}$$
$$\Psi_2(x, y, t) = (k_1/p_1) A_1 e^{i(k_1 x - p_1 y - \omega_1 t)} + (k_1/p_1) A_1 e^{i(k_1 x + p_1 y - \omega_1 t)}.$$

The displacement u_z, due to $(10.94)_1$ and (10.111), assumes the form

$$u_z = p_1^2 + k_1^2 p_1 [A_1 e^{i(k_1 x - p_1 y - \omega_1 t)} + A_1 e^{i(k_1 x + p_1 y - \omega_1 t)}],$$

and the following conclusion is proved: *an SH wave incident on the surface $y = 0$ is reflected as an SH wave with the same amplitude and with the angle of reflection equal to the angle of incidence. In this case there is no mode conversion.*

10.11 Rayleigh Waves

In the previous sections we proved that, in an elastic unbounded medium, there are two waves: a longitudinal, or dilational, wave and a transverse, or distorsional, wave. When dealing with a half-space, it can be proved that there exists an additional wave, called a **Rayleigh wave**, which propagates confined only in the neighborhood of the surface of the medium, with less speed than the body waves; its displacement decays exponentially with the distance from the surface.

10.11. Rayleigh Waves

Consider a wave propagating in the x direction, parallel to the surface $y = 0$, attenuated in the y direction. Since the displacements occur in the plane Oxy, only the scalar potential ϕ and the component Ψ_3 of the vector potential $\boldsymbol{\Psi}$ have to be considered.

According to these requirements, we refer to $(10.82)_1$ and $(10.84)_1$ by taking the positive sign for the functions $Y(y)$ and $Y_\alpha(y)$, $\alpha = 3$, in the method of separation of variables (see Section 10.9). Accordingly, it holds that

$$Y(y) = A_y e^{-\bar{p}y} + B_y e^{\bar{p}y},$$
$$Y_3(y) = A_3 e^{-\bar{p}_3 y} + B_3 e^{\bar{p}_3 y}, \tag{10.111}$$

where

$$\bar{p} = -p = \sqrt{k^2 - \frac{\omega^2}{c_P^2}},$$
$$\bar{p}_3 = -p_3 = \sqrt{k_3^2 - \frac{\omega_3^2}{c_P^2}}. \tag{10.112}$$

In (10.111), since the wave is supposed to be confined to the surface, it follows that $B_y = B_3 = 0$ and the potentials become

$$\phi_3 = A e^{-\bar{p}y} e^{i(kx - \omega t)},$$
$$\Psi = B e^{-\bar{p}_3 y} e^{i(k_3 x - \omega_3 t)}. \tag{10.113}$$

Constants A and B must satisfy the boundary condition $\mathbf{TN} = \mathbf{0}$; in scalar terms, they can be written as

$$T_{12} = T_{22} = 0 \quad \text{if } y = 0. \tag{10.114}$$

By introducing the potentials (10.113) into $(10.90)_{2,4}$, the first part of (10.114) gives

$$-2i\bar{p}k A e^{ikx} + (\bar{p}_3^2 + k_3^2) B e^{ik_3 x} = 0,$$

and, since the x-axis is arbitrary, it follows that

$$k = k_3, \quad -2i\bar{p}k A + (\bar{p}_3^2 + k^2) B = 0. \tag{10.115}$$

The second part of (10.114) gives

$$[(\lambda + 2\mu)\bar{p}^2 - \lambda k^2] A + 2\mu i \bar{p}_3 k B = 0. \tag{10.116}$$

Moreover, taking into account (10.112) with $k = k_3$ and considering the relation $c_P^2/c_S^2 = (\lambda + 2\mu)/\mu$, we also find that

$$\bar{p}_3^2 + k^2 = 2k^2 - \frac{\omega^2}{c_S^2},$$
$$\bar{p}^2(\lambda + 2\mu) - \lambda k^2 = \left(k^2 - \frac{\omega^2}{c_P^2}\right)(\lambda + 2\mu) - \lambda k^2 = \mu \left(2k^2 - \frac{\omega^2}{c_S^2}\right).$$

By introducing these relations into (10.116) and (10.115)$_2$, we obtain the following system:

$$\begin{pmatrix} 2k^2 - \dfrac{\omega^2}{c_S^2} & 2ik\sqrt{k^2 - \dfrac{\omega^2}{c_S^2}} \\ -2ik\sqrt{k^2 - \dfrac{\omega^2}{c_P^2}} & 2k^2 - \dfrac{\omega^2}{c_S^2} \end{pmatrix} \begin{pmatrix} A \\ B \end{pmatrix} = 0, \qquad (10.117)$$

which admits a nontrivial solution only if the determinant of the coefficients vanishes. By imposing this condition, we obtain the ***dispersion relation***, which relates the circular frequency ω to the wave number k:

$$\left(2k^2 - \dfrac{\omega^2}{c_S^2}\right)^2 - 4k^2 \sqrt{k^2 - \dfrac{\omega^2}{c_S^2}} \sqrt{k^2 - \dfrac{\omega^2}{c_P^2}} = 0. \qquad (10.118)$$

Finally, by using the relation

$$c_R = \dfrac{\omega}{k}, \qquad (10.119)$$

we derive the ***Rayleigh equation*** (1887):

$$\left(2 - \dfrac{c_R^2}{c_S^2}\right)^2 - 4\left(1 - \dfrac{c_R^2}{c_S^2}\right)^{1/2} \left(1 - \dfrac{c_S^2 \, c_R^2}{c_P^2 \, c_S^2}\right)^{1/2} = 0. \qquad (10.120)$$

If the potentials (10.113) are introduced into (10.72) with $k = k_3$, then the displacement components can be written as

$$\begin{aligned} u_x &= (ikAe^{-\bar{p}y} - \bar{p}Be^{-\bar{p}_3 y}) \, e^{i(kx - \omega t)}, \\ u_y &= -(\bar{p}Ae^{-\bar{p}y} + ikBe^{-\bar{p}_3 y}) \, e^{i(kx - \omega t)}. \end{aligned} \qquad (10.121)$$

The previous results allow us to make the following remarks:

1. First, from (10.82)$_1$ and (10.84)$_1$, when considering the positive sign as well as $\alpha = 3$, $k = k_3$, and recalling (10.119), we get

$$c_R < c_S < c_P, \qquad (10.122)$$

   i.e., the Rayleigh wave propagates more slowly than the longitudinal and shear waves. Furthermore, from (10.120) it follows that the ratio c_R/c_S depends only on the ratio c_P/c_S or, equivalently, on Poisson's ratio; since usually $0.15 \leq v \leq 0.3$, it follows that $0.90 \leq c_R/c_S \leq 0.93$.

2. In the relation (10.120) the wave number k does not appear, so that in an elastic homogeneous medium, surface waves are not dispersive (also see the next section). To see this, consider a wave which is obtained by superimposing two waves of different circular frequency ω. Due to the linear character of the problem we are dealing with, all the wave components have the same phase velocity and the wave form does not change in time.

3. Since the determinant of the matrix in (10.117) vanishes, the constant B can be derived as a function of A:

$$B = \frac{2iA}{2 - c_R^2/c_S^2} \sqrt{1 - c_R^2/c_P^2}, \qquad (10.123)$$

where (10.119) has also been taken into account, so that the displacement components (10.121) depend on the arbitrary constant A. Because of the imaginary unit, these components exhibit a phase difference of 90 degrees. For this reason, the motion $\mathbf{X} = (x, y)$ of a particle near to the boundary is composed of two harmonic motions along the axes Ox and Oy, characterized by the same circular frequency and amplitude A and B. The trajectory of $\mathbf{X}$ is an ellipse, the major semi-axis B is normal to the free surface, and the motion is counterclockwise with respect to the normal at the surface. At the boundary, the normal component is about 1.5 times the tangential component.

4. Because the motion is attenuated with depth, it is confined to a narrow zone of about 2 times the wavelength, and this aspect justifies the definition of a surface wave. This circumstance, as well as the fact that along the propagation direction the surface waves are less attenuated than the body waves, renders the analysis of this problem very attractive when dealing with geological sites.

10.12 Reflection and Refraction of SH Waves

If the boundary surface is an interface $y = 0$ separating two media in contact, we need to find the displacement and stress continuity across the interface.[6] It can be shown that an SH wave arriving at the surface $y = 0$

---

[6] If the media are bounded together, the transverse slip can also occur at lubricated interfaces, a case not considered here but of interest in ultrasonics, when transducers on lubricated plates are used to generate and receive waves from solids.

from one medium generates only reflected and refracted SH waves. In the following discussion, the superscript a characterizes properties of the first medium ($y > 0$) and the superscript b characterizes those of the second medium ($y < 0$). The boundary conditions at the interface are

$$u_z^a = u_z^b, \quad T_{23}^a = T_{23}^b. \tag{10.124}$$

Applying Helmholtz's condition to both media, we can write the governing potentials in the following way:

$$\begin{aligned}\Psi_1^\tau &= A_1^\tau e^{i(k_1^\tau x - p_1^\tau y - \omega_1^\tau t)} + B_1^\tau e^{i(k_1^\tau x + p_1^\tau y - \omega_1^\tau t)}, \\ \Psi_2^\tau &= A_2^\tau e^{i(k_1^\tau x - p_1^\tau y - \omega_1^\tau t)} + B_2^\tau e^{i(k_1^\tau x + p_1^\tau y - \omega_1^\tau t)}, \quad \tau = a, b.\end{aligned} \tag{10.125}$$

By inserting these relations into $(10.72)_3$, we can derive the displacement component u_z in both media:

$$\begin{aligned}u_z^a &= i(k_1^a A_2^a + p_1^a A_1^a) e^{i(k_1^a x - p_1^a y - \omega_1^a t)} \\ &\quad + i(k_1^a B_2^a - p_1^a B_1^a) e^{i(k_1^a x + p_1^a y - \omega_1^a t)}, \\ u_z^b &= i(k_1^b A_2^b + p_1^b A_1^b) e^{i(k_1^b x - p_1^b y - \omega_1^b t)} \\ &\quad + i(k_1^b B_2^b - p_1^b B_1^b) e^{i(k_1^b x + p_1^b y - \omega_1^b t)},\end{aligned} \tag{10.126}$$

and they describe four waves directed along the vectors $(k_1^a, -p_1^a)$, (k_1^a, p_1^a), $(k_1^b, -p_1^b)$, (k_1^b, p_1^b). Since the SH wave has been supposed to propagate in the a-medium, we can write

$$\begin{aligned}u_z^a &= iA(I) e^{i(k_1^a x - p_1^a y - \omega_1^a t)} + iA(Rfl) e^{i(k_1^a x + p_1^a y - \omega_1^a t)}, \\ u_z^b &= iA(Rfr)^{i(k_1^b x - p_1^b y - \omega_1^b t)},\end{aligned} \tag{10.127}$$

where

$$\begin{aligned}A(I) &= k_1^a A_2^a + p_1^a A_1^a, \\ A(Rfl) &= k_1^a B_2^a - p_1^a B_1^a, \\ A(Rfr) &= k_1^b A_2^b + p_1^b A_1^b,\end{aligned} \tag{10.128}$$

denote the amplitudes of the incident, reflected, and refracted waves, respectively.

Helmholtz's condition introduces the following relations between the quantities $A_1^\tau, A_2^\tau, B_1^\tau$, and B_2^τ, $\tau = a, b$:

$$k_1^\tau(A_1^\tau + B_1^\tau) + p_1^\tau(B_2^\tau - A_2^\tau) = 0, \tag{10.129}$$

so that the six constants $A_1^a, A_2^a, B_1^a, B_2^a, A_1^b, A_2^b$, which appear in (10.127), have to satisfy the four conditions (10.124) and (10.129). Therefore, if the

10.12. Reflection and Refraction of SH Waves

characteristics of the incident wave are given, the solution is completely determined.

If θ^a and θ^b denote the angles between $\mathbf{N}$ and the propagation vectors of the incident and refracted waves, respectively, we obtain

$$k_1^a = k^a \sin\theta^a, \; p_1^a = k^a \cos\theta^a,$$
$$k_1^b = k^b \sin\theta^b, \; p_1^b = k^b \cos\theta^b, \tag{10.130}$$

where k^a and k^b are the wave numbers. Inserting the displacements (10.127) into (7.18), we derive

$$T_{23}^a = -\mu^a p_1^a[-A(I) + A(Rfl)]e^{i(k_1^a x - \omega_1^a t)},$$
$$T_{23}^b = \mu^b p_1^b A(Rfr) e^{i(k_1^b x - \omega_1^b t)}. \tag{10.131}$$

Taking into account $(10.124)_1$ and (10.130), we derive the equation

$$[A(I) + A(Rfl)]e^{i(k^a x \sin\theta^a - \omega_1^a t)} = A(Rfr)e^{i(k^b x \sin\theta^b - \omega_1^b t)}. \tag{10.132}$$

The arbitrariness of x and t leads to

$$\omega_1^a = \omega_1^b, \; \frac{\sin\theta^a}{\sin\theta^b} = \frac{k_b}{k_a} = \frac{c_S^a}{c_S^b}, \tag{10.133}$$

$$A(I) + A(Rfl) = A(Rfr).$$

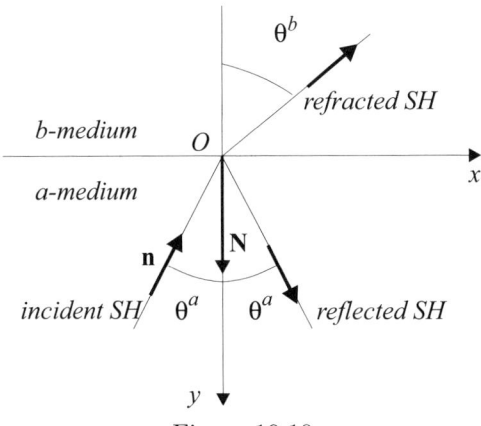

Figure 10.10

In turn, the continuity condition $(10.124)_2$, when $(10.133)_{1,2}$ are taken into account, implies that

$$\mu^a k^a \cos\theta^a[-A(I) + A(Rfl)] = -\mu^b k^b \cos\theta^b A(Rfr). \tag{10.134}$$

This condition, together with (10.133)$_3$, gives the wave amplitude ratios

$$\frac{A(Rfr)}{A(I)} = \frac{2\mu^a \cos\theta^a}{\mu^a \cos\theta^a + \mu^b(c_S^a/c_S^b)\cos\theta^b},$$
$$\frac{A(Rfl)}{A(I)} = \frac{\mu^a \cos\theta^a - \mu^b(c_S^a/c_S^b)\cos\theta^b}{\mu^a \cos\theta^a + \mu^b(c_S^a/c_S^b)\cos\theta^b}.$$
(10.135)

The previous formulae contain the following results (see Figure 10.10): Let Σ be an SH wave which propagates in the a-medium and is incident on the surface $y = 0$. If θ^a is the angle which Σ forms with the vector $\mathbf{N}$ normal to this surface, two waves are generated: a reflected one propagating in the a-medium and forming with $\mathbf{N}$ the same angle θ^a, and a refracted wave propagating in the b-medium and forming with $\mathbf{N}$ an angle θ^b, which is related to θ^a by (10.133)$_2$ Snell's law. In particular, in the limit case

$$\mu^a \cos\theta^a = \mu^b(c_S^a/c_S^b)\cos\theta^b,$$

Σ is completely refracted.

10.13 Harmonic Waves in a Layer

This section deals with the propagation of P and SV waves in an elastic layer Δ of finite thickness $2h$, having free boundary surfaces. We introduce a coordinate system whose origin O is on the plane of symmetry π of Δ, parallel to the boundary surfaces, with the axes Ox, Oy on π and Oz orthogonal to π. The analysis of this problem is relevant as it allows us to introduce the concepts of *dispersion, group velocity,* and *frequency spectrum.*

The displacement field is represented by the following functions:

$$\begin{aligned} u_x &= u_x(x,y,t), \\ u_y &= u_x(x,y,t), \\ u_x &= 0. \end{aligned}$$
(10.136)

From (10.93), the scalar potential ϕ and the component Ψ_3 of the vector potential are relevant, and, by considering the solutions obtained with the method of separation of variables (see Section 10.9) and in particular (10.85), we see that they assume the following form:

$$\begin{aligned} \phi &= [A\cos(py) + B\sin(py)]\,e^{i(kx-\omega t)}, \\ \Psi_3 &= [A_3\cos(p_3 y) + B_3\sin(p_3 y)]\,e^{i(k_3 x - \omega_3 t)}, \end{aligned}$$
(10.137)

10.13. Harmonic Waves in a Layer

where

$$p = \sqrt{\frac{\omega^2}{c_P^2} - k^2},$$

$$p_3 = \sqrt{\frac{\omega_3^2}{c_S^2} - k_3^2}. \tag{10.138}$$

By substituting (10.137) into (10.72)$_{1,2}$, we obtain the displacement components

$$\begin{aligned}
u_x &= ik\,[A\cos(py) + B\sin(py)]\,e^{i(kx-\omega t)} \\
&\quad + p_3\,[-A_3\sin(p_3 y) + B_3\cos(p_3 y)]\,e^{i(k_3 x - \omega_3 t)}, \\
u_y &= p\,[-A\sin(py) + B\cos(py)]\,e^{i(kx-\omega t)} \\
&\quad - ik_3\,[A_3\cos(p_3 y) + B_3\sin(p_3 y)]\,e^{i(k_3 x - \omega_3 t)}.
\end{aligned} \tag{10.139}$$

By imposing the boundary conditions

$$\mathbf{Tn} = \mathbf{0} \iff T_{12} = T_{22} = 0 \quad \text{if } y = \pm h, \tag{10.140}$$

since x and t are arbitrary, we get

$$\omega = \omega_3,\ k = k_3, \tag{10.141}$$

together with the following relations:

$$\begin{aligned}
&2ikp\,[-A\sin(ph) + B\cos(ph)] \\
&\quad - (p_3^2 - k^2)\,[A_3\cos(p_3 h) + B_3\sin(p_3 h)] = 0, \\
&- [p^2(\lambda + 2\mu) + \lambda k^2]\,[A\cos(ph) + B\sin(ph)] \\
&\quad - 2\mu ikp^3\,[-A_3\sin(p_3 h) + B_3\cos(p_3 h)] = 0, \\
&2ikp\,[A\sin(ph) + B\cos(ph)] \\
&\quad - (p_3^2 - k^2)\,[A_3\cos(p_3 h) - B_3\sin(p_3 h)] = 0, \\
&- [p^2(\lambda + 2\mu) + \lambda k^2]\,[A\cos(ph) - B\sin(ph)] \\
&\quad - 2\mu ikp^3\,[A_3\sin(p_3 h) + B_3\cos(p_3 h)] = 0,
\end{aligned} \tag{10.142}$$

where (10.141) have been taken into account. Furthermore, recalling (10.138) and the relation $c_P^2/c_S^2 = (\lambda + 2\mu)/\mu$ leads to

$$p^2(\lambda + 2\mu) + \lambda k^2 = \mu(p_3^2 - k^2). \tag{10.143}$$

Now, add and subtract the first and the third and add and subtract the second and the fourth of (10.142), where (10.143) has been substituted. We obtain

$$2ikpB\cos(ph) - (p_3^2 - k^2) A_3 \cos(p_3 h) = 0,$$
$$2ikpA\sin(ph) + (p_3^2 - k^2) B_3 \sin(p_3 h) = 0,$$
$$(p_3^2 - k^2) A \cos(ph) + 2ikp_3 B_3 \cos(p_3 h) = 0,$$
$$-(p_3^2 - k^2) B \sin(ph) + 2ikp_3 A_3 \sin(p_3 h) = 0, \quad (10.144)$$

and rearranging gives

$$\begin{pmatrix} 2ikp\sin(ph) & (p_3^2 - k^2)\sin(p_3 h) \\ (p_3^2 - k^2)\cos(ph) & 2ikp_3 \cos(p_3 h) \end{pmatrix} \begin{pmatrix} A \\ B_3 \end{pmatrix} = 0, \quad (10.145)$$

$$\begin{pmatrix} 2ikp\cos(ph) & -(p_3^2 - k^2)\cos(p_3 h) \\ -(p_3^2 - k^2)\sin(ph) & 2ikp_3 \sin(p_3 h) \end{pmatrix} \begin{pmatrix} B \\ A_3 \end{pmatrix} = 0. \quad (10.146)$$

The reason for rearranging in this way is now clear: A and B_3 represent displacements symmetric with respect to the axis Ox, whereas B and A_3 represent skew-symmetric displacements.

If we assume that displacements are symmetric, then B and A_3 must vanish, and by setting to zero the determinant of the matrix in (10.145), we derive the **Rayleigh–Lamb frequency equation**

$$\frac{4k^2 p p_3}{(p_3^2 - k^2)^2} = -\frac{\tan(p_3 h)}{\tan(ph)}. \quad (10.147)$$

By similar arguments, from (10.146) we obtain the frequency equation for antisymmetric waves:

$$\frac{4k^2 p p_3}{(p_3^2 - k^2)^2} = -\frac{\tan(ph)}{\tan(p_3 h)}. \quad (10.148)$$

Recalling (10.138), we see that these equations give the wave number k or the phase velocity $c = \omega/k$ *for any selected value of the circular frequency* ω, i.e., (10.147) and (10.148) represent the relation between the wave velocity and the wave number, called the **frequency spectrum**.

If the velocity changes with ω, the propagation is said to be *dispersive*, because any wave obtained as a superposition of waves of different frequencies will change its shape in time, since single components tend to disperse. In this case, we introduce the concept of **group velocity** v_G, and it can be shown, by making reference to two monochromatic waves characterized by the same amplitude but slightly different frequencies, that the following relation holds:

$$v_G = \frac{d\omega}{dk} = c + k\frac{dc}{dk} = c - \lambda\frac{dc}{d\lambda}. \quad (10.149)$$

10.14. Exercises

The above argument is kinematic in its essence. From a dynamic point of view, it can be proved that the group velocity is the velocity of the energy propagation. The proof is in this case rather long, but we can rely on intuition and observe that energy cannot be transferred through nodes, which move with group velocity.

A final remark is that usually $dc/dk < 0$, so that $v_G < c$; this case is called *normal dispersion*. When $v_G > c$, the case is called *inverse dispersion*.

10.14 Exercises

1. Suppose that a uniform layer of isotropic elastic material of thickness H is subjected at the lower boundary to a planar steady-state motion represented as
$$u_1 = Ae^{-i\omega t}. \tag{10.150}$$
Determine the displacement field of the layer and show that the layer response depends on the frequency of the boundary motion.

Due to reflection at the upper boundary, there will also be a wave traveling in the negative x_3 direction, so that for the displacement field we can assume a steady-state solution of the form
$$u_1(x_3, t) = Be^{i(kx_3 - \omega t)} + Ce^{i(-kx_3 - \omega t)}. \tag{10.151}$$
Since the surface of the layer is stress free, it must hold that
$$\frac{\partial u_1(H,t)}{\partial x_3} = ikB \exp[i(kH - \omega t)] - ikC \exp[i(kH + \omega t)] = 0,$$
which implies that
$$Be^{ikH} - Ce^{-ikH} = 0. \tag{10.152}$$
At the lower boundary the motion must be equal to the prescribed value, so that
$$B + C = A, \tag{10.153}$$
and both these conditions allow us to find the two constants
$$B = \frac{e^{-ikH}}{2\cos kH} A,$$
$$C = \frac{e^{ikH}}{2\cos kH} A. \tag{10.154}$$
By introducing these constants into (10.151), we obtain the solution for the displacement field:
$$u_1 = \frac{A\cos(kx_3 - kH)}{\cos kH} e^{-i\omega t}. \tag{10.155}$$

This relatively simple example allows us to introduce an aspect of relevance in earthquake engineering known as soil amplification. Let us define as an amplification factor the ratio between the motion at the upper boundary (which is the relevant one for buildings interacting with the soil) and the one imposed at the lower boundary (which can be assumed to be the motion of the bedrock). This ratio is

$$\text{amplification ratio} = \frac{u_1(H,t)}{u_1(0,t)} = \frac{1}{\cos kH} = \frac{1}{\cos(\omega H/v_s)}, \quad (10.156)$$

and it can be observed that a resonance phenomenon occurs (i.e., the motion tends to infinity) when the ratio has the value

$$\frac{\omega H}{v_S} = \frac{\pi}{2} + n\pi. \quad (10.157)$$

We note that, even if an infinite amplification cannot occur (because the present analysis assumes no dissipation of energy), the major aspects of this phenomenon are qualitatively accounted for: the amplification factor depends on the layer properties as well as the frequency of the motion at its base.

2. Assuming a Kelvin–Voigt solid

$$T_{ij} = 2G\varepsilon_{ij} + 2\eta\dot{\varepsilon}_{ij},$$

where η is a damping coefficient, show, with reference to the layer of Exercise 1, that the amplification factor is equal to

$$\frac{1}{\sqrt{\cos^2 kH + \sinh^2 \xi kH}}.$$

Chapter 11

Other Approaches to Thermodynamics

11.1 Basic Thermodynamics

In Chapter 5 we introduced Truesdell, Coleman, and Noll's approach (see, for instance, [23], [24], [25]) to thermodynamics of continua, together with some of its applications. This approach has the following main advantages:

- It suggests a formulation of thermodynamic principles that we can use in continuum mechanics.

- It suggests that we use the second principle as a restriction for the constitutive equations, rather than as a constraint on processes.

Despite these advantages, some authors have noted shortcomings or flaws about this approach. Starting from these objections, we now present a concise review of various approaches. To do this, we first recall some concepts of basic thermodynamics.

Essentially, elementary thermodynamics (see [33]) deals with **homogeneous** and **steady-state processes**; i.e., it considers processes evolving in such a way that, at any instant, the state of a system S is always very near equilibrium. It follows that the state of S can be described by a convenient and finite number of real parameters $\mathbf{x} \equiv (x_i)$, $i = 1, \ldots, n$, and its evolution is given by the functions $x_i(t)$ of the time t.

As an example, the equilibrium state of a gas in a container is macroscopically described by its volume V and the pressure p. If we consider a binary mixture of gas, then in addition to the parameters V and p, the concentration c of one of them must be introduced. Accordingly, the steady-state evolution is described through the functions $p(t)$, $V(t)$, and $c(t)$.

As a second step, the **empirical temperature** τ of the system S in a given state π is defined by the **zero postulate of thermodynamics**: If two systems are at equilibrium when they are in contact with a third one, they will also be at equilibrium when kept in contact with each other.

According to this postulate, it can be proved that the equilibrium of two systems requires the existence of at least one state function, called the *empirical temperature*, which is equal for both of them. The state of a system can then be regarded as an *index* of this equilibrium. The reference system can be assumed to be a thermometer and, if its state is expressed by its volume, this latter parameter is an index of the empirical temperature. This is what happens in common thermometers, where the temperature scales refer to volume changes.

In the following discussion, instead of the pressure, the empirical temperature $\tau = x_1$ is taken as one of the state variables.

Moreover, we suppose that the mechanical power w and the thermal power q can be represented by the following relationships:

$$\begin{cases} w = \sum_{k=1}^{n} \pi_k(\mathbf{x}) \dot{x}_k, \\ q = c(\mathbf{x})\dot{\tau} + \sum_{k=2}^{n} \nu_k(\mathbf{x}) \dot{x}_k, \end{cases} \qquad (11.1)$$

where the state functions $\pi_k(\mathbf{x})$ are called *partial pressures*, the function $c(\mathbf{x})$ is the *specific heat*, and $\nu_k(\mathbf{x})$ are the *latent heats*.

When dealing with fundamentals of elementary thermodynamics, it is important to recall that the first and second laws have been formulated for **cyclic** processes, at the end of which the system returns to its initial state.

Furthermore, these laws only involve the notion of mechanical and thermal power. By using the previous notations, we represent these cycles by closed curves γ_c of $\Re^n$ of the equation $\mathbf{x} = \mathbf{x}(t)$.

First law of thermodynamics: In any cyclic process γ_c the mechanical work $L^{(e)}$ done on the system is equal to the absorbed heat Q:

$$L^{(e)} = Q. \qquad (11.2)$$

Since $L^{(e)}$ and Q are written as

$$L^{(e)} = \int_{\gamma_c} w(t)\, dt, \qquad Q = \int_{\gamma_c} q(t)\, dt,$$

the differential form

$$w = (\pi_1(\mathbf{x}) - c(\mathbf{x}))\, d\tau + \sum_{k=2}^{n} (\pi_k(\mathbf{x}) - \nu_k(\mathbf{x}))\, dx_k$$

11.1. Basic Thermodynamics

is integrable, so that we conclude that there exists a state function $E(\mathbf{x})$, called the *internal energy*, such that, in any process which is not a cycle, the change ΔE is given by

$$\Delta E = L^{(e)} - Q = \int_{\gamma_c} \omega. \qquad (11.3)$$

It is well known that the first law states that the transformation of one form of energy into another is completely symmetric. On the other hand, since in physical processes such a symmetry does not hold (i.e., the transformation of nonthermal energy into thermal energy is the preferred process), the second law imposes severe limitations on the transformation of heat into work. In fact, its formulation given by Lord Kelvin is as follows:

Second law of thermodynamics (Lord Kelvin, 1851): *It is impossible to perform a cyclic transformation whose only result is to convert into work a given heat quantity Q, extracted from a source at uniform temperature.*

It can be also proved that the above formulation is equivalent to the following one, due to Clausius (1854):

It is impossible to realize a cyclic transformation whose only result is the heat transfer from a body, having a given temperature, to another one at a higher temperature.

Starting from the second law, we can prove the existence of an universal function θ (the term *universal* highlights its applicability to any body), which is positive and depends only on the empirical temperature τ. This function θ is called the **absolute temperature**. Furthermore, it can be proved that, in any cyclic steady process in the interval (t_0, t_1), if $\theta(t)$ is the uniform temperature at which the heat exchange takes place, the integral relation

$$\int_{t_0}^{t_1} \frac{q(t)}{\theta(t)} dt = 0 \qquad (11.4)$$

holds provided that the cyclic transformation is **reversible**.

The above result can then be used to prove the existence of a state function η, called **entropy**, such that, in any *reversible but not necessarily cyclic process*, it holds that

$$\Delta \eta = \eta(t_1) - \eta(t_0) = \int_{t_0}^{t_1} \frac{q(t)}{\theta(t)} dt. \qquad (11.5)$$

In elementary thermodynamics the previous concepts are also usually extended to *irreversible processes* and, in such cases, (11.4) and (11.5) become

$$\int_{t_0}^{t_1} \frac{q(t)}{\theta(t)} dt \leq 0, \qquad \eta(t_1) - \eta(t_0) \geq \int_{t_0}^{t_1} \frac{q(t)}{\theta(t)} dt. \qquad (11.6)$$

But these extensions are not so straightforward for the following reasons:

- an irreversible transformation of a thermodynamic system is not *analytically* defined;
- it is far from being clear how to compute q and w, as well as the integrals in (11.3) and (11.6).

These shortcomings call for a new formulation of elementary thermodynamics, in order to describe nonhomogeneous and irreversible processes. However, it is relevant to observe that in this theory the concepts of energy and entropy are derived from properties of cyclic processes, rather than being introduced as primitive concepts, as is done in the formulation of Truesdell and Coleman.

In the remaining sections, we present **extended thermodynamics** (see [34]) and the formulation suggested by J. Serrin [35].[1]

11.2 Extended Thermodynamics

Let S be a rigid body of uniform density ρ. The energy balance equation in this case is
$$\rho \dot{\epsilon} = -\nabla \cdot \mathbf{h},$$
and, by introducing the constitutive relationships
$$\epsilon = \epsilon(\theta), \qquad \mathbf{h} = -K \nabla \theta, \tag{11.7}$$
we obtain
$$\rho c \dot{\theta} = K \Delta \theta,$$
where the thermal conductivity K is a positive quantity and $c = \partial \epsilon / \partial \theta$ is the heat capacity under constant volume. The equation obtained is a second-order partial differential equation of parabolic type, for the unknown field of temperature $\theta(\mathbf{x}, t)$. As it is well recognized, this equation describes a diffusion phenomenon, and accordingly any perturbation appears instantaneously at any spatial point. Therefore, in contrast with the experimental result, it cannot foresee a finite velocity of the heat propagation at low temperature. The need to predict such an experimental result was the starting point for extended thermodynamics.

---
[1] In the papers of A. Day [36], [37], [38], the notions of internal energy and temperature are accepted and, by conveniently rewriting the second principle for cyclic processes, the existence of the entropy for a large class of materials is proved.

11.2. Extended Thermodynamics

The assumptions of this approach can be summarized as follows. Let S be a continuous system described by n fields $\mathbf{u} = (u_{(1)}(\mathbf{x}_\alpha), \ldots, \mathbf{u}_{(n)}(\mathbf{x}_\alpha))$, where $(x_\alpha) = (t, \mathbf{x})$, $\alpha = 0, 1, 2, 3$, which satisfy n balance equations

$$\frac{\partial \mathbf{F}_{(i)\alpha}(\mathbf{u})}{\partial x_\alpha} = \mathbf{P}_{(i)}(\mathbf{u}), \ i = 1, \ldots, n. \tag{11.8}$$

Quantities like $\mathbf{F}_{(i)0}$ are *densities*, whereas quantities like $\mathbf{F}_{(i)j}$, $j = 1, 2, 3$, are *fluxes*; on the right-hand side, quantities like $\mathbf{P}_{(i)}$ represent *production terms*. If the *constitutive equations* $\mathbf{F}_{(i)\alpha}(\mathbf{u})$ and $\mathbf{P}_{(i)}(\mathbf{u})$ are specified, then the quasi-linear system (11.8) allows us to determine the fields $\mathbf{u}_{(i)}(\mathbf{x}_\alpha)$.

In order to reduce the generality of the constitutive equations, they must satisfy

- the entropy inequality;

- the convexity;

- the principle of relativity; i.e., they must be covariant for a Galileian transformation.

The entropy inequality is expressed as

$$\frac{\partial h_\alpha(\mathbf{u})}{\partial x_\alpha} \equiv \Sigma(\mathbf{u}) \geq 0,$$

where *this inequality must be satisfied by all the thermodynamic processes compatible with the balance equations* (11.8).

It is relevant to remark that, even if the entropy principle is accepted as a constraint for the constitutive equations, this principle must be satisfied when considering processes compatible with balance equations, which play the role of constraints for the processes.

As proved in [39], this requirement is equivalent to requiring that the inequality

$$\frac{\partial h_\alpha(\mathbf{u})}{\partial x_\alpha} - \sum_{i=1}^{n} \mathbf{\Lambda}_{(i)} \cdot \left(\frac{\partial \mathbf{F}_{(i)\alpha}(\mathbf{u})}{\partial x_\alpha} - \mathbf{P}_{(i)}(\mathbf{u}) \right) \geq 0$$

is satisfied for *any* differentiable process $\mathbf{u}_{(i)}(x_\alpha)$, where the quantities $\mathbf{\Lambda}_{(i)}$ are unknown Lagrangian multipliers.

Since the previous inequality can also be written as

$$\sum_{j=1}^{n} \left(\frac{\partial h_\alpha(\mathbf{u})}{\partial \mathbf{u}_{(j)}} - \sum_{i=1}^{n} \mathbf{\Lambda}_{(i)} \cdot \frac{\partial \mathbf{F}_{(i)\alpha}(\mathbf{u})}{\partial \mathbf{u}_{(j)}} \right) \frac{\partial \mathbf{u}_{(j)}}{\partial x_\alpha} + \sum_{i=1}^{n} \mathbf{\Lambda}_{(i)} \cdot \mathbf{P}_{(i)} \geq 0,$$

and the quantities $\partial \mathbf{u}_{(i)}/\partial x_\alpha$ are arbitrary, we derive

$$\frac{\partial h_\alpha(\mathbf{u})}{\partial \mathbf{u}_{(j)}} - \sum_{i=1}^{n} \mathbf{\Lambda}_{(i)} \cdot \frac{\partial \mathbf{F}_{(i)\alpha}(\mathbf{u})}{\partial \mathbf{u}_{(j)}} \geq 0, \quad \sum_{i=1}^{n} \mathbf{\Lambda}_{(i)} \cdot \mathbf{P}_{(i)} \geq 0. \qquad (11.9)$$

Finally it can be proved that the $n \times n$ matrices

$$\frac{\partial \mathbf{F}_{(i)\alpha}(\mathbf{u})}{\partial \mathbf{\Lambda}_{(j)}}$$

are symmetric and that this imposes severe constraints on the functions $\mathbf{F}_{(i)\alpha}(\mathbf{u})$.

This approach leads to quasi-linear systems of hyperbolic differential equations, gives a generalized form for the entropy flux, and proves the existence of the absolute temperature. The reader interested in applications of extended thermodynamics to fluids, mixtures, and other fields is referred to [34].

11.3 Serrin's Approach

Serrin's approach is based on the notion of an *ideal material* and provides a mathematically correct formulation of all the results of the elementary thermodynamics of homogeneous processes. In the context of this volume, we limit ourselves to outlining that this approach allows us

1. to introduce the **absolute temperature** θ as a positive function of the empirical temperature, measured by thermometers constituted by ideal materials;

2. to introduce the **accumulation function** $\mathcal{Q}(\theta, P)$, dependent on the absolute temperature θ as well as on the thermodynamic process P of the system S. This function expresses the heat quantity exchanged by S with the surroundings at a *temperature less than or equal to* θ, during the process P;

3. to formulate the thermodynamic principles for cyclic processes, without assuming the existence of energy and entropy.

The two fundamental laws of thermodynamics are formulated by Serrin in the following way.

First Law. *In any cyclic process P, the balance of energy is given by*

$$L^{(i)}(P) = Q(P),$$

11.3. Serrin's Approach

where $L^{(i)}$ denotes the mechanical work of the system on the exterior world and Q is the quantity of heat received from it.

Second Law. *There is no cyclic process P in which the accumulation function assumes positive values.*

This formulation states that in any cyclic process there are temperatures at which the system provides heat to the surroundings. It can be further proved that the above formulation, when applied to homogeneous systems, leads to Kelvin and Clausius's postulate.

Moreover, starting from the second law and the existence of ideal materials, Serrin proves that the second law is equivalent to the following mathematical condition in any cyclic process:

$$\int_0^\infty \frac{Q(\theta, P)}{\theta^2} d\theta \leq 0. \tag{11.10}$$

It remains now to find the expressions of $L^{(i)}$, Q and $Q(\theta, P)$ for a continuous system S in order to write explicitly the previous thermodynamic laws. To this end, we introduce the following definitions.

A *process of duration* $J = [0, t_1]$ is defined by a couple of functions

$$\mathbf{x}(\mathbf{X}, t), \quad \theta(\mathbf{X}, t),$$

conveniently regular on $C_* \times J$, which give a motion and an absolute temperature field at any instant $t \in J$. The set $\mathbb{P}(S)$ of the *cyclic processes* of duration J is given by all processes which satisfy the conditions

$$\begin{cases} \mathbf{x}(\mathbf{X}, 0) = \mathbf{x}(\mathbf{X}, t_1), \\ \theta(\mathbf{X}, 0) = \theta(\mathbf{X}, t_1), \end{cases} \tag{11.11}$$

as well as the balance equations of mass, momentum, and angular momentum.

The change of kinetic energy ΔT of $c(t)$, in the time interval J and along the process P, is written as

$$\Delta T = \int_{c(t_1)} \frac{1}{2}\rho v^2 \, dc - \int_{c(0)} \frac{1}{2}\rho v^2 \, dc. \tag{11.12}$$

In addition, the work $L^{(e)}(P)$ done by the exterior of $c(t)$ during the same process P is given by

$$L^{(e)}(P) = \int_J dt \int_{\partial c(t)} \mathbf{v} \cdot \mathbf{Tn} \, d\sigma + \int_J dt \int_{c(t)} \mathbf{v} \cdot \rho \mathbf{b} \, dc. \tag{11.13}$$

But, $L^{(i)}(P) = \Delta T - L^{(e)}(P)$, so that the previous relation implies that

$$L^{(i)}(P) = \Delta T - \int_J dt \int_{\partial c(t)} \mathbf{v} \cdot \mathbf{Tn} \, d\sigma + \int_J dt \int_{c(t)} \mathbf{v} \cdot \rho \mathbf{b} \, dc. \tag{11.14}$$

Taking the scalar product of the local balance equation (5.30) with $\mathbf{v}$ and integrating the result over $c(t)$, we obtain

$$\frac{d}{dt}\int_{c(t)} \frac{1}{2}\rho v^2 \, dc = \int_{\partial c(t)} \mathbf{v} \cdot \mathbf{T}\mathbf{n} \, d\sigma + \int_{c(t)} \mathbf{v} \cdot \rho \mathbf{b} \, dc$$
$$- \int_{c(t)} \mathbf{T} : \nabla \mathbf{v} \, dc. \qquad (11.15)$$

By integrating this over the time interval J and by recalling (11.13), we have the relation

$$\Delta T = L^{(e)}(P) - \int_J dt \int_{c(t)} \mathbf{T} : \nabla \mathbf{v} \, dc,$$

which, compared with (11.14), leads to

$$L^{(i)} = -\int_J dt \int_{c(t)} \mathbf{T} : \nabla \mathbf{v} \, dc. \qquad (11.16)$$

In a similar way, assuming that the heat exchanged by $c(t)$ with the surroundings is given (see Chapter 5) by a source term r and a flux through the boundary $\partial c(t)$ of density $\mathbf{h} \cdot \mathbf{n}$, we get

$$Q(P) = \int_J dt \left(\int_{\partial c(t)} \mathbf{h} \cdot \mathbf{n} \, d\sigma + \int_{c(t)} \rho r \, dc \right). \qquad (11.17)$$

Finally, by introducing the notations

$$\begin{cases} \partial \tilde{c}_t(\theta) = \{\mathbf{x} \in \partial c(t) | \vartheta(\mathbf{x}, t) < \theta\}, \\ \tilde{c}_t(\theta) = \{\mathbf{x} \in c(t) | \vartheta(\mathbf{x}, t) < \theta\}, \end{cases}$$

where $\vartheta(\mathbf{x}, t)$ is the field of absolute temperature and θ is any arbitrary but fixed value of temperature, the accumulation function is written as

$$\mathcal{Q}(\theta, P) = \int_J dt \left(\int_{\partial \tilde{c}_t(\theta)} \mathbf{h} \cdot \mathbf{n} \, d\sigma + \int_{\tilde{c}_t(\theta)} \rho r \, dc \right). \qquad (11.18)$$

At this stage we are in a position to prove that the previous relations allow us to formulate the first and second laws of thermodynamics in a manner which is completely analogous to that of elementary thermodynamics.

First, by combining (11.16) and (11.17), the first law can be written as

$$\int_J dt \int_{c(t)} (\mathbf{T} : \nabla \mathbf{v} + \nabla \cdot \mathbf{h} + \rho r) \, dc = 0, \qquad (11.19)$$

11.4. An Application to Viscous Fluids

whereas (11.18) allows us to put the second law (11.10) in the form

$$\int_0^\infty \frac{1}{\theta^2} \left(\int_J dt \left(\int_{\partial \widetilde{c}_t(\theta)} \mathbf{h} \cdot \mathbf{n} \, d\sigma + \int_{\widetilde{c}_t(\theta)} \rho r \, dc \right) \right) d\theta \leq 0. \quad (11.20)$$

Equations (11.19) and (11.20) are valid for any cyclic process of duration J.

To continue, it is more convenient to rewrite the previous formulae in the Lagrangian formalism (see Section 5.6) as follows:

$$\int_J dt \int_{c_*(t)} \left(\mathbf{T}_* : \dot{\mathbf{F}} + \nabla \cdot \mathbf{h}_* + \rho_* r \right) dc_* = 0, \quad (11.21)$$

$$\int_0^\infty \frac{1}{\theta^2} \left(\int_J dt \left(\int_{\partial \widetilde{c}_{*t}(\theta)} \mathbf{h}_* \cdot \mathbf{n}_* \, d\sigma_* + \int_{\widetilde{c}_{*t}(\theta)} \rho_* r \, dc_* \right) \right) d\theta \leq 0, \quad (11.22)$$

where $\mathbf{T}_*$ is the Piola–Kirchhoff stress tensor, $\mathbf{h}_*$ is the material heat flux vector, and $\mathbf{F}$ is the deformation gradient.

The application of Fubini's theorem leads to the relation

$$\int_0^\infty \frac{1}{\theta^2} d\theta \int_{\partial \widetilde{c}_{*t}(\theta)} \mathbf{h}_* \cdot \mathbf{n}_* \, d\sigma_* = \int_{\partial c_*} \mathbf{h}_* \cdot \mathbf{n}_* \, d\sigma_* \int_\theta^\infty \frac{1}{\theta^2} d\theta. \quad (11.23)$$

The second of the previous relations can also be written as

$$\int_J \left(\int_{\partial c_*} \frac{\mathbf{h}_* \cdot \mathbf{n}_*}{\theta} d\sigma_* + \int_{c_*} \rho_* \frac{r}{\theta} dc_* \right) dt \leq 0. \quad (11.24)$$

Finally, by applying Gauss's theorem and using the arbitrariness of the material region c_*, we see that the fundamental laws of thermodynamics assume the form

$$\int_J \left(\mathbf{T}_* : \dot{\mathbf{F}} + q \right) dt = 0, \quad (11.25)$$

$$\int_J \frac{1}{\theta} \left(q - \frac{\mathbf{h}_* \cdot \nabla \theta}{\theta} \right) dt \leq 0, \quad (11.26)$$

where

$$q = \nabla \cdot \mathbf{h}_* + \rho_* r. \quad (11.27)$$

11.4 An Application to Viscous Fluids

In order to exploit the content of the thermodynamic laws (11.25) and (11.26), they are applied to the viscous fluids (see Section 7.3). The results discussed here are contained in ([42]); for an extension to materials with fading memory see ([43]).

The constitutive equations of the material are supposed to be expressed by the differentiable functions

$$\mathbf{T}_* = \mathbf{T}_*(\mathbf{F}, \dot{\mathbf{F}}, \theta, \dot{\theta}, \nabla\theta),$$
$$q = q(\mathbf{F}, \dot{\mathbf{F}}, \theta, \dot{\theta}, \nabla\theta),$$
$$\mathbf{h}_* = \mathbf{h}_*(\mathbf{F}, \dot{\mathbf{F}}, \theta, \dot{\theta}, \nabla\theta). \tag{11.28}$$

In order to make explicit the restrictions on the constitutive equations (11.28) deriving from the two thermodynamic laws, we start by noting that in any equilibrium process

$$\mathbf{x}(\mathbf{Y}, t) = \mathbf{F} \cdot (\mathbf{Y} - \mathbf{X}), \; \theta(\mathbf{Y}, t) = \theta(\mathbf{X}), \tag{11.29}$$

the thermodynamic laws lead to the conditions

$$q(\mathbf{F}, \mathbf{0}, \theta, \dot{\theta}, \mathbf{0}) = 0,$$
$$\mathbf{h}_*(\mathbf{F}, \mathbf{0}, \theta, \dot{\theta}, \mathbf{0}) \geq 0, \tag{11.30}$$

so that the function $\mathbf{f}(\mathbf{G}) = \mathbf{h}_*(\mathbf{F}, \mathbf{0}, \theta, \dot{\theta}, \mathbf{0}) \cdot \nabla\theta$ has a minimum at $\nabla\theta = \mathbf{0}$; consequently,

$$(\nabla_{\nabla\theta}\mathbf{f})_{\nabla\theta=\mathbf{0}} = \mathbf{h}_*(\mathbf{F}, \mathbf{0}, \theta, \dot{\theta}, \mathbf{0}) = \mathbf{0}, \tag{11.31}$$

and there is no heat conduction in the absence of a temperature gradient.

Moreover, by expanding q in power series of $\dot{\mathbf{F}}$ and $\dot{\theta}$ and recalling $(11.31)_1$, we get

$$q = c(\mathbf{F}, \theta)\dot{\theta} + \nu(\mathbf{F}, \theta) \cdot \dot{\mathbf{F}} - S(\mathbf{F}, \dot{\mathbf{F}}, \theta, \dot{\theta}, \nabla\theta), \tag{11.32}$$

where S is a second-order function of $\dot{\mathbf{F}}$ and $\dot{\theta}$. If we introduce the notations

$$\mathbf{T}^e_*(\mathbf{F}, \theta) = \mathbf{T}_*(\mathbf{F}, \mathbf{0}, \theta, 0, \mathbf{0}),$$
$$\mathbf{T}^D_* = \mathbf{T}_* - \mathbf{T}^e_*,$$
$$R = \mathbf{T}^D_* : \dot{\mathbf{F}}, \tag{11.33}$$

then the fundamental laws (11.25), (11.26) reduce to the form

$$\int_J ((\rho_*\nu + \mathbf{T}^e_*) \cdot \dot{\mathbf{F}} + \rho_* c\dot{\theta})\, dt + \int_J (R - \rho_* S)\, dt = 0,$$

$$\int_J \rho_* \frac{1}{\theta}\left(\nu \cdot \dot{\mathbf{F}} + c\dot{\theta}\right) dt - \int_J \rho_* \frac{1}{\theta}\left(S + \frac{\mathbf{h}_* \cdot \nabla\theta}{\theta}\right) dt \leq 0, \tag{11.34}$$

for any cycle at the point $\mathbf{X} \in C_*$.

11.4. An Application to Viscous Fluids

Now it is possible to prove that

Theorem 11.1
For materials described by the constitutive equations (11.28), there exists a function $u(\mathbf{F}, \theta)$, called the specific internal energy, such that for any process of duration J, the following relations are satisfied:

$$u(\mathbf{F}, \theta)_{t=J} - u(\mathbf{F}, \theta)_{t=0} = \int_0^J (\rho_* q - \mathbf{T}_* : \dot{\mathbf{F}}) \, dt, \qquad (11.35)$$

$$\frac{\partial u}{\partial \mathbf{F}} = \rho_* \nu + \mathbf{T}_*, \quad \frac{\partial u}{\partial \theta} = \rho_* c, \qquad (11.36)$$

$$R = \rho_* S. \qquad (11.37)$$

PROOF Let P be a cyclic process defined by the motion $\mathbf{x}(\mathbf{X}, t)$ and the temperature field $\theta(\mathbf{X}, t)$, where $\mathbf{X}, t) \in C_* \times [0, J]$. Let P_ϵ denote a new cyclic process characterized by the new fields

$$\mathbf{x}_\epsilon(\mathbf{X}, t) = \mathbf{x}(\mathbf{X}, \epsilon t), \; \theta_\epsilon(\mathbf{X}, t) = \theta(\mathbf{X}_0 + \epsilon(\mathbf{X} - \mathbf{X}_0), \epsilon t), \qquad (11.38)$$

where $\mathbf{X}_0$ is an arbitrary point of C_*. In the process P_ϵ, any particle $\mathbf{X}$ of the system assumes, at the instant t, the same position and temperature than it assumes at the instant ϵt during the process P. For $\epsilon \to 0$, the system tends to equilibrium at uniform and constant temperature.

It is easy to verify the following relations:

$$\mathbf{F}_\epsilon(\mathbf{X}_0, t) = \mathbf{F}(\mathbf{X}_0, \epsilon t), \; \dot{\mathbf{F}}_\epsilon(\mathbf{X}_0, t) = \epsilon \dot{\mathbf{F}}(\mathbf{X}_0, \epsilon t)$$
$$\theta_\epsilon(\mathbf{X}_0, t) = \theta(\mathbf{X}_0, \epsilon t), \; \dot{\theta}_\epsilon(\mathbf{X}_0, t) = \epsilon \dot{\theta}(\mathbf{X}_0, \epsilon t)$$
$$\nabla \theta_\epsilon = \epsilon \nabla \theta. \qquad (11.39)$$

The first fundamental law $(11.34)_1$, applied to the process P_ϵ, leads to the equality

$$\int_0^{J/\epsilon} ((\rho_* \nu_\epsilon + \mathbf{T}^e_{*\epsilon}) \cdot \epsilon \dot{\mathbf{F}} + \rho_* c \epsilon \dot{\theta}) \, dt + \int_0^{J/\epsilon} (R_\epsilon - \rho_* S_\epsilon) \, dt = 0, \qquad (11.40)$$

which, by the variable change $t' = \epsilon t$, can be written as

$$\int_0^J ((\rho_* \nu + \mathbf{T}^e_*) \cdot \dot{\mathbf{F}} + \rho_* c \dot{\theta}) \, dt' + \frac{1}{\epsilon} \int_0^{J/\epsilon} (R_\epsilon - \rho_* S_\epsilon) \, dt' = 0. \qquad (11.41)$$

In the limit $\epsilon \to 0$, we have that

$$\int_0^J ((\rho_* \nu + \mathbf{T}^e_*) \cdot \dot{\mathbf{F}} + \rho_* c \dot{\theta}) \, dt' = 0 \qquad (11.42)$$

in any cyclic process. Owing to a well-known theorem of differential forms, from (11.42) the conditions (11.36) are derived.

Let t be any instant in the interval $[0, J]$ of the cyclic process P and let P' be the process bringing the system to the initial state and having a duration $[t, t + J']$. Finally, P'_ϵ will denote the process defined by (11.38). Applying $(11.34)_1$ to the cyclic composed process P_c constituted by P in the time interval $[0, t]$ and by P'_ϵ in the time interval $[t, t + J'/\epsilon]$, and taking into account (11.42), we derive

$$\int_0^t (R - S)\, dt' + \int_t^{t+J'/t} (R'_\epsilon - S'_\epsilon)\, dt' = 0. \qquad (11.43)$$

By the variable change $t'' = \epsilon(t' - t)$ in the second integral, the previous relation becomes

$$\int_0^t (R - S)\, dt' + \frac{1}{\epsilon} \int_0^{J''} (R''_\epsilon - S''_\epsilon)\, dt'' = 0 \qquad (11.44)$$

so that, for $\epsilon \to 0$, we conclude that

$$\int_0^t (R - S)\, dt = 0 \quad \forall t \in [0, J], \qquad (11.45)$$

and the proof is complete. ∎

In a similar way, the following theorem can be proved.

Theorem 11.2
For materials described by the constitutive equations (11.28), there exists a function $\eta(\mathbf{F}, \theta)$, called the specific entropy, such that in any process of duration J the following relations are satisfied:

$$\eta(\mathbf{F}, \theta)_{t=J} - \eta(\mathbf{F}, \theta)_{t=0} = \int_0^J \frac{1}{\theta}\left(\rho_* q - \frac{\mathbf{h}_* \cdot \nabla \theta}{\theta}\right) dt, \qquad (11.46)$$

$$\theta \frac{\partial \eta}{\partial \mathbf{F}} = \rho_* \nu, \quad \theta \frac{\partial \eta}{\partial \theta} = \rho_* c, \qquad (11.47)$$

$$\rho_* S\theta + \mathbf{h}_* \cdot \nabla \theta \geq 0. \qquad (11.48)$$

References

[1] L. D. Landau, E. M. Lifshitz, *Fluid Mechanics*, Butterworth-Heinemann, 2000.

[2] C. Truesdell, *A First Course in Rational Continuum Mechanics*, Vol. 1, Second Edition, Academic Press, 1991.

[3] J. Marsden, T. Hughes, *Mathematical Foundations of Elasticity*, Dover, 1993.

[4] I-Shih Liu, *Continuum Mechanics*, Springer, 2002.

[5] N. Bellomo, L. Preziosi, A. Romano, *Mechanics and Dynamical Systems with Mathematica®*, Birkhäuser, 2000.

[6] A. Marasco, A. Romano, *Scientific Computing with Mathematica®: Mathematical Problems for ODEs*, Birkhäuser, 2001.

[7] S. Wolfram, *Mathematica®. A System for Doing Mathematics by Computer*, Addison-Wesley Publishing Company, 1991.

[8] C. Tolotti, *Lezioni di Meccanica Razionale*, Liguori Editore, 1965.

[9] A. Romano, *Meccanica Razionale con Elementi di Meccanica Statistica*, Liguori Editore, 1996.

[10] G. Mostov, J. Sampson, and J. Meyer, *Fundamental Structures of Algebra*, Birkhäuser, 2000.

[11] A. Lichnerowicz, *Algèbra et Analyse Linéaires*, Birkhäuser, 2000.

[12] A. Romano, *Elementi di Algebra Lineare e Geometria Differenziale*, Liguori Editore, 1996.

[13] G. P. Moeckel, Thermodynamics of an interface, *Arch. Rat. Mech. Anal.*, 57, 1974.

[14] M. Gurtin, On the two-phase Stefan problem with interfacial entropy and energy, *Arch. Rat. Mech. Anal.*, 96, 1986.

[15] F. Dell'Isola, A. Romano, On a general balance law for continua with an interface, *Ricerche Mat.*, 2, XXXV, 1986.

[16] A. Marasco, A. Romano, Balance laws for charged continuous systems with an interface, *Math. Models Methods Appl. Sci. (MAS)*, 12, no.1, 2002.

[17] J. Serrin, J. Dunn, On the thermodynamics of interstitial working, *Arch. Rat. Mech. Anal.*, 88, 1985.

[18] A. Romano, A macroscopic theory of thermoelastic dielectrics, *Ric. Mat. Univ. Parma*, 5, 1979.

[19] A. Romano, A macroscopic non-linear theory of magnetothermoelastic continua, *Arch. Rat. Mech. Anal.*, 65, 1977.

[20] C. Wang, A new representation theorem for isotropic functions. Part I and Part II, *Arch. Rat. Mech. Anal.*, 36, 1971.

[21] R. Rivlin, Further remarks on the stress-deformation relations for isotropic materials, *J. Rational Mech. Anal.*, 4, 1955.

[22] R. Rivlin and J. Ericksen, Stress-deformation relations for isotropic materials, *J. Rational Mech. Anal.*, 4, 1955.

[23] B. Coleman and V. Mizel, Existence of caloric equations of state in thermodynamics, *J. Chem. Phys.*, 40, 1964.

[24] C. Truesdell and W. Noll, *The Nonlinear Field Theories of Mechanics*, Handbuch der Physik, Vol. III/3, Springer-Verlag, Berlin New York, 1965.

[25] A. Eringen and E. Suhubi, *Elastodynamics*, Vol. I, Academic Press, New York and London, 1974.

[26] G. Smith and R. Rivlin, Integrity bases for vectors. The crystal classes, *Arch. Rat. Mech. Anal.*, 15, 1964.

[27] M. E. Gurtin, *The Linear Theory of Elasticity*, Handbuch der Physik, Vol. VIa/2, Springer-Verlag, Berlin New York, 1972.

[28] A. E. H. Love, *Mathematical Theory of Elasticity*, Dover, New York 1944.

[29] A. Clebsch, *Thèorie de l'èlasticitè des corpes solides*, Dunod, Paris, 1883.

[30] R. Marcolongo, *Teoria Matematica dell'Equilibrio dei Corpi Elastici*, Manuali Hoepli, Serie Scientifica, Milano, 1904.

[31] R. Toupin, Saint-Venant's Principle, *Arch. Rat. Mech. Anal.*, 18, 1965.

References

[32] G. Fichera, *Problemi Analitici Nuovi nella Fisica Matematica Classica*, Quaderni del G. N. F. M., 1985.

[33] E. Fermi, *Termodinamica*, Boringhieri, Torino, 1958.

[34] I. Müller and T. Ruggeri, *Rational Extended Thermodynamics*, Second Edition, Springer Tracts in Natural Philosophy, Vol 37, Springer-Verlag, New York, 1998.

[35] J. Serrin, *Foundations of Thermodynamics*, Lecture notes of a course held in Naples, 1977.

[36] W. A. Day, Thermodynamics based on a work axiom, *Arch. Rat. Mech. Anal.*, 31, 1968.

[37] W. A. Day, Entropy and Hidden variables in continuum thermodynamics, *Arch. Rat. Mech. Anal.*, 62, 1976.

[38] W. A. Day, *The Thermodynamics of Simple Materials with Fading Memory*, Second Edition, Springer Tracts in Natural Philosophy, Vol. 22, Springer-Verlag, New York, 1972.

[39] I-Shih Liu, Method of Lagrange multipliers for exploitation of the entropy principle, *Arch. Rat. Mech. Anal.*, 46, 1972.

[40] W. Noll, Lectures on the foundations of continuum mechanics and thermodynamics, *Arch. Rat. Mech. Anal.*, 52, 1973.

[41] I. Müller, The coldness, a universal function in thermoelastic bodies, *Arch. Rat. Mech. Anal.*, 41, 1971.

[42] D. Iannece, G. Starita, A new proof of the existence of the internal energy and entropy for rate type materials, *Ric. di Matematica*, XXIX, 1981.

[43] D. Iannece, A. Romano, D. Starita, The coldness, a universal function in thermoelastic bodies, *Arch. Rat. Mech. Anal.*, 75, 1981.

[44] P. D. Lax, Hyperbolic systems of conservation laws, *Comm. Pure Appl. Math.*, 10, Wiley, New York, 1957.

[45] A. Jeffrey, *Quasilinear Hyperbolic Systems and Waves*, Pitman Publishing, London–San Francisco–Melbourne, 1976.

[46] V. I. Smirnov, A Course of Higher Mathematics, Pergamon Press, Vol. V, 1964.

Index

2-tensor, 14

absolute temperature, 369
acceleration, 110
accumulation function, 372
acoustic tensor, 289, 339
adiabatic shock equation, 292
adjoint, 88
affine space, 30
algebraic multiplicity, 21
angle of attack, 268
anisotropic solid, 173, 181
Archimedes' principle, 248
axial vector, 12

balance of the angular
 momentum, 140
basis, 3
basis, dual or reciprocal, 8
bending, nonuniform, 337
bending, uniform, 336
Bernoulli's theorem, 251
bipolar coordinates, 68
Blasius equation, 280
Blasius formula, 267
boundary layer, 276
Boussinesq–Papkovich–Neuber
 method, 329

Cauchy's hypothesis, 132
Cauchy's polar decomposition
 theorem, 28

Cauchy's problem, 190, 200
Cauchy's stress tensor, 139
Cauchy's theorem, 132
Cauchy–Green tensor, left, 81
Cauchy–Green tensor, right, 80
Cauchy–Kovalevskaya
 theorem, 192
Cayley–Hamilton theorem, 92
center of buoyancy, 249
central axis, 36
characteristic equation, 21
characteristic space, 20
characteristic surface, 190, 194,
 201
Christoffel symbols, 51
Clausius–Duhem inequality, 145
Clausius–Planck inequality, 144
coefficient of kinematic viscosity,
 272
compatibility conditions, 93
complex potential, 257
complex velocity, 257
components, contravariant, 3
components, covariant, 9
compressible, 245
configuration, actual or
 current, 77
configuration, reference, 77
conjugate, 173
constitutive axioms, 156
constitutive equations, 155

Index

contact forces, 137
convective derivative, 111
coordinate curves, 46
coordinates, spatial and material, 77
covariant derivative, 52
curl, 54
curvilinear coordinates, 46
cylindrical coordinates, 64

D'Alembert's equation, 196
D'Alembert's paradox, 266
deformation gradient, 78
Deformation program, 101
deformation, homogeneous, 327
derivative, weak or generalized, 380
dilational wave, 340
dilational waves, 289
dimension, 3
direct sum, 4
directional derivative, 54
Dirichlet boundary value problem, 383
dispersion relation, 358
displacement boundary value problem, 319
displacement field, 84
displacement gradient, 84
divergence, 53
doublet, 260
dynamic process, 156

EigenSystemAG program, 41
eigenvalue, 20
eigenvalue equation, 20
eigenvector, 20
eikonal equation, 209
elastic behavior, 160
elliptic coordinates, 66
elliptic equation, 195
elliptic system, 202
empirical temperature, 367
endomorphism, 14

energy balance, 141
energy density, 321
entropy, 143, 369
entropy principle, 143, 144
equivalent to zero, 37
equivalent vector system, 36
Euclidean point space, 31
Euclidean tensor, 14, 29
Euclidean vector space, 5
Euler fluid, 245
Euler–Cauchy postulate, 138
Eulerian coordinates, 109
Eulerian form, 110
extended thermodynamics, 370

finite deformation, 77
first Helmholtz theorem, 253
first law of Thermodynamics, 368
first law of thermodynamics, 141
first Piola–Kirchhof tensor, 148
first-order singular surface, 56
Fourier inequality, 145
frame of reference, 31
frame, natural, or holonomic, 46
free vector set, 2
frequency spectrum, 364

Gauss coordinates, 190
Gauss's theorem, 55
generalized polar coordinates, 63
geometric compatibility condition, 59
geometric multiplicity, 20
grade n, 157
gradient, 51
gradient of velocity, 112
Gram–Schmidt procedure, 6
Green–St.Venant tensor, 84
group velocity, 364

Hadamard's theorem, 58, 209, 210
harmonic function, 115, 256
heat equation, 197

Index

heat flux vector, 142
heat source, specific, 142
Helmholtz free energy, 145
Helmholtz theorem, 343
homogeneous material, 173
Hugoniot's equation, 292
hyperbolic equation, 196
hyperbolic system, 202

image, 17
incompressible, 245
incompressible fluid, 180
incompressible material, 165
infinitesimal strain tensor, 85
integral curves, 54
invariant, first, second, and third, 22
inverse dispersion, 365
irrotational motion, 114
isochoric motion, 114
isotropic function, 174
isovolumic motion, 114

Joukowsky program, 309
Joukowsky's transformation, 269
JoukowskyMap, 312
jump conditions, 56, 131
jump system, 208

kernel, 17
kinetic energy, 141
kinetic energy theorem, 141
kinetic field, 111
kinetic potential, 114
Kutta–Joukowsky theorem, 267

Lagrange's theorem, 250
Lagrangian coordinates, 109
Lagrangian form, 110
Lagrangian mass conservation, 147
Lamé coefficients, 176
Laplace's equation, 115, 196, 256
Laplacian, 54
Lax condition, 218

length, 6
Levi–Civita symbol, 11
lift, 267
linear combination, 2
linear isotropic solid, 176
linear mapping, 14
linear PDE, 190
linearly dependent vector set, 3
linearly independent vector set , 2
LinElasticityTensor program, 185
local balance equation, 131
local speed of propagation, 117
locally equivalent processes, 156
longitudinal unit extension, 80
longitudinal wave, 340

Mach number, 289
mass conservation principle, 136
mass density, 136
mass forces, 137
material coordinates, 109
material derivative, 110
material frame-indifference, 157
material objectivity, 157
material response, 156
material volume, 114, 131
metric coefficients, 48
Minkowski's inequality, 6
mixed boundary value problem, 319
mixed product, 13
mode conversion, 355
momentum balance principle, 137

Navier–Stokes behavior, 180
Navier–Stokes equation, 272
Neumann boundary value problem, 383
normal dispersion, 365

normal speed, 117
normal stress, 138

objective tensor, 116
objective vector, 116
Operator program, 70
order, 2
ordinary wave, 208
orientation, 11
orthogonal vectors, 6
orthonormal system, 6

P wave, 350
parabolic coordinates, 67
parabolic equation, 195
parabolic system, 202
paraboloidal coordinates, 69
particle path, 111
Pascal's principle, 245
PdeEqClass program, 221
PdeSysClass program, 227
perfect fluid, 179, 245
perfect gas, 247
perturbed region, 208
phase velocity, 345
Poincaré inequality, 382
Poisson's condition, 139
Poisson's ratio, 177
polar continuum, 138
polar vector, 12
Potential program, 298
potential, stream or Stokes, 257
potential,velocity or kinetic, 257
Prandtl's equations, 279
pressure, 245
principal direction of stress, 139
principal stress, 139
principal stretching, 79
principle of determinism, 156
principle of dissipation, 158
principle of equipresence, 159
principle of local action, 157
principle of virtual work, 151
product, 2, 14

product, inner or scalar, 5
product, vector or cross, 11
prolate spheroidal
        coordinates, 68
proper rotation, 25
pseudovector, 12
pure compression, 335
pure extension, 335

quasi-linear PDE, 190

Rankine–Hugoniot jump
        conditions, 215
rate of deformation, 112
Rayleigh equation, 358
Rayleigh wave, 356
Rayleigh–Lamb frequency
        equation, 364
Rayleigh–Lamb relation, 346
rectilinear coordinates, 31
reduced dissipation
        inequality, 145
reflection, 345
refraction, 346
Reynolds number, 276
Riemann–Christoffel tensor, 93
rotation, 25
rotation axis, 27
rotation tensor, 79

Saint-Venant conjecture, 335
Saint-Venant principle, 335
scalar invariant, 36
Schwarz's inequality, 6
second Helmholtz theorem, 254
second law of thermodynamics,
        143, 144
second Piola–Kirchhoff
        tensor, 148
second principle of
        thermodynamics, 369
semi-inverse method, 332
semilinear PDE, 190
SH wave, 350

Index 387

shear, 80
shear stress, 138
shear wave, 340
shock intensity, 217
shock wave, 208
simple continuum, 138
simple material, 157
singular perturbation, 277
singular surface, 56, 118
sink, 260
Sobolev space, 381
solenoidal vector field, 252
sound velocity, 289
source, 260
spatial coordinates, 109
specific entropy, 144
specific force, 137
specific internal energy, 142
spectral decomposition, 23
spectrum, 20
spherical coordinates, 65
spin, 112
stagnation points, 263
stationary motion, 111
Stevino's law, 247
Stokes's theorem, 55
stream tube, 252
streamline, 111
strength of a source or sink, 260
stress, 137
stress boundary conditions, 139
stress boundary value
  problem, 319
stretch ratio, 80
stretching, 112
stretching tensor, left, 79
stretching tensor, right, 79
subsonic motion, 289
subspace, 4
sum, 2
supersonic motion, 289
SV wave, 350
symmetric tensor, 14
symmetry, 172

symmetry group, 173

temperature, absolute, 143
tensor of linear elasticity, 167
tensor of thermal
  conductivity, 167
tensor product, 14, 29
tensor, acoustic, 339
tensor, first Piola–Kirchhof, 148
tensor, Green–St.Venant, 84
tensor, infinitesimal stress, 85
tensor, left Cauchy–Green, 81
tensor, orthogonal, 24
tensor, Riemann–Christoffel, 93
tensor, right Cauchy–Green, 80
tensor, rotation, 79
tensor, second Piola–Kirchhoff,
  148
tensor, skew-symmetric, 14
tensor, symmetric, 14
tensor, two-point, 89
thermal power flux, 142
thermoelastic behavior, 160
thermoelastic isotropic solid, 173
thermokinetic process, 156
thermometer, 143
thermoviscoelasticity, 160
thermoviscous fluid, 177
Thomson–Kelvin theorem, 250
Torricelli's theorem, 253
torsion, 337
torsional stiffness, 334
totally hyperbolic system, 202
trace, 382, 383
traction, 137
transpose, 14
transverse wave, 340
Tricomi's equation, 198
two-point tensor, 89

undisturbed region, 208

vector, 2
vector field, 50

vector space, 2
vector, unit or normal, 6
Vectorsys program, 36
velocity, 110
velocity or kinetic potential, 256
Velocity program, 126
vortex line, 252
vortex potential, 258
vortex tube, 252
vortex vector, 252
vorticity tensor, 112

wave, dilational, 340
wave, longitudinal, 340

wave, P, SH, and SV, 350
wave, Rayleigh, 356
wavefront, 208
WavesI program, 233
WavesII program, 239
weak solution, 384
Wing program, 306

Young's modulus, 177

zero postulate of
    thermodynamics, 367

QA 808.2 .R66 2004

Romano, Antonio, 1941-

Continuum mechanics using
 Mathematica